D1617334

Uwe Krey · Anthony Owen

Basic Theoretical Physics

A Concise Overview

With 31 Figures

 Springer

Prof. Dr. Uwe Krey
University of Regensburg (retired)
FB Physik
Universitätsstraße 31
93053 Regensburg, Germany
E-mail: uwe.krey@physik.uni-regensburg.de

Dr. *rer nat habil* Anthony Owen
University of Regensburg (retired)
FB Physik
Universitätsstraße 31
93053 Regensburg, Germany
E-mail: anthony.owen@physik.uni-regensburg.de

Library of Congress Control Number: 2007930646

ISBN 978-3-540-36804-5 Springer Berlin Heidelberg New York

Springer is a part of Springer Science+Business Media

springer.com

© Springer-Verlag Berlin Heidelberg 2007

Typesetting and production: LE-TeX Jelonek, Schmidt & Vöckler GbR, Leipzig
Cover design: eStudio Calamar S.L., F. Steinen-Broo, Pau/Girona, Spain

Printed on acid-free paper   SPIN 11492665   57/3180/YL - 5 4 3 2 1 0

Preface

This textbook on theoretical physics (I-IV) is based on lectures held by one of the authors at the University of Regensburg in Germany. The four 'canonical' parts of the subject have been condensed here into a single volume with the following main sections:

I = Mechanics and Basic Relativity;
II = Electrodynamics and Aspects of Optics;
III = Quantum Mechanics (non-relativistic theory), and
IV = Thermodynamics and Statistical Physics.

Our compendium is intended primarily for revision purposes and/or to aid in a deeper understanding of the subject. For an introduction to theoretical physics many standard series of textbooks, often containing seven or more volumes, are already available (see, for example, [1]).

Exercises closely adapted to the book can be found on one of the authors websites [2], and these may be an additional help.

We have laid emphasis on relativity and other contributions by Einstein, since the year 2005 commemorated the centenary of three of his ground-breaking theories.

In Part II (Electrodynamics) we have also treated some aspects with which every physics student should be familiar, but which are usually neglected in textbooks, e.g., the principles behind cellular (or mobile) phone technology, synchrotron radiation and holography. Similarly, Part III (Quantum Mechanics) additionally covers aspects of quantum computing and quantum cryptography.

We have been economical with figures and often stimulate the reader to sketch his or her own diagrams. The frequent use of italics and quotation marks throughout the text is to indicate to the reader where a term is used in a specialized way. The Index contains useful keywords for ease of reference.

Finally we are indebted to the students and colleagues who have read parts of the manuscript and to our respective wives for their considerable support.

Regensburg,   *Uwe Krey*
May 2007   *Anthony Owen*

Contents

Part I

Mechanics and Basic Relativity

1 Space and Time

1.1 Preliminaries to Part I

This book begins in an elementary way, before progressing to the topic of *analytical mechanics*.[1] Nonlinear phenomena such as "chaos" are treated briefly in a separate chapter (Chap. 12). As far as possible, only elementary formulae have been used in the presentation of relativity.

1.2 General Remarks on Space and Time

a) Physics is based on *experience* and *experiment*, from which axioms or generally accepted principles or laws of nature are developed. However, an axiomatic approach, used for the purposes of reasoning in order to establish a formal deductive system, is potentially dangerous and inadequate, since axioms do not constitute a necessary truth, experimentally.

b) Most theories are only approximate, preliminary, and limited in scope. Furthermore, they cannot be *proved* rigorously in every circumstance (i.e., *verified*), only shown to be untrue in certain circumstances (i.e., *falsified; Popper*).[2] For example, it transpires that Newtonian mechanics only applies as long as the magnitudes of the velocities of the objects considered are very small compared to the velocity c of light *in vacuo*.

c) Theoretical physics develops (and continues to develop) in "phases" (*Kuhn*[3], *changes of paradigm*). The following list gives examples.

1. From $\sim 1680-1860$: *classical Newtonian mechanics*, falsified by experiments of those such as Michelson and Morley (1887). This falsification was ground-breaking since it led Einstein in 1905 to the insight that the perceptions of space and time, which were the basis of Newtonian theory, had to be modified.

2. From $\sim 1860-1900$: *electrodynamics (Maxwell)*. The full consequences of Maxwell's theory were only later understood by Einstein through his *special theory of relativity* (1905), which concerns both Newtonian

---

[1] See, for example, [3].
[2] Here we recommend an internet search for *Karl Popper*.
[3] For more information we suggest an internet search for *Thomas Samuel Kuhn*.

mechanics (Part I) and Maxwell's electrodynamics (Part II). In the
same year, through his hypothesis of *quanta of electromagnetic waves
(photons)*, Einstein also contributed fundamentally to the developing
field of *quantum mechanics* (Part III).

3. 1905: Einstein's *special theory of relativity*, and 1916: his *general theory
   of relativity*.
4. From 1900: Planck, Bohr, Heisenberg, de Broglie, Schrödinger: *quantum mechanics; atomic and molecular physics*.
5. From ~ 1945: *relativistic quantum field theories, quantum electrodynamics, quantum chromodynamics, nuclear and particle physics*.
6. From ~ 1980: *geometry (spacetime) and cosmology: supersymmetric theories, so-called 'string' and 'brane' theories; astrophysics; strange matter*.
7. From ~ 1980: *complex systems and chaos; nonlinear phenomena in mechanics related to quantum mechanics; cooperative phenomena*.

Theoretical physics is thus a discipline which is open to change. Even in
mechanics, which is apparently old-fashioned, there are many unsolved problems.

1.3 Space and Time in Classical Mechanics

Within classical mechanics it is implicitly assumed – from relatively inaccurate measurements based on everyday experience – that

a) *physics* takes place in a *three-dimensional Euclidean space* that is *not*
   influenced by material properties and physical events. It is also assumed
   that
b) *time* runs separately as an *absolute* quantity; i.e., it is assumed that all
   clocks can be *synchronized* by a signal transmitted at a speed $v \to \infty$.

Again, the underlying experiences are only approximate, e.g., that

α) measurements of lengths and angles can be performed by translation and
   rotation of rigid bodies such as rods or yardsticks;
β) the sum of the interior angles of a triangle is $180°$, as Gauss showed in
   his famous geodesic triangulation of 1831.

Thus, according to the laws of classical mechanics, rays of light travel in
straight lines (rectilinear behavior). Einstein's prediction that, instead, light
could travel in curved paths became evident as a result of very accurate astronomical measurements when in 1919 during a solar eclipse rays of light
traveling near the surface of the sun were observed showing that stellar bodies under the influence of gravitation give rise to a curvature of spacetime
(*general theory of relativity*), a phenomenon which was not measurable in
Gauss's time.

Assumption b) was also shown to be incorrect by Einstein (see below).

2 Force and Mass

2.1 Galileo's Principle (Newton's First Axiom)

Galileo's principle, which forms the starting point of theoretical mechanics, states that in an *inertial frame of reference* all bodies not acted upon by any force move rectilinearly and homogeneously at constant velocity v.

The main difficulty arising here lies in the realization of an *inertial frame*, which is only possible by iteration: to a zeroth degree of approximation an inertial frame is a system of Cartesian coordinates, which is rigidly rotating with the *surface* of the earth, to which its axes are attached; to the next approximation they are attached to the *center* of the earth; in the following approximation they are attached to the center of the sun, to a third approximation to the center of our galaxy, and so on. According to *Mach* an *inertial frame* can thus only be defined by the *overall* distribution of the stars. The final difficulties were only resolved later by Einstein, who proposed that inertial frames can only be defined *locally*, since gravitation and acceleration are equivalent quantities (see Chap. 14).

Galileo's principle is essentially equivalent to Newton's First Axiom (or Newton's First Law of Motion).

2.2 Newton's Second Axiom: Inertia; Newton's Equation of Motion

This axiom constitutes an essential widening and accentuation of Galileo's principle through the introduction of the notions of *force*, F, and *inertial mass*, $m_t \equiv m$. (This is the *inertial aspect* of the central notion of *mass, m.*)

Newton's second law was originally stated in terms of momentum. The rate of change of momentum of a body is proportional to the force acting on the body and is in the same direction. where the momentum of a body of inertial mass m_t is quantified by the vector $p := m_t \cdot v$.[1] Thus

$$F = \frac{\mathrm{d}p}{\mathrm{d}t} . \tag{2.1}$$

---

[1] Here we consider only bodies with infinitesimal volume: so-called *point masses*.

The notion of *mass* also has a *gravitational aspect*, m_s (see below), where $m_t = m_s(\equiv m)$. However, primarily a body possesses 'inertial' mass m_t, which is a quantitative measure of its inertia or resistance to being moved[2]. (Note: In the above form, (2.1) also holds in the *special theory of relativity*, see Sect. 15 below, according to which the momentum is given by

$$\boldsymbol{p} = \frac{m_0 \boldsymbol{v}}{\sqrt{1 - \frac{v^2}{c^2}}} \; ;$$

m_0 is the *rest mass*, which only agrees with m_t in the Newtonian approximation $v^2 \ll c^2$, where c is the velocity of light *in vacuo*.)

Equation (2.1) can be considered to be essentially a definition of *force* involving (inertial) mass and velocity, or equivalently a definition of *mass* in terms of force (see below).

As already mentioned, a body with (inertial) mass also produces a gravitational force proportional to its gravitational mass m_s. Astonishingly, in the conventional units, i.e., apart from a universal constant, one has the well-known identity $m_s \equiv m_t$, which becomes still more astonishing, if one simply changes the name and thinks of m_s as a "gravitational charge" instead of "gravitational mass". This remarkable identity, to which we shall return later, provided Einstein with strong motivation for developing his *general theory of relativity*.

2.3 Basic and Derived Quantities; Gravitational Force

The basic quantities underlying all physical measurements of motion are

- *time*: defined from multiples of the period of a so-called '*atomic clock*', and
- *distance*: measurements of which are nowadays performed using radar signals.

The conventional units of time (e.g., *second*, *hour*, *year*) and length (e.g., *kilometre*, *mile*, etc.) are arbitrary. They have been introduced historically, often from astronomical observations, and can easily be transformed from one to the other. In this context, the so-called "archive metre" (in French: "*mètre des archives*") was adopted historically as the universal prototype for a standard length or distance: 1 metre (1 m).

Similarly, the "archive kilogram" or international prototype kilogram in Paris is the universal standard for the unit of *mass*: 1 kilogram (1 kg).

---

[2] in German: *inertial mass* = *träge Masse* as opposed to *gravitational mass* = *schwere Masse* m_s. The fact that in principle one should distinguish between the two quantities was already noted by the German physicist H. Hertz in 1884; see [4].

However, the problem as to whether the archive kilogram should be used as a definition of *(inertial) mass* or a definition of *force* produced a dilemma. In the nineteen-fifties the "kilopond (kp)" (or kilogram-force (kgf)) was adopted as a standard quantity in many countries. This quantity is defined as the gravitational force acting on a 1 kg mass in standard earth gravity (in Paris where the archive kilogram was deposited). At that time the quantity *force* was considered to be a "basic" quantity, while *mass* was (only) a "derived" one. More recently, even the above countries have reverted to using *length, time*, and *(inertial) mass* as base quantities and *force* as a derived quantity. In this book we shall generally use the international system (SI) of units, which has 7 dimensionally independent base units: metre, kilogram, second, ampere, kelvin, mole and candela. All other physical units can be derived from these base units.

What can be learnt from this? Whether a quantity is *basic* or (only) *derived*, is a *matter of convention*. Even the *number* of base quantities is not fixed; e.g., some physicists use the 'cgs' system, which has three base quantities, *length* in centimetres (cm), *time* in seconds (s) and *(inertial) mass* in grams (g), or multiples thereof; or the mksA system, which has four base quantities, corresponding to the standard units: metre (m), kilogram (kg), second (s) and ampere (A) (which only comes into play in electrodynamics). Finally one may adopt a system with only one basic quantity, as preferred by high-energy physicists, who like to express everything in terms of a fundamental unit of energy, the electron-volt eV: e.g., *lengths* are expressed in units of $\hbar \cdot c/(\text{eV})$, where $\hbar$ is Planck's constant divided by 2π, which is a universal quantity with the physical dimension $action = energy \times time$, while c is the velocity of light *in vacuo*; *masses* are expressed in units of eV/c^2, which is the "rest mass" corresponding to the energy $1\,\text{eV}$. (Powers of $\hbar$ and c are usually *replaced* by unity[3]).

As a consequence, writing Newton's equation of motion in the form

$$m \cdot \boldsymbol{a} = \boldsymbol{F} \tag{2.2}$$

(relating acceleration $\boldsymbol{a} := \frac{\mathrm{d}^2 \boldsymbol{r}}{\mathrm{d}t^2}$ and force $\boldsymbol{F}$), it follows that one can equally well say that in this equation the *force* (e.g., calibrated by a certain spring) is the 'basic' quantity, as opposed to the different viewpoint that the *mass* is 'basic' with the force being a *derived quantity*, which is 'derived' by the above equation. (This arbitariness or dichotomy of viewpoints reminds us of the question: "Which came first, the chicken or the egg?!"). In a more modern didactical framework based on current densities one could, for example, write the left-hand side of (2.2) as the time-derivative of the momentum, $\frac{\mathrm{d}\boldsymbol{p}}{\mathrm{d}t} \equiv \boldsymbol{F}$, thereby using the force as a secondary quantity. However, as already

---

[3] One should avoid using the semantically different formulation "set to 1" for the quantities with non-vanishing *physical dimension* such as $c(= 2.998 \cdot 10^8\,\text{m/s})$, etc.

mentioned, a different viewpoint is also possible, and it is better to keep an open mind on these matters than to *fix* our ideas unnecessarily.

Finally, the problem of planetary motion dating back to the time of Newton where one must in principle distinguish between the *inertial mass* m_t entering (2.2) and a *gravitational mass* m_s, which is numerically identical to m_t (apart from a universal constant, which is usually replaced by unity), is far from being trivial; m_s is defined by the gravitational law:

$$F(r) = -\gamma \frac{M_s \cdot m_s}{|r - R|^2} \cdot \frac{r - R}{|r - R|},$$

where r and m_s refer to the planet, and R and M_s to the central star ("sun"), while γ is the gravitational constant. Here the [gravitational] masses play the role of gravitational charges, similar to the case of Coulomb's law in electromagnetism. In particular, as in Coulomb's law, the proportionality of the gravitational force to M_s and m_s can be considered as representing an *active* and a *passive* aspect of gravitation.[4] The fact that inertial and gravitational mass are indeed equal was first proved experimentally by Eötvös (Budapest, 1911 [6]); thus we may write $m_s = m_t \equiv m$.

2.4 Newton's Third Axiom ("Action and Reaction ...")

Newton's third axiom states that action and reaction are equal in magnitude and opposite in direction.[5] This implies *inter alia* that the "active" and "passive" gravitational masses are equal (see the end of the preceding section), i.e., on the one hand, a body with an (active) gravitational charge M_s generates a gravitational field

$$G(r) = -\gamma \frac{M_s}{|r - R|^2} \cdot \frac{r - R}{|r - R|},$$

in which, on the other hand, a different body with a (passive) gravitational charge m_s is acted upon by a force, i.e., $F = m_s \cdot G(r)$. The relations are analogous to the electrical case (Coulomb's law). The equality of active and passive gravitational charge is again not self-evident, but in the considered context it is implied that no *torque* arises (see also Sect. 5.2). Newton also recognized the general importance of his third axiom, e.g., with regard to the application of tensile stresses or compression forces between two bodies.

Three additional consequences of this and the preceding sections will now be discussed.

---

[4] If one only considers the relative motion, active and passive aspects cannot be distinguished.

[5] In some countries this is described by the abbreviation in Latin "actio=reactio".

a) As a consequence of equating the inertial and gravitational masses in Newton's equation $\boldsymbol{F}(\boldsymbol{r}) = m_{\mathrm{s}} \cdot \boldsymbol{G}(\boldsymbol{r})$ it follows that *all bodies fall equally fast* (if only gravitational forces are considered), i.e.: $\boldsymbol{a}(t) = \boldsymbol{G}(\boldsymbol{r}(t))$. This corresponds to Galileo's experiment[6], or rather *thought experiment*, of dropping different masses simultanously from the top of the Leaning Tower of Pisa.

b) The *principle of superposition* applies with respect to gravitational forces:

$$\boldsymbol{G}(\boldsymbol{r}) = -\gamma \sum_k \frac{(\Delta M_{\mathrm{s}})_k}{|\boldsymbol{r} - \boldsymbol{R}_k|^2} \cdot \frac{\boldsymbol{r} - \boldsymbol{R}_k}{|\boldsymbol{r} - \boldsymbol{R}_k|} \ .$$

Here $(\Delta M_{\mathrm{s}})_k := \varrho_k \Delta V_k$ is the mass of a small volume element ΔV_k, and ϱ_k is the mass density. An analogous "superposition principle" also applies for electrostatic forces, but, e.g., not to nuclear forces. For the principle of superposition to apply, the equations of motion must be linear.

c) Gravitational (and Coulomb) forces *act in the direction of the line joining* the point masses i and k. This implies a different emphasis on the meaning of Newton's third axiom. In its weak form, the postulate means that $\boldsymbol{F}_{i,k} = -\boldsymbol{F}_{k,i}$; in an intensified or "strong" form it means that $\boldsymbol{F}_{i,k} = (\boldsymbol{r}_i - \boldsymbol{r}_k) \cdot f(r_{i,k})$, where $f(r_{ik})$ is a scalar function of the distance $r_{i,k} := |\boldsymbol{r}_i - \boldsymbol{r}_k|$.

As we will see below, the above intensification yields a sufficient condition that Newton's third axiom not only implies $\boldsymbol{F}_{i,k} = -\boldsymbol{F}_{k,i}$, but also $\boldsymbol{\mathcal{D}}_{i,k} = -\boldsymbol{\mathcal{D}}_{k,i}$, where $\boldsymbol{\mathcal{D}}_{i,k}$ is the *torque* acting on a particle at $\boldsymbol{r}_i$ by a particle at $\boldsymbol{r}_k$.

---

[6] In essence, the early statement of Galileo already contained the basis not only of the later equation $m_{\mathrm{s}} = m_{\mathrm{t}}$, but also of the Eötvös experiment, [6] (see also [4]), and of Einstein's equivalence principle (see below).

3 Basic Mechanics of Motion in One Dimension

3.1 Geometrical Relations for Curves in Space

In this section, motion is considered to take place on a *fixed* curve in three-dimensional Euclidean space. This means that it is essentially one-dimensional; motion in a straight line is a special case of this.

For such trajectories we assume they are described by the radius vector $\boldsymbol{r}(t)$, which is assumed to be continuously differentiable at least twice, for $t \in [t_a, t_b]$ (where t_a and t_b correspond to the beginning and end of the motion, respectively). The instantaneous *velocity* is

$$\boldsymbol{v}(t) := \frac{\mathrm{d}\boldsymbol{r}}{\mathrm{d}t} \,,$$

and the instantaneous *acceleration* is

$$\boldsymbol{a}(t) := \frac{\mathrm{d}\boldsymbol{v}}{\mathrm{d}t} = \frac{\mathrm{d}^2\boldsymbol{r}}{\mathrm{d}t^2} \,,$$

where for convenience we differentiate all three components, $x(t)$, $y(t)$ and $z(t)$ in a fixed Cartesian coordinate system,

$$\boldsymbol{r}(t) = x(t)\boldsymbol{e}_x + y(t)\boldsymbol{e}_y + z(t)\boldsymbol{e}_z : \frac{\mathrm{d}\boldsymbol{r}(t)}{\mathrm{d}t} = \dot{x}(t)\boldsymbol{e}_x + \dot{y}(t)\boldsymbol{e}_y + \dot{z}(t)\boldsymbol{e}_z \,.$$

For the *velocity* vector we can thus simply write: $\boldsymbol{v}(t) = v(t)\boldsymbol{\tau}(t)$, where

$$v(t) = \sqrt{(\dot{x}(t))^2 + (\dot{y}(t))^2 + (\dot{z}(t))^2}$$

is the *magnitude* of the velocity and

$$\boldsymbol{\tau}(t) := \frac{\boldsymbol{v}(t)}{|\boldsymbol{v}(t)|}$$

the *tangential unit vector* to the curve (assuming $\boldsymbol{v} \neq 0$).

$v(t)$ and $\boldsymbol{\tau}(t)$ are thus *dynamical* and *geometrical* quantities, respectively, with an absolute meaning, i.e., independent of the coordinates used.

In the following we assume that $\boldsymbol{\tau}(t)$ is not constant; as we will show, the acceleration can then be decomposed into, (i), a tangential component, and, (ii), a normal component (typically: radially inwards), which has the direction of a so-called *osculating normal* $\boldsymbol{n}$ to the curve, where the unit vector $\boldsymbol{n}$ is proportional to $\frac{d\boldsymbol{\tau}}{dt}$, and the magnitude of the force (ii) corresponds to the well-known "centripetal" expression $\frac{v^2}{R}$ (see below); the quantity R in this formula is the (instantaneous) so-called *radius of curvature* (or *osculating radius*) and can be evaluated as follows:

$$\frac{1}{R} = \left|\frac{d\boldsymbol{\tau}}{v \cdot dt}\right| .^1$$

Only the *tangential* force, (i), is relevant at all, whereas the *centripetal* expression (ii) is compensated for by *forces of constraint*[2], which keep the motion on the considered curve, and need no evaluation except in special instances.

The quantity

$$\int_{t_a}^{t} v(t)\,dt$$

is called the *arc length* $s(t)$, with the differential $ds := v(t)\,dt$. As already mentioned, the *centripetal acceleration*, directed towards the center of the *osculating circle*, is given by

$$\boldsymbol{a}_{\text{centrip.}}(t) = \boldsymbol{n}(t)\frac{v^2(t)}{R(t)} .$$

We thus have

$$\boldsymbol{a}(t) \equiv \boldsymbol{\tau}(t) \cdot \frac{dv(t)}{dt} + \boldsymbol{a}_{\text{centrip.}}(t) .$$

The validity of these general statements can be illustrated simply by considering the special case of circular motion at constant angular velocity, i.e.,

$$\boldsymbol{r}(t) := R \cdot (\cos(\omega t)\boldsymbol{e}_x + \sin(\omega t)\boldsymbol{e}_y) .$$

The tangential vector is

$$\boldsymbol{\tau}(t) = -\sin(\omega t)\boldsymbol{e}_x + \cos(\omega t)\boldsymbol{e}_y ,$$

and the *osculating normal* is

$$\boldsymbol{n}(t) = -(\cos(\omega t)\boldsymbol{e}_x + \sin(\omega t)\boldsymbol{e}_y) ,$$

i.e., directed towards the center. The *radius of curvature* $R(t)$ is of course identical with the radius of the circle. The acceleration has the above-mentioned magnitude, $R\omega^2 = v^2/R$, directed inwards.

---

[1] It is strongly recommended that the reader should produce a sketch illustrating these relations.

[2] This is a special case of *d'Alembert's principle*, which is described later.

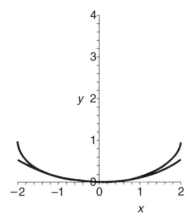

Fig. 3.1. Osculating circle and radius of curvature. The figure shows as a typical example the lower part of an ellipse (described by the equation $\frac{y_1}{b} := 1 \pm \sqrt{1 - \frac{x^2}{a^2}}$, with $a := 2$ and $b := 1$) and a segment (the lower of the two curves!) of the *osculating circle* at $x = 0$ (with $R := \frac{a^2}{b} \equiv 4$). Usually a one-dimensional treatment suffices, since an infinity of lines have the same osculating circle at a given point, and since usually one does *not* require the *radial* component $\boldsymbol{n} \cdot \frac{mv^2}{R}$ of the force (which is compensated by forces of constraint), as opposed to the *tangential* component $\boldsymbol{F}_{\text{tangent}} := m \cdot \dot{v}(t) \cdot \boldsymbol{\tau}(t)$ (where t is the time, m the inertia and v the magnitude of the considered point mass); $\boldsymbol{n}$ and $\boldsymbol{\tau}$ are the *osculating normal* and the *tangential* unit vectors, respectively. A one-dimensional treatment follows

Finally, as already alluded to above, the quanties $\boldsymbol{\tau}(t)$, $\boldsymbol{n}(t)$, $R(t)$, and $s(t)$ have purely geometrical meaning; i.e., they do not change due to the kinematics of the motion, but only depend on geometrical properties of the curve on which the particle moves. The kinematics are determined by Newton's equations (2.2), and we can specialize these equations to a one-dimensional problem, i.e., for $\sim \boldsymbol{\tau}(t)$, since (as mentioned) the transverse forces, $\boldsymbol{n}(t) \cdot \frac{v^2}{R}$, are compensated for by *constraining forces*.

The preceding arguments are supported by Fig. 3.1 above.

Thus, for simplicity we shall write $x(t)$ instead of $s(t)$ in the following, and we have

$$v(t) = \frac{dx(t)}{dt} \quad \text{and} \quad m \cdot a(t) = m \cdot \frac{d^2x(t)}{dt^2} = F(t, v(t), a(t)),$$

where F is the tangential component of the force, i.e., $F \equiv \boldsymbol{F} \cdot \boldsymbol{\tau}$.

3.2 One-dimensional Standard Problems

In the following, for simplicity, instead of F we consider the reduced quantity $f := \frac{F}{m}$. If $f(t, v, x)$ depends on only one of the three variables t, v or x, the equations of motion can be solved *analytically*. The most simple case is where f is a function of t. By direct integration of

$$\frac{\mathrm{d}^2 x(t)}{\mathrm{d}t^2} = f(t)$$

one obtains:

$$v(t) = v_0 + \int_{t_0}^{t} \mathrm{d}\tilde{t} f(\tilde{t}) \quad \text{and} \quad x(t) = x_0 + v_0 \cdot (t - t_0) + \int_{t_0}^{t} \mathrm{d}\tilde{t} v(\tilde{t}) \,.$$

(x_0, v_0 and t_0 are the real initial values of position, velocity and time.)

The next most simple case is where f is given as an explicit function of v. In this case a standard method is to use *separation of variables* ($\hat{=}$ transition to the inverse function), if possible: Instead of

$$\frac{\mathrm{d}v}{\mathrm{d}t} = f(v)$$

one considers

$$\mathrm{d}t = \frac{\mathrm{d}v}{f(v)} \,, \quad \text{or} \quad t - t_0 = \int_{v_0}^{v} \frac{\mathrm{d}\tilde{v}}{f(\tilde{v})} \,.$$

One obtains t as a function of v, and can thus, at least implicitly, calculate $v(t)$ and subsequently $x(t)$.

The third case is where $f \equiv f(x)$. In this case, for one-dimensional problems, one always proceeds using the principle of *conservation of energy*, i.e., from the equation of motion,

$$m \frac{\mathrm{d}v}{\mathrm{d}t} = F(x) \,,$$

by multiplication with

$$v = \frac{\mathrm{d}x}{\mathrm{d}t}$$

and subsequent integration, with the substitution $v\,\mathrm{d}t = \mathrm{d}x$, it follows that

$$\frac{v^2}{2m} + V(x) \equiv E$$

is constant, with a *potential energy*

$$V(x) := - \int_{x_0}^{x} \mathrm{d}\tilde{x} F(\tilde{x}) \,.$$

Therefore,

$$v(x) = \sqrt{\frac{2}{m}(E - V(x))} \,,$$

or $\quad dt = \dfrac{dx}{\sqrt{\frac{2}{m}(E - V(x))}} \quad$, and finally

$$t - t_0 = \int_{x_0}^{x} \frac{d\tilde{x}}{\sqrt{\frac{2}{m}(E - V(\tilde{x}))}} \ .$$

This relation is very useful, and we shall return to it often later.

If f depends on two or more variables, one can only make analytical progress in certain cases, e.g., for the driven harmonic oscillator, with damping proportional to the magnitude of the velocity. In this important case, which is treated below, one has a *linear* equation of motion, which makes the problem solvable; i.e.,

$$\ddot{x} = -\omega_0^2 x - \frac{2}{\tau}v + f(t) \ ,$$

where useful general statements can be made (see below). (The above-mentioned ordinary differential equation applies to harmonic springs with a spring constant k and mass m, corresponding to the Hookean force $F_H := -k \cdot x$, where $\omega_0^2 = k/m$, plus a *linear* frictional force $F_R := -m\frac{2}{\tau} \cdot v$, plus a driving force $F_A := m \cdot f(t)$.)

There are cases where the frictional force depends *quadratically* on the velocity (so-called *Newtonian friction*),

$$F_R := -\alpha \cdot \frac{mv^2}{2} \ ,$$

i.e., with a so-called *technical friction factor* α, and a driving force depending mainly, i.e., explicitly, on x, and only implicitly on t, e.g., in motor racing, where the acceleration may be very high in certain places, $F_a = mf(x)$. The equation of motion,

$$m\dot{v} = -\alpha \frac{mv^2}{2} + mf(x) \ ,$$

can then be solved by multiplying by

$$\frac{dt}{dx} \left(\equiv \frac{1}{v} \right) \ :$$

One thus obtains the ordinary first-order differential equation

$$\frac{dv}{dx} + \frac{\alpha v}{2} = \frac{f(x)}{v} \ ,$$

which can be solved by iteration. On the r.h.s. of this equation, one uses, for example, an approximate expression for $v(x)$ and obtains a refinement on the l.h.s., which is then substituted into the r.h.s., etc., until one obtains convergence. In almost all other cases one has to solve an ordinary second-order differential equation *numerically*. Many computer programs are available for solving such problems, so that it is not necessary to go into details here.

4 Mechanics of the Damped and Driven Harmonic Oscillator

In this section the potential energy $V(x)$ for the motion of a one-dimensional system is considered, where it is assumed that $V(x)$ is smooth everywhere and differentiable an arbitrarily often number of times, and that for $x = 0$, $V(x)$ has a parabolic local minimum. In the vicinity of $x = 0$ one then obtains the following Taylor expansion, with $V''(0) > 0$:

$$V(x) = V(0) + \frac{1}{2}V''(0)x^2 + \frac{1}{3!}V'''(0)\,x^3 + \dots \, ,$$

i.e.,

$$V(x) = V(0) + \frac{m\omega_0^2}{2}x^2 + \mathcal{O}\left(x^3\right) \, ,$$

with $\omega_0^2 := V''(0)/m$, neglecting terms of third or higher order. For small oscillation amplitudes we thus have the differential equation of a free harmonic oscillator of angular frequency ω_0:

$$m\ddot{x} = -\frac{\mathrm{d}V}{\mathrm{d}x} \, , \quad \ddot{x} = -\omega^2 x \, ,$$

whose general solution is: $x(t) = x_0 \cdot \cos(\omega_0 t - \alpha)$, with arbitrary real quantities x_0 and α.

In close enough proximity to a parabolic local potential energy minimum, one always obtains a harmonic oscillation (whose frequency is given by $\omega_0 := \frac{V''(0)}{m}$).

If one now includes (i) a *frictional force*, $F_R := -\gamma v$, which can be characterized by a so-called "relaxation time" τ (i.e., $\gamma =: m \cdot 2/\tau$), and (ii) a *driving force* $F_A(t) = m \cdot f(t)$, then one obtains the ordinary differential equation

$$\ddot{x} + \frac{2}{\tau}\dot{x} + \omega_0^2 x = f(t) \, .$$

This is a *linear* ordinary differential equation of second order ($n \equiv 2$) with constant coefficients. For $f(t) \equiv 0$ this differential equation is *homogeneous*, otherwise it is called *inhomogeneous*. (For arbitrary $n = 1, 2, \dots$ the general inhomogeneous form is: $\left(\frac{\mathrm{d}^n}{\mathrm{d}^n t} + \sum_{\nu=0}^{n-1} a_\nu \frac{\mathrm{d}^\nu}{\mathrm{d}^\nu t} \right) x(t) = f(t)$).

For such differential equations, or for *linear* equations in general, the *principle of superposition applies: The sum of two solutions of the homogeneous equation, possibly weighted with real or complex coefficients, is also a solution of the homogeneous equation; the sum of a "particular solution" of the inhomogeneous equation plus an "arbitrary solution" of the homogeneous equation yields another solution of the inhomogeneous equation for the same inhomogeneity; the sum of two particular solutions of the inhomogeneous equation for different inhomogeneities yields a particular solution of the inhomogeneous differential equation, i.e., for the sum of the inhomogeneities.*

The general solution of the inhomogeneous differential equation is therefore obtained by adding a relevant *particular* solution of the inhomogeneous (i.e., *"driven"*) equation of motion to the *general* solution of the homogeneous equation, i.e., the general *"free oscillation"*.

As a consequence, in what follows we shall firstly treat a "general free oscillation", and afterwards the seemingly rather special, but actually quite general "periodically driven oscillation", and also the seemingly very special, but actually equally general so-called "ballistically driven oscillation".

The general solution of the equation for a free oscillation, i.e., the general solution of

$$\left(\frac{d^n}{d^n t} + \sum_{\nu=0}^{n-1} a_\nu \frac{d^\nu}{d^\nu t} \right) x(t) = 0 \quad \text{for} \quad n = 2 \,,$$

is obtained by linear combination of solutions of the form $x(t) \propto e^{\lambda \cdot t}$. After elementary calculations we obtain:

$$x(t) = \exp\left(-\frac{t}{\tau} \right) \cdot \left\{ x_0 \cos\left(\sqrt{\omega_0^2 - \frac{1}{\tau^2}} \cdot t \right) \right.$$

$$\left. + \left(v_0 + \frac{x_0}{\tau} \right) \cdot \frac{\sin\left(\sqrt{\omega_0^2 - \frac{1}{\tau^2}} \cdot t \right)}{\sqrt{\omega_0^2 - \frac{1}{\tau^2}}} \right\} . \tag{4.1}$$

This expression only looks daunting at first glance, until one realizes that the bracketed expression converges for $t \to 0$ to

$$x_0 + \left(v_0 + \frac{x_0}{\tau} \right) \cdot t \,,$$

as it must do.

Equation (4.1) applies not only for real

$$\varepsilon = \sqrt{\omega_0^2 - \frac{1}{\tau^2}}$$

but also for imaginary values, because in the limit $t \to 0$ not only

$$\frac{\sin(\varepsilon t)}{\varepsilon} \ ,$$

but also

$$\frac{\sin i\varepsilon t}{i\varepsilon} \to t \ ;$$

x_0 and v_0 are the initial position and the initial velocity, respectively. Thus the above-mentioned formula applies

a) not only for *damped oscillations*, i.e., for $\varepsilon > 0$, or

$$\omega_0^2 > \frac{1}{\tau^2} \ ,$$

i.e., sine or cosine oscillations with frequency

$$\omega_1 := \sqrt{\omega_0^2 - \frac{1}{\tau^2}}$$

and the damping factor $e^{-\lambda_1 t}$, with $\lambda_1 := \frac{1}{\tau}$,

b) but also for the *aperiodic case*,

$$\omega_0^2 < \frac{1}{\tau^2} \ ,$$

since for real

$$\varepsilon :$$

$$\frac{\sin(i\varepsilon \cdot t)}{i\varepsilon} \equiv \sinh \varepsilon t \ , \quad \text{with}$$

$$\sinh(x) := \frac{1}{2}(e^x - e^{-x}) \ , \quad \text{and}$$

$$\cos(i\varepsilon t) \equiv \cosh(\varepsilon t) \ , \quad \text{with}$$

$$\cosh(x) := \frac{1}{2}(e^x + e^{-x}) \ :$$

In the aperiodic case, therefore, an exponential behavior with two characteristic decay frequencies results ("relaxation frequencies"),

$$\lambda_\pm := \frac{1}{\tau} \pm \sqrt{\frac{1}{\tau^2} - \frac{1}{\omega_0^2}} \ .$$

Of these two relaxation frequencies the first is large, while the second is small.

c) Exactly in the *limiting case*, $\varepsilon \equiv 0$, the second expression on the r.h.s. of equation (4.1) is simply $(v_0 + \frac{x_0}{\tau}) \cdot t$ for all t, i.e., one finds the fastest decay.

Thus far we have only dealt with the *free* harmonic oscillator.

1) We now consider a *forced* harmonic oscillator, restricting ourselves at first to the simple *periodic* "driving force"

$$f(t) := f_A e^{i\omega_A t} .$$

However, due to the superposition principle this is no real restriction, since almost every driving force $f(t)$ can be written by Fourier integration as the sum or integral of such terms:

$$f(t) = \int_{-\infty}^{+\infty} d\omega_A \tilde{f}(\omega_A) e^{i\omega_A \cdot t} , \quad \text{with} \quad \tilde{f}(\omega_A) = (2\pi)^{-1} \int_{-\infty}^{+\infty} dt f(t) e^{i\omega_A \cdot t} .$$

For $x(t)$ in the above-mentioned case, after relaxation of a *transient process*, the following stationary solution results:

$$x(t) = x_A \cdot e^{i\omega_A t} ,$$

with (complex) amplitude

$$x_A = \frac{-f_A}{\omega_A^2 - \omega_0^2 + 2i\frac{\omega_A}{\tau}} . \tag{4.2}$$

If one now plots, for the case of weak friction, i.e., for $\tau\omega_0 \gg 1$, the fraction

$$\left| \frac{x_A}{f_A} \right|$$

as a function of the driving frequency ω_A, the following typical *amplitude resonance curve* is obtained:

$$\left| \frac{x_A}{f_A} \right| (\omega_A) = \frac{1}{\sqrt{(\omega_0^2 - \omega_A^2)^2 + \frac{4\omega_A^2}{\tau^2}}} . \tag{4.3}$$

This curve has a very sharp maximum (i.e., a *spike*) of height $\frac{\tau}{2\omega_0}$ at the resonance frequency, $\omega_A = \omega_0$; for *slight deviations* (positive or negative) from resonance, i.e., for

$$\omega_A \approx \omega_0 \pm \frac{1}{\tau} ,$$

the amplitude almost immediately becomes smaller by a factor of $\frac{1}{\sqrt{2}}$, compared to the maximum. For small frequencies, i.e., for $\omega_A \ll \omega_0$, the amplitude x_A is *in phase* with the driving force; for high frequencies, i.e., for $\omega_A \gg \omega_0$, they are of opposing phase (out of phase by 180° or π); at resonance the motion x_A is exactly 90° (or $\frac{\pi}{2}$) behind the driving force f_A. The transition from *in phase* to *opposite phase* behavior is very rapid, occurring in the narrow interval given by the values

$$\omega_A \approx \omega_0 \mp \frac{1}{\tau} \ .$$

The dimensionless ratio $\tau\omega_0 (\gg 1)$ is called the quality factor of the resonance. It can be of the order of 1.000 or 10.000, or even higher.

2) In the *ballistic case*, there is again no restriction. For a sequence of *ultra-short and ultra-strong* pulses $\sim \delta(t-t')$ with t' between $t' = t_0$ and $t' = t$, i.e., for the formal case

$$f(t) = \int\limits_{t_0}^{t} \mathrm{d}t' g(t')\delta(t-t') \ ,$$

with the *Dirac δ-function* $\delta(x)$ (a very high and very narrow 'bell-shaped function' of *height* $\propto \frac{1}{\varepsilon}(\to \infty)$ and *width* $\propto \varepsilon \ (\to 0)$, but where the integral should always $\overset{!}{=} 1$, e.g., $\delta(x) := \frac{e^{-x^2/(2\varepsilon^2)}}{\sqrt{2\pi\varepsilon^2}})$ we have quite generally:

$$x(t) = \int\limits_{t_0}^{t} \mathrm{d}t' g(t') \, \mathcal{G}(t-t') \ . \tag{4.4}$$

The so-called *Green's function* $\mathcal{G}(t-t')$ is thus identical with the response to the specific pulse $\delta(t-t')$, but actually it depends only on the system considered (not on the driving force itself), and the response agrees with the general principles of *causality* (i.e., $t' \le t$) and of *linearity* (i.e., the superposition principle)[1] analogous to (4.1); i.e., independent of $g(t')$, we have in the limit of $\varepsilon \to 0$:

$$\mathcal{G}(t-t') \equiv e^{-\frac{t-t'}{\tau}} \cdot \frac{\sin\left[(t-t')\sqrt{\omega_0^2 - \frac{1}{\tau^2}}\right]}{\sqrt{\omega_0^2 - \frac{1}{\tau^2}}} \ . \tag{4.5}$$

---

[1] Due to every pulse (at time $t = t'$) the velocity increases, $v(t'+\varepsilon) - v(t'-\varepsilon) \equiv g(t')$.

5 The Three Classical Conservation Laws; Two-particle Problems

We shall now deal with the three classical conservation laws in mechanics, *viz* for momentum, angular momentum, and energy. We shall formulate and prove them for N *point masses* at positions $\boldsymbol{r}_i$, for which

a) *internal forces* $\boldsymbol{F}_{ik}$ $(i, k = 1, \ldots, N)$ exist; these are the forces by which the k-th point mass acts on the i-th point mass. These internal forces are assumed to obey Newton's third axiom in its strong version, see above, i.e.,

$$\boldsymbol{F}_{i,k} = -\boldsymbol{F}_{k,i} \propto (\boldsymbol{r}_i - \boldsymbol{r}_k) \ .$$

This implies that the internal forces are not only mutually opposite to each other but also act in the direction of the vector joining the mutal positions, as applies to Newtonian gravitational forces and electrostatic Coulomb forces;

b) *external* forces $\boldsymbol{F}_i^{\text{ext}}$ are also assumed to act on every point mass ($i = 1, \ldots, N$).

Newton's equation for each point mass can therefore be written

$$m_i \dot{\boldsymbol{v}}_i = \left(\sum_{k=1}^{N} \boldsymbol{F}_{i,k} \right) + \boldsymbol{F}_i^{\text{ext}} \ . \tag{5.1}$$

Next we shall state a theorem on the time-derivative of the total momentum. (This is identical to a well-known theorem on the motion of the center of mass).

5.1 Theorem for the Total Momentum (or for the Motion of the Center of Mass)

Carrying out the summation in (5.1) for $i = 1, \ldots, N$ and using Newton's third axiom, since

$$\sum_{i,k=1}^{N} \boldsymbol{F}_{i,k} \equiv 0 \ ,$$

one obtains that the time-derivative of the *total momentum,*

$$\boldsymbol{P} := \sum_{i=1}^{N} m_i \boldsymbol{v}_i \ ,$$

is identical to the sum of (only) the *external* forces, whereas the *internal forces* compensate each other:

$$\dot{\boldsymbol{P}} = \sum_{i=1}^{N} \boldsymbol{F}_i^{\mathrm{ext}} \ . \tag{5.2}$$

At the same time, this becomes a well-known theorem on the motion of the center of mass, because it is easily shown that the total momentum $\boldsymbol{P}$ is identical with the expression $M\boldsymbol{v}_{\mathrm{s}}$, where

$$M \left(= \sum_i m_i \right)$$

is the total mass of the system considered, while

$$\boldsymbol{v}_{\mathrm{s}} := M^{-1} \sum_i m_i \boldsymbol{v}_i$$

is the velocity of the center of mass.

5.2 Theorem for the Total Angular Momentum

Similarly the time derivative of the total angular momentum $\boldsymbol{L}$ is equal to the sum of the external torques, where both quantities, $\boldsymbol{L}$ and the sum of the external torques, are related to the center of mass,

$$\boldsymbol{R}_{\mathrm{s}} := M^{-1} \sum_{i=1}^{N} m_i \boldsymbol{r}_i \ ,$$

even if this is in motion:

$$\dot{\boldsymbol{L}} = \sum_i (\boldsymbol{r}_i - \boldsymbol{R}_{\mathrm{s}}(t)) \times m_i \dot{\boldsymbol{v}}_i = \sum_{i=1}^{N} \boldsymbol{\mathcal{D}}_i^{\mathrm{ext}} \ . \tag{5.3}$$

Here $\boldsymbol{\mathcal{D}}_i^{\mathrm{ext}}$ is the external torque about the (resting or moving) center of mass, i.e., for gravitational forces, or for the forces of electrostatic monopoles (Coulomb forces):

$$\boldsymbol{\mathcal{D}}_i = (\boldsymbol{r}_i - \boldsymbol{R}_{\mathrm{s}}) \times \boldsymbol{F}_i^{\mathrm{ext}} \ .$$

Thus, the theorem states that under the above-mentioned conditions the time-derivative of the total angular momentum $\boldsymbol{L}$, i.e., related to the center of mass, is identical to the sum of external torques. We write consciously "*sum of external torques*", and *not* "sum of the torques generated by the external forces", because the external torques are not always identical to the torques of the external forces: A 'counter-example' is given in the case of *electric dipoles*, $\boldsymbol{p}_i$: Here the forces and torques, through which they interact, are complicated (see below), but even in this case Newton's third axiom applies in its strong sense,

$$\boldsymbol{F}_{i,k} = -\boldsymbol{F}_{k,i} \quad \text{plus} \quad \boldsymbol{\mathcal{D}}_{i,k} = -\boldsymbol{\mathcal{D}}_{k,i} \ .$$

To be specific: on the one hand one has for the force $\boldsymbol{F}_i$ on an electric dipole

$$\boldsymbol{F}_i = (\boldsymbol{p}_i \cdot \nabla) \left(\sum_{k(\neq i)} \boldsymbol{E}^{(k)}(\boldsymbol{r}_i) + \boldsymbol{E}^{\text{ext}} \right) ,$$

where $\boldsymbol{E}^{(k)}(\boldsymbol{r}_i)$ is the electric field generated by a dipole k at the position $\boldsymbol{r}_i$ and $\boldsymbol{E}^{\text{ext}}$ the external electric field. But on the other hand, for these forces, one does not have as usual

$$\boldsymbol{\mathcal{D}}_i = (\boldsymbol{r}_i - \boldsymbol{R}_{\text{s}}(t)) \times \left(\sum_{k(\neq i)} \boldsymbol{F}_{i,k} + \boldsymbol{F}_i^{\text{ext}} \right) ,$$

but instead

$$\boldsymbol{\mathcal{D}}_i = \boldsymbol{p}_i \times \left(\sum_{k(\neq i)} \boldsymbol{E}^{(k)}(\boldsymbol{r}_i) + \boldsymbol{E}^{\text{ext}} \right)$$

(see Part II of this volume, Sect. 17.2.6).

Even in this more complicated case the theorem on the time-derivative of the total angular momentum applies, basically because dipole forces and torques can be derived as limiting cases from monopole forces, i.e., by opposite point charges at slightly different places (as will be shown below, see Sect. 17.2.6).

What can be learnt from this?

a) The time-derivative of the *total linear momentum* $\boldsymbol{P}$ or *total angular momentum* $\boldsymbol{L}$ is equal to the sum of the *external* forces ($\sum_i \boldsymbol{F}_i^{\text{ext}}$) or the sum of the *external* torques ($\sum_i \boldsymbol{\mathcal{D}}_i^{\text{ext}}$), respectively. Due to Newton's third axiom the *internal* forces or torques compensate in a pairwise manner. These statements of course embody the principle of conservation of total momentum (or total angular momentum) in the special case that the sum of the external forces (or torques) vanish. In any case, Newton's third axiom includes both forces and torques: it includes both conservation theorems separately (i.e., the second theorem is not just a consequence of the first one, or *vice versa*).

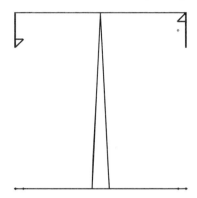

Fig. 5.1. A lever with two opposite forces producing a torque

b) Sometimes a good question can decisively help our understanding. For example the following question aids in understanding the separation of forces and torques in connection with Newton's third axiom: *Does the sum of the internal torques vanish if the sum of the internal forces vanishes?* The answer is of course "No". A simple drawing helps. Consider the special case of a horizontal *lever*, where two opposite external forces are applied at the ends, the first one upwards to the r.h.s. and the second one downwards to the l.h.s. of the lever, generating a torque, as shown in Fig. 5.1.

In this example the sum of the forces vanishes, but the sum of the torques does not vanish as an automatic consequence, unless the forces have the direction of the vector $\boldsymbol{r}_i - \boldsymbol{r}_k$.

We shall now consider the energy theorem.

5.3 The Energy Theorem; Conservative Forces

In this case too we shall be somewhat more general, i.e., we will not simply derive the *energy conservation law*, but look in detail at the necessary prerequisites. For this purpose, we shall consider the totality of the forces, $\boldsymbol{F}_i$, acting on N point masses, i.e., with $3N$ components. These correspond to $3N$ coordinates x_α, $\alpha = 1, \ldots, 3N$, for the N point masses. For the components F_α we assume

$$F_\alpha = -\frac{\partial \mathcal{V}(x_1, \ldots, x_{3N}, t)}{\partial x_\alpha} - v_\alpha \cdot f_R^{(\alpha)}(x_1, \ldots, v_1, \ldots, t) + \sum_{\beta, \lambda} \varepsilon_{\alpha, \beta, \lambda} v_\beta C_\lambda \;,$$

$$(5.4)$$

with non-negative frictional forces $f_R^{(\alpha)}(x_1, \ldots, v_1, \ldots, t)$.

In the first term on the r.h.s. of (5.4), $\mathcal{V}(x_1, \ldots, x_{3N}, t)$ describes a *potential energy*, which we assume to exist. This is a strong assumption (see below) and typically it includes the most important contribution to the considered forces. (We stress here that the existence of $\mathcal{V}$ is far from being self-evident).

In contrast, the second term on the r.h.s. describes *frictional forces*, which are often less important, and the third term, which looks rather formal, but is very strong in certain cases, (with the vector C_λ ($\lambda = 1, 2, 3$) and an antisymmetric tensor $\varepsilon_{\alpha,\beta,\lambda} = -\varepsilon_{\alpha,\lambda,\beta}$) describes some kind of *Lorentz force* or *Coriolis force*, i.e.,

$$F_{\text{Lorentz}} = q v \times B \ ,$$

with electric charge q and magnetic induction B, or

$$F_{\text{Coriolis}} = -2 m_{\text{t}} \omega \times v \ ,$$

where ω is the rotation of the coordinate system and v the velocity in this system. We thus have $C = qB$ or $C = -2 m_{\text{t}} \omega$, respectively. We shall return to the Coriolis force below, see Sect. 14.2, in connection with so-called "fictitious (or inertial) forces", which are always multiplied by a factor m_{t}.

A characteristic property of the Lorentz and Coriolis forces is that, in contrast to frictional forces, they *do no work*, *viz* in the following sense:

$$\delta A_B := \mathrm{d} r \cdot [v \times B] \equiv 0 \ ,$$

as $\mathrm{d} r = v \, \mathrm{d} t$. In fact one obtains from Newton's equation (5.4), multiplying with $v_\alpha = \dot{x}_\alpha$,

$$\frac{\mathrm{d}}{\mathrm{d} t} \sum_{\alpha=1}^{3N} \frac{m_\alpha v_\alpha^2}{2} = -\frac{\mathrm{d} \mathcal{V}}{\mathrm{d} t} + \frac{\partial \mathcal{V}}{\partial t} - \sum_\alpha v_\alpha^2 f_R^{(\alpha)} \ . \tag{5.5}$$

Here we have used the fact that

$$\frac{\mathrm{d} \mathcal{V}(x_\alpha, t)}{\mathrm{d} t} = \dot{x}_\alpha \frac{\partial \mathcal{V}}{\partial x_\alpha} + \frac{\partial \mathcal{V}}{\partial t} \ .$$

As a consequence we obtain the following *energy theorem*

$$\frac{\mathrm{d}}{\mathrm{d} t} \left(\sum_{\alpha=1}^{3N} \frac{m_\alpha v_\alpha^2}{2} + \mathcal{V}(x_1, \dots, x_{3N}, t) \right) = \frac{\partial \mathcal{V}}{\partial t} - \sum_{\alpha=1}^{3N} v_\alpha^2 f_R^{(\alpha)} \ . \tag{5.6}$$

Thus the total mechanical energy, i.e., the sum of the kinetic energy $\mathcal{T}$ and potential energy $\mathcal{V}$, can change, but only as follows:

a) due to *frictional forces* the total mechanical energy always *decreases*,
b) whereas Lorentz and Coriolis forces alone do not influence the energy although they do influence the motion itself,
c) while the total mechanical energy can *increase* or *decrease* according to the sign of the *explicit* change with time, $\frac{\partial \mathcal{V}}{\partial t}$, of the potential energy $\mathcal{V}$.

For example, if a load attached to the end of a rope, which is oscillating transversally, is continuously *pulled upwards* by an external force during the motion, then the potential energy of the load increases with time, and this has consequences for the oscillation.

We shall return later to the possibility of enhancing the energy of an oscillation by systematically influencing its potential energy (*parametric amplification, parametric resonance*, building up of the amplitude).

Finally some conceptual remarks should be made. Forces depending explicitly on *velocity* and/or *time* are called non-conservative forces; therefore, both *Lorentz forces* and *frictional forces* are *non-conservative*, although the Lorentz forces do not "hamper" the conservation of energy. On the other hand, for so-called *conservative forces*, one requires that they can be derived solely from a potential energy $\mathcal{V}$, where $\mathcal{V}$ should be a function of position only.

In short, for conservative forces a function $\mathcal{V}(x_1, \ldots, x_{3N})$ is assumed to exist such that for $\alpha = 1, \ldots, 3N$:

$$F_\alpha \equiv -\frac{\partial}{\partial x_\alpha}\mathcal{V}(x_1, \ldots, x_{3N}) . \tag{5.7}$$

Thus for conservative forces (but also for a sum of conservative forces plus Lorentz and Coriolis forces) the *energy theorem* can be simplified to the well-known *theorem on the conservation of mechanical energy* $\mathcal{T} + \mathcal{V}$.

The postulate that a potential energy should exist, implies that the $3N \cdot (3N-1)/2$ conditions

$$\frac{\partial F_\alpha}{\partial x_\beta} = \frac{\partial F_\beta}{\partial x_\alpha}$$

are satisfied (so-called *generalized irrotational behavior*).

This criterion results from the equality of the mixed derivatives of $\mathcal{V}$, i.e.,

$$\frac{\partial \mathcal{V}}{\partial x_i \partial x_j} \equiv \frac{\partial \mathcal{V}}{\partial x_j \partial x_i} ,$$

and it can be checked in finite time, whereas the other criteria mentioned in this context, e.g., the criterion that for every closed loop $\mathcal{W}$ the work-integral

$$\oint_{\mathcal{W}} \sum_{\alpha=1}^{N} F_\alpha \cdot \mathrm{d}x_\alpha$$

should vanish, *cannot* be checked directly in finite time, although it follows from the above-mentioned *generalized irrotational behavior*.

In any case, if the above-mentioned $3N \cdot (3N-1)/2$ generalized irrotational conditions are satisfied and no explicit time dependence arises, one can define the potential energy $\mathcal{V}$ (apart from an arbitrary constant) by the following expression:

$$\mathcal{V}(x_1, \ldots, x_{3N}) = -\int_{r_0}^{r} \sum_{\alpha=1}^{3N} \mathrm{d}x_\alpha \cdot F_\alpha(x_1, \ldots, x_{3N}) . \tag{5.8}$$

Here $\boldsymbol{r} := (x_1, \ldots, x_{3N})$; $\boldsymbol{r}_0$ is an arbitrary starting point, and the integration path from $\boldsymbol{r}_0$ to $\boldsymbol{r}$ is also arbitrary.

(With regard to the mathematics: on the r.h.s. of (5.8), under the integral, we have a differential form of the first kind, a so-called "Pfaff form", which neither depends on $\boldsymbol{v}$ nor t. We *assume* that this differential form is *total*, i.e., that it can be derived from a uniquely defined function $-\mathcal{V}$. This requires additionally that the boundary of the considered region does not consist of separate parts.)

5.4 The Two-particle Problem

With regard to the Center of Mass theorem one should mention that so-called two-particle problems can always be solved analytically by separating the motion into that of the *center of mass* and the corresponding *relative motion*:

Consider for $i = 1, 2$ two coupled vectorial equations of motion of the form

$$m_i \ddot{\boldsymbol{r}}_i = \boldsymbol{F}_i^{\text{ext}} + \boldsymbol{F}_{i,k}^{\text{int}} \,,$$

where k is complementary to i, i.e.,

$$\boldsymbol{F}_{1,2}^{\text{int}} \quad \text{and} \quad \boldsymbol{F}_{2,1}^{\text{int}} = -\boldsymbol{F}_{1,2}^{\text{int}} \,.$$

Adding the equations of motion one obtains

$$M \dot{\boldsymbol{v}}_{\text{s}} = (\boldsymbol{F}_1^{\text{ext}} + \boldsymbol{F}_2^{\text{ext}})$$

($\to$ motion of the center of mass). Similarly a weighted *difference*,

$$\frac{\boldsymbol{F}_{1,2}^{\text{int}}}{m_1} - \frac{\boldsymbol{F}_{2,1}^{\text{int}}}{m_2} \,,$$

yields

$$\frac{\mathrm{d}^2}{\mathrm{d}t^2} (\boldsymbol{r}_1 - \boldsymbol{r}_2) = \frac{1}{m_{\text{red}}} \boldsymbol{F}_{1,2}^{\text{int}}$$

($\to$ *relative motion*). Here we assume that

$$\frac{\boldsymbol{F}_1^{\text{ext}}}{m_1} - \frac{\boldsymbol{F}_2^{\text{ext}}}{m_2}$$

vanishes (no external forces at all).

This means that in the Newtonian equations for the relative quantity $\boldsymbol{r} := \boldsymbol{r}_1 - \boldsymbol{r}_2$ not only the internal force $\boldsymbol{F}_{1,2}^{int}$ has to be used, but also the so-called reduced mass m_{red},

$$\frac{1}{m_{\text{red}}} := \frac{1}{m_1} + \frac{1}{m_2} \,.$$

6 Motion in a Central Force Field; Kepler's Problem

Kepler's problem concerning the motion of the planets in our solar system is one of the main problems in Classical Mechanics, similar to that of the *hydrogen atom* in Quantum Mechanics. Incidentally they are both closely related (A typical examination question might read: *"In what respect are these problems and their solutions analogous or different?"*).

6.1 Equations of Motion in Planar Polar Coordinates

We have

$$\boldsymbol{r}(t) = r(t) \cdot (\cos\varphi(t), \sin\varphi(t)) \equiv r(t)\boldsymbol{e}_r(t) \ ,$$

$$\frac{\mathrm{d}\boldsymbol{r}}{\mathrm{d}t} = \boldsymbol{v} = \dot{r}(t)\boldsymbol{e}_r + r(t)\dot{\boldsymbol{e}}_r = \dot{r}(t)\boldsymbol{e}_r(t) + r(t)\dot{\varphi}(t)\boldsymbol{e}_\varphi(t) \ .$$

Thereby $\boldsymbol{e}_r = (\cos\varphi(t), \sin\varphi(t))$, and

$$\boldsymbol{e}_\varphi(t) := (-\sin\varphi(t), \cos\varphi(t)) \ .$$

But $\dot{\boldsymbol{e}}_\varphi = -\dot{\varphi}\boldsymbol{e}_r$.

As a consequence, one obtains for the acceleration:

$$\boldsymbol{a}(t) = \frac{\mathrm{d}^2\boldsymbol{r}}{\mathrm{d}t^2} = \{\ddot{r}(t) - r(t)\dot{\varphi}(t)^2\}\boldsymbol{e}_r(t) + \{2\dot{r}\dot{\varphi} + r\ddot{\varphi}\}\,\boldsymbol{e}_\varphi(t) \ . \tag{6.1}$$

For a purely radial force, $\boldsymbol{F} = F_r(r,\varphi)\boldsymbol{e}_r$, we thus have $F_\varphi \equiv 0$; therefore

$$2\dot{r}\dot{\varphi} + r\ddot{\varphi} \equiv \frac{1}{r}\frac{\mathrm{d}(r^2\dot{\varphi})}{\mathrm{d}t} \equiv 0 \ ,$$

or

$$\frac{1}{2}r^2\dot{\varphi} = \text{constant} \ .$$

This is Kepler's second law, or "law of equal areas", since

$$\frac{1}{2}r^2\dot{\varphi}\,\mathrm{d}t$$

is the triangular area covered by the vector from the center of the sun to the planet in the time interval $\mathrm{d}t$.

At this point we should remind ourselves of Kepler's three laws.

6.2 Kepler's Three Laws of Planetary Motion

They are:

1) The planets orbit the central star, e.g., the sun, on an *elliptical path*, where the sun is at one of the two *foci* of the ellipse.
2) The vector from the center of the sun to the planet covers equal areas in equal time intervals.
3) The ratio T^2/a^3, where T is the time period and a the major principal axis of the ellipse, is *constant* for all planets (of the solar system). (In his famous interpretation of the motion of the *moon* as a planet orbiting the earth, i.e., the earth was considered as the "central star", Newton concluded that this *constant* parameter is not just a universal number, but proportional to the mass M of the respective central star.)

As already mentioned, Kepler's second law is also known as the *law of equal areas* and is equivalent to the angular momentum theorem for relative motion, because (for relative motion[1])

$$\frac{\mathrm{d}\boldsymbol{L}}{\mathrm{d}t} = \boldsymbol{r} \times \boldsymbol{F} \ ,$$

i.e., $\equiv 0$ for central forces, i.e., if $\boldsymbol{F} \sim \boldsymbol{r}$. In fact, we have

$$\boldsymbol{L} = \boldsymbol{r} \times \boldsymbol{p} = m \cdot r^2 \dot{\varphi} \boldsymbol{e}_z \ .$$

Here m is the *reduced mass* appearing in Newton's equation for the relative motion of a "two-particle system" (such as *planet–sun*, where the other planets are neglected); this *reduced mass*,

$$\frac{m}{1 + \frac{m}{M}} \ ,$$

is practically identical to the mass of the planet, since $m \ll M$.

The complete law of gravitation follows from Kepler's laws by further *analysis* which was first performed by Newton himself. The gravitational force $\boldsymbol{F}$, which a point mass M at position $\boldsymbol{R}_M$ exerts on another point mass m at $\boldsymbol{r}$ is given by:

$$\boldsymbol{F}(\boldsymbol{r}) \equiv -\gamma \frac{(m \cdot M) \cdot (\boldsymbol{r} - \boldsymbol{R}_M)}{|\boldsymbol{r} - \boldsymbol{R}_M|^3} \ . \tag{6.2}$$

The gravitational force, which acts in the direction of the line joining $\boldsymbol{r}$ and $\boldsymbol{R}_\mathrm{M}$, is (i) *attractive* (since the gravitional constant γ is > 0), (ii) $\propto m \cdot M$, and (iii) (as Coulomb's law in electromagnetism) inversely proportional to the square of the separation.

---

[1] We do not write down the many sub-indices $_{\mathrm{rel.}}$, which we should use in principle.

As has already been mentioned, the *principle of superposition* applies to Newton's law of gravitation with regard to summation or integration over M, i.e., Newton's theory of gravity, in contrast to *Einstein's general theory of relativity* (which contains Newton's theory as a limiting case) is *linear* with respect to the sources of the gravitational field.

Newton's systematic *analysis* of Kepler's laws (leading him to the important idea of a central gravitational force) follows below; but firstly we shall discuss the reverse path, the derivation (*synthesis*) of Kepler's laws from Newton's law of gravitation, (6.2). This *timeless* achievement of Newtonian theory was accomplished by using the newly developed (also by Newton himself[2]) mathematical tools of differential and integral calculus.

For 200 years, Newton was henceforth the ultimate authority, which makes Einstein's accomplishments look even greater (see below).

6.3 Newtonian Synthesis: From Newton's Theory of Gravitation to Kepler

Since in a central field the force possesses only a radial component, F_r, (here depending only on r, but not on φ), we just need the equation

$$m \cdot \left(\ddot{r} - r\dot{\varphi}^2 \right) = F_r(r) \ .$$

The force is trivially conservative, i.e.,

$$\boldsymbol{F} = -\mathrm{grad}V(\boldsymbol{r}) \ ,$$

with potential energy

$$V(\boldsymbol{r}) \equiv - \int_{\infty}^{r} \mathrm{d}\tilde{r} F_r(\tilde{r}) \ .$$

Thus we have *conservation of the energy*:

$$\frac{m}{2} \cdot \left(\dot{r}^2 + r^2 \dot{\varphi}^2 \right) + V(r) = E \ . \tag{6.3}$$

Further, with the conservation law for the *angular momentum* we can eliminate the variable $\dot{\varphi}$ and obtain

$$\frac{m}{2} \cdot \dot{r}^2 + \frac{L^2}{2mr^2} + V(r) = E \ . \tag{6.4}$$

Here we have used the fact that the square of the angular momentum

$$(\boldsymbol{L} = \boldsymbol{r} \times \boldsymbol{p})$$

---

[2] *Calculus* was also invented independently by the *universal genius Wilhelm Leibniz*, a philosopher from Hanover, who did not, however, engage in physics.

is given by[3] the following relation:

$$L^2 = (mr^2\dot\varphi)^2 \ .$$

Equation (6.4) corresponds to a one-dimensional motion with an *effective potential energy*

$$V_{\text{eff}}(r) := V(r) + \frac{L^2}{2mr^2} \ .$$

The one-dimensional equation can be solved using the above method based on energy conservation:

$$t - t_0 = \int_{r_0}^{r} \frac{\mathrm{d}\tilde r}{\sqrt{\frac{2}{m}(E - V_{\text{eff}}(\tilde r))}} \ .$$

Similarly we obtain from the conservation of angular momentum:

$$t - t_0 = \frac{m}{L} \int_{\varphi_0}^{\varphi} r^2(\tilde\varphi) \cdot \mathrm{d}\tilde\varphi \ .$$

Substituting

$$\mathrm{d}t = \left(\frac{\mathrm{d}\tilde\varphi}{\mathrm{d}t}\right)^{-1} \cdot \mathrm{d}\tilde\varphi$$

we obtain:

$$\varphi(r) = \varphi_0 + \frac{L}{m} \int_{r_0}^{r} \frac{\mathrm{d}\tilde r}{\tilde r^2 \sqrt{\frac{2}{m}(E - V_{\text{eff}}(\tilde r))}} \ . \tag{6.5}$$

All these results apply quite generally. In particular we have used the fact that the distance r depends (via t) uniquely on the angle φ, and *vice versa*, at least if the motion starts (with $\varphi = \varphi_0 = 0$) at the point closest to the central star, the so-called *perihelion*, and ends at the point farthest away, the so-called *aphelion*.

6.4 Perihelion Rotation

What value of φ is obtained at the *aphelion*? It is far from being trivial (see below) that this angle is exactly π, so that the planet returns to the *perihelion* exactly after 2π. In fact, this is (almost) only true for *Kepler potentials*, i.e., for $V = -A/r$, where A is a constant[4], whereas (6.4) applies for more general potentials that only depend on r. If these potentials deviate slightly from

---

[3] In quantum mechanics we have $L^2 \to \hbar^2 l \cdot (l+1)$, see Part III.
[4] We write "almost", because the statement is also true for potentials $\propto r$.

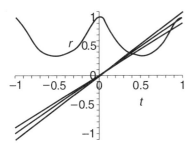

Fig. 6.1. Perihelion rotation. The function $r(t) = 0.5/(1+0.5\cos(2\pi t))$ is plotted together with the three straight lines: $\varphi(t) = (2\pi)*(1+\varepsilon)*t$, with $\varepsilon = -0.1$, 0.0 and $+0.1$. The three corresponding orbits $r(\phi)$ yield a closed ellipse only for $\varepsilon \equiv 0$; in the other two cases one obtains so-called *rosette orbits*, see the following figure

the Kepler potential, one actually observes the phenomenon of *perihelion rotation*, i.e., the aphelion position is not obtained for $\varphi = \pi$, but later (or earlier), *viz* for

$$\varphi = \pi \pm \frac{1}{2}\Delta\varphi\,,$$

and the planet returns to the *perihelion* distance only at an angle deviating from 2π, *viz* at $2\pi \pm \Delta\varphi$, see Fig. 6.1 below.

Such a *perihelion rotation* is actually observed, primarily for the planet *Mercury* which is closest to the sun. The reasons, all of them leading to tiny, but measurable deviations from the -A/r-potential, are manifold, for example

- perturbations by the other planets and/or their moons can be significant,
- also deviations from the exact spherical shape of the central star may be important,
- finally there are the *general relativistic effects* predicted by Einstein, which have of course a revolutionary influence on our concept of space and time. (Lest we forget, this even indirectly became a political issue during the dark era of the Nazi regime in Germany during the 1930s.)

How *perihelion rotation* comes about is explained in Figs. 6.1 and 6.2.

In the following section we shall perform an *analysis* of Kepler's laws, analogously to *Newtonian analysis*, in order to obtain the laws of gravitation.

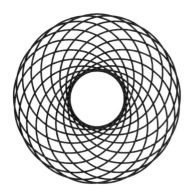

Fig. 6.2. Rosette orbits. If $r(t)$ and $\varphi(t)$ have different periods (here $r(t) = 0.5/(1 + 0.5\cos(2\pi t))$ but $\varphi(t) = 1.9\pi \cdot t$), one obtains the *rosette orbit* shown. It corresponds exactly to a potential energy of the non-Keplerian form $V(r) = -A/r - B/r^2$, and although looking more complicated, it consists of only one continuous line represented by the above function

6.5 Newtonian Analysis: From Kepler's Laws to Newtonian Gravitation

As mentioned above, Newton used a rather long but systematic route to obtain his law of gravitation,

$$\boldsymbol{F}(\boldsymbol{r}) = -\gamma \frac{\tilde{m}M}{r^2} \frac{\boldsymbol{r}}{r} \ ,$$

from Kepler's laws.

6.5.1 Newtonian Analysis I: Law of Force from Given Orbits

If the *orbits* are of the form $\frac{1}{r} = f(\varphi)$, then one obtains by straightforward differentiation:

$$-\frac{\dot{r}}{r^2} = \dot{\varphi} \cdot \frac{\mathrm{d}}{\mathrm{d}\varphi} f(\varphi) = \left(\frac{L}{mr^2}\right) \cdot \frac{\mathrm{d}f}{\mathrm{d}\varphi} \ , \quad \text{or}$$

$$\dot{r} = -\frac{L}{m} \frac{\mathrm{d}f}{\mathrm{d}\varphi} \ , \quad \text{i.e.,}$$

$$\ddot{r} = -\dot{\varphi} \frac{L}{m} \frac{\mathrm{d}^2 f}{\mathrm{d}\varphi^2} = -\frac{L^2}{m^2 r^2} \frac{\mathrm{d}^2 f}{\mathrm{d}\varphi^2} \ ,$$

or finally the law of force:

$$\frac{F_r}{m} \equiv \ddot{r} - r\dot{\varphi}^2 = -\frac{L^2}{m^2 r^2} \left(\frac{\mathrm{d}^2 f}{\mathrm{d}\varphi^2} + \frac{1}{r}\right) \ . \tag{6.6}$$

Equation (6.6) will be used later.

6.5.2 Newtonian Analysis II: From the String Loop Construction of an Ellipse to the Law $F_r = -A/r^2$

Reminding ourselves of the elementary method for drawing an ellipse using a loop of string, we can translate this into the mathematical expression

$$r + r' = r + \sqrt{r^2 + (2a)^2 - 2r \cdot 2e \cdot \cos \varphi} \overset{!}{=} 2a \ .$$

Here $2a$ is the length of the *major axis* of the ellipse (which extends from $x = -a$ to $x = +a$ for $y \equiv 0$); r and r' are the distances from the two *foci* (the ends of the loop of string) which are situated at $x = \pm e$ on the *major axis*, the x-axis, and φ is the *azimuthal angle*, as measured e.g., from the left *focus*.[5] From this we obtain the *parametric representation* of the ellipse, which was already well-known to Newton:

$$r = \frac{p}{1 - \varepsilon \cos \varphi} \ ,$$

or

$$\frac{1}{r} = \frac{1}{p} \cdot (1 - \varepsilon \cos \varphi) \ . \tag{6.7}$$

Here $a^2 - e^2 =: b^2$, and $\frac{b^2}{a} = p$; b is the length of the *minor semiaxis* of the ellipse. The parameter p is the distance from the left focus to the point of the ellipse corresponding to the azimuthal angle $\varphi = \frac{\pi}{2}$, and

$$\varepsilon := \frac{a^2 - b^2}{a^2}$$

is the *ellipticity*, $0 \leq \varepsilon < 1$.

As a result of the above relations we already have the following inverse-square law of force:

$$F_r(r) = \frac{-A}{r^2} \ , \quad \text{where} \quad A > 0$$

(attractive interaction), while the parameter p and the angular momentum L are related to A by:

$$p = \frac{L^2}{A \cdot m} \ .$$

6.5.3 Hyperbolas; Comets

Equation (6.7) also applies where $\varepsilon \geq 1$. In this case the orbits are no longer *ellipses* (or circles, as a limiting case), but *hyperbolas* (or parabolas, as a limiting case).[6]

Hyperbolic orbits in the solar system apply to the case of *nonreturning comets*, where the sun is the central point of the hyperbola, i.e., the *perihelion* exists, but the *aphelion* is replaced by the limit $r \to \infty$. For repulsive interactions, $A < 0$, one would only have hyperbolas.

---

[5] Here we recommend that the reader makes a *sketch*.

[6] For the hydrogen atom the quantum mechanical case of *continuum states at E > 0* corresponds to the *hyperbolas* of the Newtonian theory, whereas the *ellipses* in that theory correspond to the *bound states* of the quantum mechanical problem; see Part III of this volume.

6.5.4 Newtonian Analysis III: Kepler's Third Law and Newton's Third Axiom

Up till now we have not used Kepler's third law; but have already derived an attractive force of the correct form:

$$F_r(r) = -\frac{A}{r^2} \ .$$

We shall now add Kepler's third law, starting with the so-called *area velocity* V_F, see below. Due to the lack of any *perihelion rotation*, as noted above, Newton at first concluded from Kepler's laws that the *"time for a round trip"* T must fulfil the equation

$$T = \frac{\pi a \cdot b}{V_F} \ ,$$

since the expression in the numerator is the area of the ellipse; hence

$$T^2 = \frac{\pi^2 a^2 b^2}{V_F^2} = \frac{\pi^2 a^3 p}{V_F^2} \ .$$

However, according to Kepler's third law we have

$$\frac{T^2}{a^3} = \frac{\pi^2 p}{V_F^2} = \frac{\pi^2 p}{L^2/(4m^2)} = \frac{4\pi^2 p}{L^2} m^2 = C \ ,$$

i.e., this quantity must be the same for all planets of the planetary system considered. The parameter A appearing in the force

$$F_r(r) = -\frac{A}{r^2} \ , \quad \text{i.e.,} \quad A = \frac{L^2}{p^2 m} \ ,$$

is therefore given by the relation:

$$F_r(r) = -\frac{A}{r^2} = -\frac{4\pi^2}{C} \cdot \frac{m}{r^2} \ ,$$

i.e., it is proportional to the mass m of the planet.

In view of the principle of action and reaction being equal in magnitude and opposite in direction, Newton concluded that the prefactor

$$\frac{4\pi^2}{C}$$

should be proportional to the mass M of the central star,

$$\frac{4\pi^2}{C} = \gamma \cdot M \ , \quad \text{where} \quad \gamma$$

is the gravitational constant.

By systematic analysis of Kepler's three laws Newton was thus able to derive his general gravitational law from his three axioms under the implicit proposition of a fixed Euclidean (or preferably Galilean) space-time structure.

It is obvious that an inverse approach would also be possible; i.e., for given gravitational force, Kepler's laws follow from Newton's equations of motion. This has the didactic virtue, again as mentioned above, that the (approximate) nonexistence of any *perihelion rotation*, which otherwise would be easily overlooked, or erroneously taken as self-evident, is now explicitly recognized as *exceptional*[7].

We omit at this point any additional calculations that would be necessary to perform the above task. In fact this so-called *synthesis* of Kepler's laws from Newton's equations can be found in most of the relevant textbooks; it is essentially a systematic exercise in *integral calculus*.

For the purposes of school physics many of these calculations may be simplified, for example, by replacing the ellipses by circles and making systematic use of the compensation of *gravitational forces* and so-called *centrifugal forces*. However, we shall refrain from going into further details here.

6.6 The Runge-Lenz Vector as an Additional Conserved Quantity

The so-called *Runge-Lenz vector*, $\mathcal{L}_e$, is an *additional* conserved quantity, independent of the usual three conservation laws for energy, angular momentum, and linear momentum for a planetary system. The additional conservation law only applies for potentials of the form $\mp A/r$ (as well as $\mp A \cdot r$ potentials), corresponding to the fact that for these potentials the orbits are *ellipses*, i.e., they close exactly, in contrast to *rosettes*, see above.

The Runge-Lenz vector is given by

$$\mathcal{L}_e := \frac{\boldsymbol{v} \times \boldsymbol{L}}{A} - \boldsymbol{e}_r \,, \quad \text{where} \quad \boldsymbol{L}$$

is the angular momentum. It is not difficult to show that $\mathcal{L}_e$ is conserved:

$$\frac{\mathrm{d}}{\mathrm{d}t}(\boldsymbol{v} \times \boldsymbol{L}) = \boldsymbol{a} \times \boldsymbol{L} = \boldsymbol{a} \times \left(mr^2\dot{\varphi}\boldsymbol{e}_z\right)$$

$$= -\frac{A}{r^2}\boldsymbol{e}_r \cdot r^2\dot{\varphi} \times \boldsymbol{e}_z$$

$$= -A\dot{\varphi}\boldsymbol{e}_r \times \boldsymbol{e}_z = A\dot{\varphi}\boldsymbol{e}_\varphi = A\dot{\boldsymbol{e}}_r$$

---

[7] The (not explicitly stated) *non*-existence of any *perihelion rotation* in Kepler's laws corresponds quantum mechanically to the (seemingly) *incidental* degeneracy of orthogonal energy eigenstates $\psi_{n,l}$ of the hydrogen atom, see Part III, i.e., states with the *same* value of the main quantum number n but *different* angular quantum numbers l.

i.e., $\dot{\mathcal{L}}_e = 0$. The geometrical meaning of $\mathcal{L}_e$ is seen from the identity

$$\mathcal{L}_e \equiv \frac{e}{a} \,, \quad \text{where} \quad 2e$$

is the vector joining the two *foci* of the ellipse, and a is the length of its principal axis: $a\mathcal{L}_e$ is thus equal to e. This can be shown as follows.

From a string loop construction of the ellipse we have

$$r + r' = r + \sqrt{(2e - r)^2} = 2a \,,$$

i.e.,

$$r^2 + 4e^2 - 2e \cdot r = (2a - r)^2 = r^2 - 4ar + 4a^2 \,,$$

hence on the one hand

$$r = \frac{a^2 - e^2}{a} - \frac{e \cdot r}{a} \equiv p - \frac{e}{a} \cdot r \,.$$

On the other hand we have

$$r \cdot \mathcal{L}_e = r \cdot \frac{v \times L}{A} - r \cdot e_r = \frac{[r \times v] \cdot L}{A} - r = \frac{L \cdot L}{mA} - r = p - r \,,$$

hence

$$r \equiv p - \mathcal{L}_e \cdot r \,.$$

We thus have

$$\mathcal{L}_e = \frac{e}{a} \,,$$

as stated above.

7 The Rutherford Scattering Cross-section

We assume in the following that we are dealing with a radially symmetric potential energy $\mathcal{V}(r)$, see also section 6.3, which is either attractive (as in the preceding subsections) or repulsive.

Consider a *projectile* (e.g., a comet) approaching a *target* (e.g, the sun) which, without restriction of generality, is a point at the origin of coordinates. The projectile approaches from infinity with an initial velocity v_∞ parallel to the x-axis with a *perpendicular* distance b from this axis. The quantity b is called the *impact parameter*[1].

Under the influence of $\mathcal{V}(r)$ the projectile will be deflected from its original path. The *scattering angle* ϑ describing this deflection can be calculated by (6.5) in Sect. 6.3, where $\varphi(r)$ describes the path with

$$\varphi(-\infty) = 0 , \quad \text{and} \quad \frac{\vartheta}{2} = \varphi_0 \quad \text{while} \quad \vartheta = \varphi(+\infty) ;$$

r_0 is the shortest distance from the target. It corresponds to the *perihelion point* $\boldsymbol{r}_0$.

The main problem in evaluating ϑ from (6.5) is the calculation of $\boldsymbol{r}_0$. For this purpose we shall use the conservation of angular momentum L and energy E. Firstly we may write

$$L = m \cdot b \cdot v_\infty \equiv m \cdot r_0 \cdot v_0 , \quad \text{where} \quad v_0$$

is the velocity at the perihelion. In addition, since the potential energy vanishes at infinity, conservation of energy implies:

$$V(r_0) = \frac{m}{2} v_\infty^2 - \frac{m}{2} v_0^2 ,$$

where the second term can be expressed in terms of L and r_0. In this way the *perihelion* can be determined together with the corresponding orbit $r(\vartheta)$ and the scattering angle $\vartheta_\infty = \vartheta(r \to \infty)$. As a consequence, there is a unique relation between the impact parameter p and the scattering angle ϑ. Furthermore we define an "element of area"

$$\mathrm{d}^{(2)}\sigma := 2\pi b \,\mathrm{d}b \equiv 2\pi b(\vartheta) \frac{\mathrm{d}b}{2\pi \sin \vartheta \mathrm{d}\vartheta} \cdot \mathrm{d}\Omega ,$$

---

[1] a sketch is recommended (this task is purposely left to the reader), but see Fig. 7.1.

where we note the fact that a solid-angle element in spherical coordinates can be written as

$$d\Omega = 2\pi \sin \vartheta d\vartheta .$$

The *differential cross-section* is defined as the ratio $\frac{d^{(2)}\sigma}{d\Omega}$:

$$\frac{d^{(2)}\sigma}{d\Omega} = \frac{b(\vartheta)\, db(\vartheta)}{\sin \vartheta\, d\vartheta} . \qquad (7.1)$$

This expression is often complicated, but its meaning can be visualized, as follows. Consider a stream of particles with current density j_0 per cross-sectional area flowing towards the target and being scattered by the potential $\mathcal{V}$. At a large distance beyond the target, a fraction of the particles enters a counter, where they are recorded. The number of counts in a time Δt is given by

$$\Delta N = \frac{d^{(2)}\sigma}{d\Omega} \cdot j_0 \cdot \Delta t \cdot \Delta\Omega .$$

The aperture of the counter corresponds to scattering angles in the interval $(\vartheta, \vartheta + d\vartheta)$, i.e., to the corresponding solid angle element

$$\Delta\Omega := 2\pi \sin \vartheta \Delta\vartheta .$$

The differential scattering cross-section is essentially the missing proportionality factor in the relation

$$\Delta N \propto j_0 \cdot \Delta t \cdot \Delta\Omega ,$$

and (7.1) should only be used for evaluation of this quantity..

The description of these relations is supported by Fig. 7.1.

For A/r-potentials the differential cross-section can be evaluated exactly, with the result

$$\frac{d^{(2)}\sigma}{d\Omega} = \frac{A^2}{16E^2} \frac{1}{\sin^4 \frac{\vartheta}{2}} ,$$

which is called the *Rutherford scattering cross-section*. This result was obtained by Rutherford in Cambridge, U.K., at the beginning of the twentieth century. At the same time he was able to confirm this formula, motivated by his ground-breaking scintillation experiments with α-particles. In this way he discovered that atoms consist of a negatively-charged electron shell with a radius of the order of 10^{-8} cm, and a much smaller, positively-charged nucleus with a radius of the order of 10^{-13} cm. In fact, the differential cross-sections for atomic nuclei are of the order of 10^{-26} cm^2, i.e., for α-particles the space between the nuclei is almost empty.

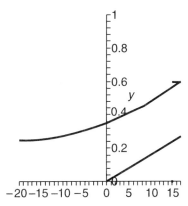

Fig. 7.1. Schematic diagram on differential scattering cross-sections. A particle enters the diagram from the left on a path parallel to the x-axis at a perpendicular distance b (the so-called "impact parameter"; here $b = 0.25$). It is then repulsively scattered by a target at the origin (here the scattering occurs for $-20 \leq x \leq 20$, where the interaction is felt) and forced to move along the path $y(x) = b + 0.00025 \cdot (x + 20)^2$, until it leaves the diagram asymptotically parallel to the inclined straight line from the origin (here $y = 0.016 \cdot x$). Finally it enters a counter at a *scattering angle* $\vartheta = \arctan 0.016$, by which the above asymptote is inclined to the x-axis. If the *impact parameter* b of the particle is slightly changed (in the (y, z)-plane) to cover an area element $\mathrm{d}^{(2)}A(b) := \mathrm{d}\varphi \cdot b \cdot \mathrm{d}b$ (where φ is the azimuthal angle in that plane), the counter covers a solid-angle element $\mathrm{d}\Omega(b) = \mathrm{d}\varphi \cdot \sin \vartheta(b) \cdot \mathrm{d}\vartheta$. The differential cross-section is the ratio $\frac{\mathrm{d}^{(2)}A}{\mathrm{d}\Omega} = \left| \frac{b \cdot \mathrm{d}b(\vartheta)}{\sin \vartheta \cdot \mathrm{d}\vartheta} \right|$.

8 Lagrange Formalism I: Lagrangian and Hamiltonian

In the present context the physical content of the *Lagrange formalism* (see below) does not essentially go beyond Newton's principles; however, *mathematically* it is much more general and of central importance for theoretical physics as a whole, not only for theoretical mechanics.

8.1 The Lagrangian Function; Lagrangian Equations of the Second Kind

Firstly we shall define the notions of "degrees of freedom", "generalized coordinates", and the "Lagrangian function" (or simply *Lagrangian*) assuming a system of N particles with $3N$ Cartesian coordinates x_α, $\alpha = 1, \ldots, 3N$:

a) The number of degrees of freedom f is the dimension of the hypersurface[1] in $3N$-dimensional space on which the system moves. This hypersurface can be *fictitious*; in particular, it may deform with time. We assume that we are only dealing with *smooth* hypersurfaces, such that f only assumes integral values $1, 2, 3, \ldots$.

b) The generalized coordinates $q_1(t), \ldots, q_f(t)$ are smooth functions that uniquely indicate the position of the system in a time interval Δt around t; i.e., in this interval $x_\alpha = f_\alpha(q_1, \ldots, q_f, t)$, for $\alpha = 1, \ldots, 3N$. The generalized coordinates (often they are angular coordinates) are *rheonomous*, if at least one of the relations f_α depends explicitly on t; otherwise they are called *skleronomous*. The *Lagrangian function* $\mathcal{L}$ is by definition equal to the *difference* (sic) between the *kinetic* and *potential* energy of the system, $\mathcal{L} = \mathcal{T} - \mathcal{V}$, expressed by $q_\alpha(t)$, $\dot{q}_\alpha(t)$ and t, where it is assumed that a potential energy exists such that for all α the relation $F_\alpha = -\frac{\partial \mathcal{V}}{\partial x_\alpha}$ holds, and that $\mathcal{V}$ can be expressed by the q_i, for $i = 1, \ldots, f$, and t.

Frictional forces and Lorentz forces (or Coriolis forces, see above) are not allowed with this definition of the Lagrangian function, but the potential energy may depend explicitly on time. However, one can generalize the definition of the Langrangian in such a way that Lorentz forces (or Coriolis forces) are

---

[1] For so-called *anholonomous constraints* (see below), f is the dimension of an *infinitesimal hypersurface element*

also included (see below). The fact that the Lagrangian contains the *difference*, and *not* the *sum* of the kinetic and potential energies, has a relativistic origin, as we shall see later.

8.2 An Important Example: The Spherical Pendulum with Variable Length

These relations are best explained by a simple example. Consider a pendulum consisting of a weightless thread with a load of mass m at its end. The length of the thread, $l(t)$, is variable, i.e., an external function. The thread hangs from the point $x_0 = y_0 = 0$, $z_0 \neq 0$, which is fixed in space, and the load can swing in all directions. In spherical coordinates we thus have (with z measured as positive downwards): $\vartheta \in [0, \pi]$, where $\vartheta = 0$ corresponds to the position of rest, and $\varphi \in [0, 2\pi)$:

$$
\begin{aligned}
x &= l(t) \cdot \sin \vartheta \cdot \cos \varphi \\
y &= l(t) \cdot \sin \vartheta \cdot \sin \varphi \\
z &= z_0 - l(t) \cdot \cos \vartheta \ .
\end{aligned}
\tag{8.1}
$$

The number of degrees of freedom is thus $f = 2$; the generalized coordinates are *rheonomous*, although this is not seen at once, since $q_1 := \vartheta$ and $q_2 = \varphi$ do not explicitly depend on t, in contrast to the relations between the cartesian and the generalized coordinates; see (8.1). Furthermore,

$$
\mathcal{V} = mgz = mg \cdot (z_0 - l(t) \cdot \cos \vartheta) \ ,
$$

whereas the kinetic energy is more complicated. A long, but elementary calculation yields

$$
\mathcal{T} = \frac{m}{2} \left(\dot{x}^2 + \dot{y}^2 + \dot{z}^2 \right) \equiv \frac{m}{2} \left\{ l^2 \cdot \left(\dot{\vartheta}^2 + \sin^2 \vartheta \dot{\varphi}^2 \right) \right.
$$
$$
\left. + 4l\dot{l} \sin \vartheta \cos \vartheta \dot{\vartheta} + \dot{l}^2 \sin^2 \vartheta \right\} \ .
$$

Apart from an additive constant the Lagrangian $\mathcal{L} = \mathcal{T} - \mathcal{V}$ is thus:

$$
\mathcal{L}(\vartheta, \dot{\vartheta}, \dot{\varphi}, t) = \frac{m}{2} \left\{ l^2 \left(\dot{\vartheta}^2 + \sin^2 \vartheta \dot{\varphi}^2 \right) + 4l\dot{l} \sin \vartheta \cos \vartheta \dot{\vartheta} \right.
$$
$$
\left. + \dot{l}^2 \sin^2 \vartheta \right\} + mg \cdot (l(t) - z_0) \cdot \cos \vartheta \ .
\tag{8.2}
$$

(In the expression for the *kinetic energy* the *inertial mass* should be used, and in the expression for the *potential energy* one should actually use the *gravitational mass*; g is the acceleration due to gravity.)

8.3 The Lagrangian Equations of the 2nd Kind

These shall now be derived, for simplicity with the special assumption $f = 1$.
Firstly we shall consider an *actual orbit* $q(t)$, i.e., following the Newtonian
equations transformed from the cartesian coordinate x to the generalized
coordinate q. At time t_1 this *actual orbit* passes through an initial point q_1,
and at t_2 through q_2. Since Newton's equation is of second order, i.e., with
two arbitrary constants, this is possible for given q_1 and q_2.

In the following, the *orbit* is varied, i.e., a set of so-called *virtual orbits*,

$$q_v(t) := q(t) + \varepsilon \cdot \delta q(t) ,$$

will be considered, where the real number $\varepsilon \in [-1, 1]$ is a so-called *variational parameter* and $\delta q(t)$ a fixed, but arbitrary function (continuously
differentiable twice) which vanishes for $t = t_1$ and $t = t_2$. Thus, the *virtual
orbits* deviate from the *actual orbit*, except at the initial point and at the end
point; naturally, the *virtual velocities* are defined as

$$\dot{q}_v(t) := \dot{q}(t) + \varepsilon \cdot \delta\dot{q}(t) , \quad \text{where} \quad \delta\dot{q}(t) = \frac{dq(t)}{dt} .$$

Using the Lagrangian function $\mathcal{L}(q, \dot{q}, t)$ one then defines the so-called
action functional

$$\mathcal{S}[q_v] := \int_{t_1}^{t_2} dt \mathcal{L}(q_v, \dot{q}_v, t) .$$

For a given function $\delta q(t)$ this functional depends on the parameter ε, which
can serve for differentiation. After differentiating with respect to ε one sets
$\varepsilon = 0$. In this way one obtains

$$\frac{d\mathcal{S}[q_v]}{d\varepsilon}\Big|_{\varepsilon=0} = \int_{t_1}^{t_2} dt \left\{ \frac{\partial\mathcal{L}}{\partial\dot{q}_v}\delta\dot{q}(t) + \frac{\partial\mathcal{L}}{\partial q_v}\delta q(t) \right\} . \tag{8.3}$$

In the first term a partial integration can be performed, so that

$$\frac{d\mathcal{S}[q_v]}{d\varepsilon}\Big|_{\varepsilon=0} = \frac{\partial\mathcal{L}}{\partial\dot{q}_v}\Big|_{t_1}\delta q(t_1) - \frac{\partial\mathcal{L}}{\partial\dot{q}_v}\Big|_{t_2}\delta q(t_2)$$
$$+ \int_{t_1}^{t_2} dt \left\{ -\frac{d}{dt}\frac{\partial\mathcal{L}}{\partial\dot{q}_v} + \frac{\partial\mathcal{L}}{\partial q_v} \right\}\delta q(t) . \tag{8.4}$$

In (8.4) the first two terms on the r.h.s. vanish, and since $\delta q(t)$ is arbitrary,
the action functional $\mathcal{S}$ becomes *extremal* for the *actual orbit*, $q_v(t) \equiv q(t)$,

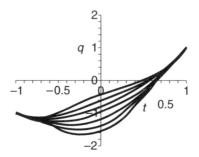

Fig. 8.1. Hamilton's variational principle. The figure shows the ε-dependent set of *virtual orbits* $q_v(t) := t + (t^2 - 1) + \varepsilon \cdot \sin(t^2 - 1)^3$, for $\varepsilon = -0.6, -0.4, \ldots, 0.6$ and times t between $t_1 = -1$ and $t_2 = 1$. The *actual orbit*, $q(t)$, corresponds to the central line ($\varepsilon = 0$) and yields an *extremum* of the action functional. The *virtual orbits* can also fan out more broadly from the initial and/or end points than in this example

iff (i.e., "if, and only if") the so-called *variational derivative*

$$\frac{\delta S}{\delta q} := \left\{ -\frac{\mathrm{d}}{\mathrm{d}t}\frac{\partial \mathcal{L}}{\partial \dot{q}_v} + \frac{\partial \mathcal{L}}{\partial q_v} \right\}$$

vanishes. An example is shown in Fig. 8.1.

The postulate that S is *extremal* for the *actual orbit* is called *Hamilton's variational principle of least[2] action*, and the equations of motion,

$$\frac{\mathrm{d}}{\mathrm{d}t}\frac{\partial \mathcal{L}}{\partial \dot{q}_v} - \frac{\partial \mathcal{L}}{\partial q_v} = 0 \,,$$

are called *Lagrangian equations of the 2nd kind* (called "2nd kind" by some authors for historical reasons). They are the so-called *Euler-Lagrange equations*[3] corresponding to Hamilton's variational principle. (The more complicated *Lagrangian equations of the 1st kind* additionally consider *constraints* and will be treated in a later section.)

For the special case where

$$\mathcal{L} = \frac{m}{2}\dot{x}^2 - V(x) \,,$$

Newton's equation results. (In fact, the Lagrangian equations of the 2nd kind can also be obtained from the Newtonian equations by a general coordinate transformation.) Thus one of the main virtues of the *Lagrangian formalism* with respect to the Newtonian equations is that the formalisms are physically equivalent; but mathematically the Lagrangian formalism has the essential

---

[2] In general, the term "least" is not true and should be replaced by "extremal".
[3] Of course any function $F(\mathcal{L})$, and also any additive modification of $\mathcal{L}$ by a total derivative $\frac{\mathrm{d}f(q(t),\dot{q}(t),t)}{\mathrm{d}t}$, would lead to the same equations of motion.

advantage of *invariance against general coordinate transformations*, whereas Newton's equations must be transformed from cartesian coordinates, where the formulation is rather simple, to the coordinates used, where the formulation at first sight may look complicated and very special.

In any case, the index v, corresponding to *virtual*, may be omitted, since finally $\varepsilon \equiv 0$.

For $f \geq 2$ the Lagrangian equations of the 2nd kind are, with $i = 1, \ldots, f$:

$$\frac{\mathrm{d}}{\mathrm{d}t} \frac{\partial \mathcal{L}}{\partial \dot{q}_i} = \frac{\partial \mathcal{L}}{\partial q_i} \ . \tag{8.5}$$

8.4 Cyclic Coordinates; Conservation of Generalized Momenta; Noether's Theorem

The quantity

$$p_i := \frac{\partial \mathcal{L}}{\partial \dot{q}_i}$$

is called the *generalized momentum* corresponding to q_i. Often p_i has the physical dimension of *angular momentum*, in the case when the corresponding generalized coordinate is an *angle*. One also calls the generalized coordinate *cyclic*[4], iff

$$\frac{\partial \mathcal{L}}{\partial q_i} = 0 \ .$$

As a consequence, from (8.5), the following theorem[5] is obtained.

If the generalized coordinate q_i is cyclic, then the related generalized momentum

$$p_i := \frac{\partial \mathcal{L}}{\partial \dot{q}_i}$$

is conserved.

As an example we again consider a *spherical pendulum* (see Sect. 8.2). In this example, the azimuthal angle φ is *cyclic* even if the length $l(t)$ of the pendulum depends explicitly on time. The corresponding generalized momentum,

$$p_\varphi = ml^2 \cdot \sin \vartheta \cdot \dot{\varphi} \ ,$$

is the z-component of the angular momentum, $p_i = L_z$. In the present case, this is in fact a *conserved quantity*, as one can also show by elementary arguments, i.e., by the vanishing of the torque $\mathcal{D}_z$.

---

[4] In *general relativity* this concept becomes enlarged by the notion of a *Killing vector*.

[5] The name *cyclic coordinate* belongs to the *canonical jargon* of many centuries and should not be altered.

Compared to the Newtonian equations of motion, the Lagrangian formalism thus:

a) not only has the decisive advantage of optimum simplicity. For suitable coordinates it is usually quite simple to write down the Lagrangian $\mathcal{L}$ of the system; then the equations of motion result almost instantly;
b) but also one sees almost immediately, because of the cyclic coordinates mentioned above, which quantities are *conserved* for the system.

For Kepler-type problems, for example, in planar polar coordinates we directly obtain the result that

$$\mathcal{L} = \frac{M}{2} v_{\mathrm{s}}^2 + \frac{m}{2} \cdot \left(\dot{r}^2 + r^2 \dot{\varphi}^2\right) - V(r) \ .$$

The *center-of-mass coordinates* and the azimuthal angle φ are therefore cyclic; thus one has the *total linear momentum* and the *orbital angular momentum* as *conserved quantities*, and because the Lagrangian does not depend on t, one additionally has *energy conservation*, as we will show immediately.

In fact, these are special cases of the basic *Noether Theorem*, named after the mathematician Emmy Noether, who was a lecturer at the University of Göttingen, Germany, immediately after World War I. We shall formulate the theorem without proof (The formulation is consciously quite sloppy):

The three conservation theorems for (i) the total momentum, (ii) the total angular momentum and (iii) the total mechanical energy correspond (i) to the homogeneity (= translational invariance) and (ii) the isotropy (rotational invariance) of space and (iii) to the homogeneity with respect to time. More generally, to any continuous n-fold global symmetry of the system there correspond n globally conserved quantities and the corresponding so-called continuity equations, as in theoretical electrodynamics (see Part II).

For the special *dynamic conserved quantities*, such as the above-mentioned Runge-Lenz vector, cyclic coordinates do not exist. The fact that these quantities are conserved for the cases considered follows only algebraically using so-called *Poisson brackets*, which we shall treat below.

8.5 The Hamiltonian

To treat the conservation of *energy*, we must enlarge our context somewhat by introducing the so-called *Hamiltonian*

$$\mathcal{H}(p_1, \ldots, p_f, q_1, \ldots, q_f, t) \ .$$

This function is a generalized and transformed version of the *Lagrangian*, i.e., the *Legendre transform* of $-\mathcal{L}$, and as mentioned below, it has many important properties. The *Hamiltonian* is obtained, as follows:

Firstly we note that the *Lagrangian* $\mathcal{L}$ depends on the generalized velocities $\dot{q}_i$, the generalized coordinates q_i and time t. Secondly we form the function

$$\tilde{\mathcal{H}}(p_1, \ldots, p_f, \dot{q}_1, \ldots, \dot{q}_f, q_1, \ldots, q_f, t) := \sum_{i=1}^{f} p_i \dot{q}_i - \mathcal{L}(\dot{q}_1, \ldots, \dot{q}_1, q_1, \ldots, q_f, t) .$$

Thirdly we assume that one can eliminate the generalized velocities $\dot{q}_i$ by replacing these quantities by functions of p_i, q_k and t with the help of the equations $p_i \equiv \frac{\partial \mathcal{L}}{\partial \dot{q}_i}$. This elimination process is almost always possible in nonrelativistic mechanics; it is a basic prerequisite of the method. After this replacement one finally obtains

$$\tilde{\mathcal{H}}(p_1, \ldots, \dot{q}_1(q_1, \ldots, p_1, \ldots, t), \ldots, q_1, \ldots, t) \equiv \mathcal{H}(p_1, \ldots, p_f, q_1, \ldots, q_f, t) .$$

As already mentioned, the final result, i.e., only after the elimination process, is called the *Hamiltonian* of the system. In a subtle way, the *Hamiltonian* is somewhat more general than the *Lagrangian*, since the variables $p_1, \ldots, p_f$ can be treated as independent and equivalent variables in addition to the variables $q_1, \ldots, q_f$, whereas in the Lagrangian formalism only the generalized coordinates q_i are independent, while the generalized velocities, $\dot{q}_i$, depend on them[6]. But above all, the Hamiltonian, and *not* the Lagrangian, will become the important quantity in the standard formulation of Quantum Mechanics (see Part III).

8.6 The Canonical Equations; Energy Conservation II; Poisson Brackets

As a result of the transformation from $\mathcal{L}$ to $\mathcal{H}$ one obtains:

$$\mathrm{d}\mathcal{H} = \sum_{i=1}^{f} \left(\mathrm{d}p_i \cdot \dot{q}_i + p_i \, \mathrm{d}\dot{q}_i - \frac{\partial \mathcal{L}}{\partial \dot{q}_i} \mathrm{d}\dot{q}_i - \frac{\partial \mathcal{L}}{\partial q_i} \mathrm{d}q_i \right) - \frac{\partial L}{\partial t} \mathrm{d}t .$$

Here the second and third terms on the r.h.s. compensate for each other, and from (8.5) the penultimate term can be written as $-\dot{p}_i \, \mathrm{d}q_i$.

Therefore, since $\mathrm{d}\mathcal{H}$ can also be written as follows:

$$\mathrm{d}\mathcal{H} = \sum_{i=1}^{f} \left(\mathrm{d}p_i \frac{\partial \mathcal{H}}{\partial p_i} + \mathrm{d}q_i \frac{\partial \mathcal{H}}{\partial q_i} \right) + \frac{\partial \mathcal{H}}{\partial t} \mathrm{d}t , \tag{8.6}$$

---

[6] Here we remind ourselves of the natural but somewhat arbitrary definition $\delta \dot{q} := \frac{\mathrm{d}(\delta q)}{\mathrm{d}t}$ in the derivation of the *principle of least action*.

one obtains by comparison of coefficients firstly the remarkable so-called *canonical equations* (here not only the signs should be noted):

$$\dot{q}_i = +\frac{\partial \mathcal{H}}{\partial p_i}, \quad \dot{p}_i = -\frac{\partial \mathcal{H}}{\partial q_i}, \quad \frac{\partial \mathcal{H}}{\partial t} = -\frac{\partial \mathcal{L}}{\partial t}. \tag{8.7}$$

Secondly, the total derivative is

$$\frac{d\mathcal{H}}{dt} = \sum_{i=1}^{f} \left(\dot{p}_i \frac{\partial \mathcal{H}}{\partial p_i} + \dot{q}_i \frac{\partial \mathcal{H}}{\partial q_i} \right) + \frac{\partial \mathcal{H}}{\partial t},$$

and for a general function

$$F(p_1(t), \ldots, p_f(t), q_1(t), \ldots, q_f(t), t): \quad \frac{dF}{dt} = \sum_{i=1}^{f} \left\{ \dot{p}_i \frac{\partial F}{\partial p_i} + \dot{q}_i \frac{\partial F}{\partial q_i} \right\} + \frac{\partial F}{\partial t}.$$

Insertion of the canonical equations reduces the previous results to:

$$\frac{d\mathcal{H}}{dt} = \frac{\partial \mathcal{H}}{\partial t} = -\frac{\partial \mathcal{L}}{\partial t}$$

and

$$\frac{dF}{dt} = \sum_{i=1}^{f} \left\{ \frac{\partial \mathcal{H}}{\partial p_i} \frac{\partial F}{\partial q_i} - \frac{\partial \mathcal{H}}{\partial q_i} \frac{\partial F}{\partial p_i} \right\} + \frac{\partial F}{\partial t},$$

respectively, where we should remember that generally the *total* and *partial* time derivatives are different!

In both cases the *energy theorem* (actually the *theorem of* $\mathcal{H}$ *conservation*) follows:

If $\mathcal{L}$ *(or* $\mathcal{H}$*) does not depend explicitly on time (e.g.,* $\frac{\partial \mathcal{H}}{\partial t} \equiv 0$*), then* $\mathcal{H}$ *is conserved during the motion (i.e.,* $\frac{d\mathcal{H}}{dt} \equiv 0$*). Usually, but not always,* $\mathcal{H}$ *equals the total mechanical energy.*

Thus some caution is in order: $\mathcal{H}$ is not always identical to the mechanical energy, and the *partial* and *total* time derivatives are also not identical; but if

$$\mathcal{L} = \mathcal{T} - \mathcal{V},$$

then (if *skleronomous generalized coordinates* are used) we automatically obtain

$$\mathcal{H} \equiv \mathcal{T} + \mathcal{V},$$

as one can easily derive by a straightforward calculation with the above definitions. Here, in the first case, i.e., with $\mathcal{L}$, one should write

$$\mathcal{T} = \frac{m\boldsymbol{v}^2}{2},$$

whereas in the second case, i.e., with $\mathcal{H}$, one should write

$$\mathcal{T} = \frac{\boldsymbol{p}^2}{2m}.$$

In the second case the sum of the braced terms yields a definition for the so-called Poisson brackets:

$$[\mathcal{H}, F]_P := \sum_{i=1}^{f} \left\{ \frac{\partial \mathcal{H}}{\partial p_i} \frac{\partial F}{\partial q_i} - \frac{\partial \mathcal{H}}{\partial q_i} \frac{\partial F}{\partial p_i} \right\} . \tag{8.8}$$

For the three components of the above-mentioned Runge-Lenz vector

$$F := (\mathcal{L}_e)_j , \quad \text{with} \quad j = x, y, z ,$$

it can be shown that with the particular (but most important) *Hamiltonian* for the Kepler problem (i.e., with -A/r potentials) the Poisson brackets $[\mathcal{H}, F]_P$ vanish, while the Poisson brackets of F with the other conserved quantities (total momentum and total angular momentum) do *not* vanish. This means that the Runge-Lenz vector is not only an additional *conserved quantity* for Kepler potentials, but is actually independent of the usual conserved quantities.

The equations of motion related to the names of Newton, Lagrange and Hamilton (i.e., the *canonical equations* in the last case) are essentially all equivalent, but ordered in ascending degree of flexibility, although the full power of the respective formalisms has not yet been (and will not be) exploited. We only mention here that there is a large class of transformations, the so-called *canonical transformations*, leading from the old (generalized) coordinates and momenta to new quantities, such that Hamilton's formalism is preserved, although generally with a new Hamiltonian. In quantum mechanics (see Part III) these transformations correspond to the important class of *unitary operations*.

Additionally we mention another relation to quantum mechanics. The *Poisson bracket* $[A, B]_P$ of two measurable quantities A and B is intimately related to the so-called *commutator* of the quantum mechanical operators $\hat{A}$ and $\hat{B}$, i.e.,

$$[A, B]_P \rightarrow \frac{\mathrm{i}}{\hbar} \left(\hat{A}\hat{B} - \hat{B}\hat{A} \right) .$$

9 Relativity I: The Principle
of Maximal Proper Time (Eigenzeit)

Obviously one could ask at this point whether Hamilton's principle *of least action*[1] is related to similar variational principles in other fields of theoretical physics, e.g., to Fermat's principle of the *shortest optical path* Δl_{opt}.[2]

The answer to this question is of course affirmative. However, we shall abstain from making the relations explicit, except for the particular interpretation and relativistic generalization of *Hamilton's principle of least-action* by *Einstein's principle of maximal proper time (or maximal eigenzeit)*.[3]

According to Einstein's principle, the motion of small particles under the influence of the effects of *special* and *general relativity* takes place in a curved four-dimensional "space-time" (three space dimensions plus one time dimension) and has the particular property that the *proper time (eigenzeit)* of the particle between the starting point and the end point under the influence of gravitational forces should be a maximum; the *eigenzeit* of the particle is simply the time measured by a conventional clock *co-moving with the particle*. As shown below, Einstein's postulate corresponds not only to Hamilton's principle, but it also yields the correct formula for the Lagrange function $\mathcal{L}$, i.e.,

$$\mathcal{L} = \mathcal{T} - \mathcal{V} \, ,$$

of course always in the limit of small velocities, i.e., for $v^2 \ll c^2$, where c is the velocity of light *in vacuo*.

---

[1] The name *least action* is *erroneous* (as already mentioned), since in fact only an *extremum* of the action functional is postulated, and only in exceptional cases (e.g., for straight-line motion without any force) is this a minimum.

[2] The differential dl_{opt} is the product of the differential $dl_{\mathrm{geom.}}$ of the geometrical path multiplied by the refractive index.

[3] As mentioned, one should be cautious with the terms "least", "shortest" or "maximal". In fact for gravitational forces in the non-relativistic limit the action, $W := \int_{t_A}^{t_B} dt \mathcal{L}$, can be identified with $-m_0 c^2 \int_{t_A}^{t_B} d\tau$ (see below) and one simply obtains complementary extrema for the proper time τ and for the Hamiltonian action W. Actually, however, in general relativity, see [7] and [8], the equations of motion of a particle with finite m_0 under the influence of gravitation are time-like geodesics in a curved Minkowski manifold, i.e., with maximal (sic) proper time. However, the distinction between "timelike" and "spacelike" does not make sense in Newtonian mechanics, where formally $c \to \infty$ (see below).

In the next section we shall firstly define some necessary concepts, such as the *Lorentz transformation*.

9.1 Galilean versus Lorentz Transformations

The perception of space and time underlying *Newtonian mechanics* corresponds to the so-called *Galilean transformation*:

All inertial frames of Newtonian mechanics are equivalent, i.e., Newton's equations have the same form in these systems. The transition between two different inertial systems is performed via a Galilean transformation: If the origin of a second inertial frame, i.e., primed system, moves with a velocity v in the x direction, then

$$x = x' + vt \quad \text{and} \quad t = t', \quad \text{and of course} \quad y = y', z = z'.$$

For the Galilean transformation, space and time are thus *decoupled*. Newton's equations of motion have the same form in both the primed and unprimed inertial frames, where for the forces we have of course

$$F_x \equiv F'_{x'}.$$

The above relation implies a simple addition of velocities, i.e., if motion occurs in the unprimed system with velocity u, and in the primed system with velocity u' (in both cases in the x direction) then

$$u = u' + v.$$

According to Newtonian mechanics, an event which took place in the unprimed coordinate system with exactly the vacuum velocity of light $u = c$ would thus have a velocity

$$u' = u - v \; (\neq c)$$

in the primed system. Thus Maxwell's theory of electrodynamics (see Part II), which describes the propagation of light with velocity

$$c = \frac{1}{\sqrt{\varepsilon_0 \mu_0}}$$

(see below), is not invariant under a Galilean transformation.

On the suggestion of Maxwell himself, the hypothesis of "additivity of velocities" was tested for light with great precision, firstly by *Michelson* (1881) and then by *Michelson and Morley* (1887), with negative result. They found that

$$u = c \Leftrightarrow u' = c,$$

implying that something fundamental was wrong.

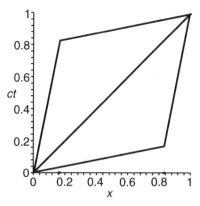

Fig. 9.1. The Lorentz transformation.

The Lorentz transformation $\left\{ x = \dfrac{x' + \frac{v}{c}ct'}{\sqrt{1 - \frac{v^2}{c^2}}}, ct = \dfrac{ct' + \frac{v}{c}x'}{\sqrt{1 - \frac{v^2}{c^2}}} \right\}$ transforms the *square* $\{0 \leq x \leq 1, 0 \leq ct \leq 1\}$ into a *rhombus* with unchanged diagonal ($x = ct \leftrightarrow x' = ct'$). The angle of inclination α between the primed and non-primed axes is given by $\alpha = \tanh \frac{v}{c}$

As mentioned previously, it had already been established before Einstein's time that the basic equations of electrodynamics (Maxwell's equations), which *inter alia* describe the propagation of light, were simply not invariant under a Galilean transformation, in contrast to Newtonian mechanics. Thus, it was concluded that with respect to electrodynamics the inertial systems were not all equivalent, i.e., there was a particular frame, the *aether*, in which Maxwell's equations had their usual form, whereas in other inertial frames they would be different.

However, Maxwellian electrodynamics can be shown to be invariant with respect to a so-called *Lorentz transformation*,[4] which transforms space and time coordinates in a similar way, as follows:

$$x = \frac{x' + \frac{v}{c}ct'}{\sqrt{1 - \frac{v^2}{c^2}}} \; ; ct = \frac{ct' + \frac{v}{c}x'}{\sqrt{1 - \frac{v^2}{c^2}}} \; .$$

(Additionally one has of course $y = y'$ and $z = z'$.)

These equations have been specifically written in such a way that the above-mentioned similarity with respect to space and time becomes obvious. (We should keep these equations in mind in this form.) Additionally, Fig. 9.1 may be useful.

---

[4] Hendryk A. Lorentz, Leiden, NL; 1904

Furthermore a nontrivial velocity transformation follows from the Lorentz transformations. Since

$$u' := \frac{\mathrm{d}x'}{\mathrm{d}t'}$$

($not: = \frac{\mathrm{d}x'}{\mathrm{d}t}$), one obtains

$$u = \frac{u' + v}{1 + \frac{u'v}{c^2}} \,,$$

which implies that $u = c \Leftrightarrow u' = c$, independently of v. Before Einstein, these relations were only considered to be *strange mathematical properties of the aether*, i.e., one believed erroneously that the Newtonian and Galilean considerations on space and time needed no modification.

Einstein's special theory of relativity (1905)[5] then changed our perception radically. It transpired that Newton's theory, not that of Maxwell, had to be modified and refined. The modifications involved our basic perception of space and time (but fortunately, since $v^2 \ll c^2$, Newtonian theory still remains valid in most practical cases of everyday life). However, an important new paradigm now entered science: a theory could be *true only under certain quantitative constraints* and could be refined or modified in other cases.

The main implications of the new theory may be restated, as follows:

*For **all** physical events all inertial frames are equivalent (i.e., there is no need for a special inertial frame called the aether). However, the transformation between different inertial frames must be made via a Lorentz transformation not a Galilean transformation. As a consequence, as already mentioned[6], with these new insights into space and time Newtonian mechanics, in contrast to Maxwell's theory of electromagnetism, had to be modified and refined, but fortunately only for very high velocities when the condition $v^2 \ll c^2$ is violated.*

9.2 Minkowski Four-vectors and Their Pseudo-lengths; Proper Time

Two years after Einstein's epochal work of 1905, the mathematician Hermann Minkowski introduced the notion of a so-called four-vector

$$\tilde{v} := (v_1, v_2, v_3, \mathrm{i} \cdot v_4) \,.$$

Here, all four variables v_α, $\alpha = 1, \ldots, 4$, are *real* quantities (i.e., the fourth component of $\tilde{v}$ is *imaginary*[7]).

---

[5] Einstein's *special theory of relativity* was published in 1905 under the title "Zur Elektrodynamik bewegter Körper" in the journal *Annalen der Physik* (see [5] or perform an internet search).

[6] Sometimes an important statement may be repeated!

[7] Many authors avoid the introduction of imaginary quantities, by using instead of $\tilde{v}$ the equivalent all-real definition $\tilde{v}' := (v^0, v^1, v^2, v^3)$, with $v^0 := v^4$; how-

Moreover, these real variables $v_1, \ldots, v_4$ *are assumed to transform by a Lorentz transformation combined with rotation in three-space, just as the variables* x, y, z *and* ct.

The union of such Minkowski four-vectors is the *Minkowski space* $\mathcal{M}_4$. A typical member is the four-differential

$$\mathrm{d}\tilde{x} := (\mathrm{d}x, \mathrm{d}y, \mathrm{d}z, \mathrm{i} \cdot c\mathrm{d}t) \,.$$

It is easy to show that under Lorentz transformations the so-called *pseudo-length*

$$\tilde{v}^2 := v_1^2 + v_2^2 + v_3^2 - v_4^2$$

of a Minkowski four-vector is invariant (e.g., the invariance of the speed of light in a Lorentz transformation results simply from the fact that for a Lorentz transformation one has: $x^2 = c^2 t^2 \Leftrightarrow x'^2 = c^2 t'^2$). In addition, the so-called *pseudo-scalar product* of two Minkowski four-vectors,

$$\tilde{v} \cdot \tilde{w} := v_1 w_1 + v_2 w_2 + v_3 w_3 - v_4 w_4 \,,$$

is also invariant for all Lorentz transformations, which means that Lorentz transformations play the role of *pseudo-rotations* in Minkowski space.

Among the invariants thus obtained is the so-called *proper time (eigenzeit)* $\mathrm{d}\tau$, which corresponds to the *pseudo-length* of the above-mentioned *Minkowski vector*

$$\mathrm{d}\tau := \sqrt{\frac{\mathrm{d}\tilde{x}^2}{-c^2}} = \mathrm{d}t \cdot \sqrt{1 - \frac{v^2}{c^2}} \,.$$

Here we have assumed $\mathrm{d}x = v_x \, \mathrm{d}t$ etc. and consider only events with $v^2 < c^2$.

A time interval $\mathrm{d}\tau$ measured with a co-moving clock, the *proper time*, is thus always shorter than the time interval $\mathrm{d}t$ measured in any other frame. This means for example that a co-moving clock transported in an aeroplane around the earth, *ticks more slowly* than an earth-based clock remaining at the airport, where the round-trip around the earth start and ends. This is a measurable effect, although very small! (More drastic effects result from the cascades of μ particles in cosmic radiation. The numerous decay products of these cascades, which have their origin at a height of ~ 30 km above the surface of the earth, have only a *proper lifetime* of $\Delta\tau \approx 10^{-6}$s. Nevertheless, showers of these particles reach the surface of the earth, even though with $v \approx c$ in 10^{-6}s they should only cover a distance of 300 m before decaying. The solution for this apparent discrepancy is the gross difference between $\mathrm{d}\tau$ and $\mathrm{d}t$ for velocities approaching the speed of light[8].)

---

ever, avoiding the imaginary unit i, one must pay some kind of *penalty*, being forced instead to distinguish between *covariant* and *contravariant* four-vector components, which is not necessary with the "imaginary" definition.

[8] Here one could introduce the terms *time dilation*, i.e., from $\mathrm{d}\tau \to \mathrm{d}t$, and *length contraction*, e.g., from 30 km to 300 m.

In a curved space the *ticking speed* of clocks is not only influenced by v, but also by gravitating bodies, as described by Einstein's *general theory of relativity* (which goes far beyond the scope of our text). Here, we only note that for *sufficiently weak gravitation potentials* $\mathcal{V}(\boldsymbol{r})$ one has the relation

$$d\tau = dt \cdot \sqrt{1 - \frac{v^2}{c^2} + \frac{2\mathcal{V}}{m_0 c^2}} \, .$$

The quantity m_0 is the *rest mass* of the considered particle, as already mentioned.

Thus, Einstein's principle of the *maximal proper time* implies that after multiplication with $(-m_0 c^2)$ the *actual path* yields an extremum of the *action*[9]

$$W := -m_0 c^2 \int_{t_1}^{t_2} dt \sqrt{1 - \frac{v^2}{c^2} + \frac{2\mathcal{V}}{m_0 c^2}} \, .$$

A Taylor expansion[10] of this expression yields (in the lowest nontrivial order w.r.t. v^2 and $\mathcal{V}$) the usual *Hamilton principle of least action*.

In addition one should note that the first three spatial components and the fourth timelike component of a Minkowski four-vector (because of the factor $i^2 = -1$) enter the final result with different *signs*. Ultimately, this different behavior of space-like and time-like components of a Minkowski four-vector (i.e., the square of the imaginary unit i appearing with the time-like component) is the genuine reason why in the formula

$$\mathcal{L} = \mathcal{T} - \mathcal{V}$$

the kinetic energy $\mathcal{T}$ and potential energy $\mathcal{V}$ enter with different signs, i.e., we are dealing here with an intrinsically relativistic phenomenon.

9.3 The Lorentz Force and its Lagrangian

If, additionally, an electric field $\boldsymbol{E}$ and a magnetic induction $\boldsymbol{B}$ are present, then the force $\boldsymbol{F}_q$ exerted by these fields on a particle with an electric charge q is given by

$$\boldsymbol{F}_q = q \cdot (\boldsymbol{E} + \boldsymbol{v} \times \boldsymbol{B}) \, ,$$

where the last term describes the Lorentz force.

The magnetic induction $\boldsymbol{B}$ can be calulated from a *vector potential* $\boldsymbol{A}$, where

$$\boldsymbol{B} = \mathrm{curl} \boldsymbol{A} \, ,$$

---

[9] Dimensional analysis: *action* := *energy* × *time*.

[10] *'Sufficiently weak'*, see above, means that a Taylor expansion w.r.t. the lowest nontrivial order makes sense.

whereas an electric field $\boldsymbol{E}$ can be mainly calculated from the *scalar potential* Φ, since

$$\boldsymbol{E} = -\text{grad}\Phi - \frac{\partial \boldsymbol{A}}{\partial t} \ .$$

The main content of the relativistic invariance of Maxwell's theory is, as we shall only mention, that Φ and $\boldsymbol{A}$ can be combined into a Minkowski four-vector

$$\tilde{A} := \left(A_x, A_y, A_z, \mathrm{i}\frac{\Phi}{c} \right) \ .$$

Furthermore, the quadruple

$$\tilde{u} := \frac{1}{\sqrt{1 - \frac{v^2}{c^2}}} (v_x, v_y, v_z, \mathrm{i}c)$$

($\equiv \frac{\mathrm{d}\tilde{x}}{\mathrm{d}\tau}$, i.e., the *Minkowski vector* $\mathrm{d}\tilde{x}$ divided by the *Minkowski scalar* $\mathrm{d}\tau$) constitutes a Minkowki four-vector, so that it is almost obvious to insert the invariant pseudo-scalar product $q\tilde{A} \cdot \tilde{u}$ into the Lagrangian, i.e.,

$$q\tilde{A} \cdot \tilde{u} = \frac{q}{\sqrt{1 - \frac{v^2}{c^2}}} \cdot (\boldsymbol{A} \cdot \boldsymbol{v} - \Phi) \ .$$

In fact, for $v^2 \ll c^2$ this expression yields an obvious addition to the Lagrangian. Thus, by the postulate of relativistic invariance, since without the Lorentz force we would have

$$\mathcal{L} \cong \mathcal{T} - \mathcal{V} - q\Phi \ ,$$

we obtain by inclusion of the Lorentz force:

$$\mathcal{L} \equiv \mathcal{T} - \mathcal{V} - \frac{q}{\sqrt{1 - \frac{v^2}{c^2}}} (\Phi - \boldsymbol{v} \cdot \boldsymbol{A}) \ , \tag{9.1}$$

and in the nonrelativistic approximation we can finally replace the complicated factor $\propto q$ simply by q itself to obtain a Lagrangian including the Lorentz force.

9.4 The Hamiltonian for the Lorentz Force; Kinetic versus Canonical Momentum; Gauge Transformations

One can also evaluate the Hamiltonian from the Lagrangian of the Lorentz force. The details are not quite trivial, but straightforward. The following result is obtained:

$$\mathcal{H} = \frac{(\boldsymbol{p} - q \cdot \boldsymbol{A}(\boldsymbol{r}, t))^2}{2m} + V(\boldsymbol{r}) + q \cdot \varPhi . \tag{9.2}$$

Here, by means of a so-called gauge transformation, not only $\boldsymbol{A}$, but simultaneously also $\boldsymbol{p}$, can be transformed as follows:

$$q \cdot \boldsymbol{A} \to q \cdot \boldsymbol{A} + \operatorname{grad} f(\boldsymbol{r}, t) ,$$
$$\boldsymbol{p} \to \boldsymbol{p} + \operatorname{grad} f(\boldsymbol{r}, t) , \quad \text{and}$$
$$q \cdot \varPhi \to q \cdot \varPhi - \partial_t f(\boldsymbol{r}, t) .$$

The *gauge function* $f(\boldsymbol{r}, t)$ is arbitrary.

What should be kept in mind here is the so-called *minimal substitution*

$$\boldsymbol{p} \to \boldsymbol{p} - q \cdot \boldsymbol{A} ,$$

which also plays a part in quantum mechanics.

From the first series of canonical equations,

$$\dot{x} = \frac{\partial \mathcal{H}}{\partial p_x}$$

etc., it follows that

$$m\boldsymbol{v} \equiv \boldsymbol{p} - q\boldsymbol{A} .$$

This quantity is called the *kinetic momentum*, in contrast to the *canonical momentum* $\boldsymbol{p}$, which, as mentioned above, must be *gauged* too, if one *gauges* the vector potential $\boldsymbol{A}$. In contrast, the kinetic momentum $m\boldsymbol{v}$ is directly measurable and gauge-invariant.

After lengthy and subtle calculations ($\to$ a typical exercise), the second series of *canonical equations*,

$$\dot{p}_x = -\frac{\partial \mathcal{H}}{\partial x}$$

etc., yields the equation corresponding to the Lorentz force,

$$m\dot{\boldsymbol{v}} = q \cdot (\boldsymbol{E} + \boldsymbol{v} \times \boldsymbol{B}) .$$

10 Coupled Small Oscillations

10.1 Definitions; Normal Frequencies (Eigenfrequencies) and Normal Modes

Let our system be described by a Lagrangian

$$\mathcal{L} = \sum_{\alpha=1}^{3N} \frac{m_\alpha}{2} \dot{x}_\alpha - V(x_1, \dots, x_{3N}) \,,$$

and let

$$\boldsymbol{x}^{(0)} := \left(x_1^{(0)}, \dots, x_{3N}^{(0)} \right)$$

be a stable second-order equilibrium configuration, i.e., $V(\boldsymbol{x}^{(0)})$ corresponds to a local minimum of 2nd order, the forces

$$F_\alpha(\boldsymbol{x}_0) := -\frac{\partial V}{\partial x_\alpha}_{|\boldsymbol{x}^{(0)}}$$

vanish and the quadratic form

$$Q := \sum_{i,k=1}^{3N} \frac{\partial^2 V}{\partial x_i \partial x_k} \Delta x_i \Delta x_k$$

is "positive definite", $Q > 0$, as long as

$$\Delta \boldsymbol{x} := \boldsymbol{x} - \boldsymbol{x}^{(0)} \neq 0 \,,$$

except for the six cases where the Δx_i correspond to a homogeneous translation or rotation of the system. In these exceptional cases the above-mentioned quadratic form should yield a vanishing result.

The $(3N) \times (3N)$-matrix

$$V_{\alpha,\beta} := \frac{\partial^2 V}{\partial x_i \partial x_k}$$

is therefore not only symmetric ($V_{\alpha,\beta} \equiv V_{\beta,\alpha}$), such that it can be diagonalized by a rotation in $\mathcal{R}^{3N}$, with real eigenvalues, but is also "positive", i.e., all eigenvalues are > 0 except for the above-mentioned six exceptional cases where they are zero (the six so-called *Goldstone modes*). Writing

$$x_i =: x_i^{(0)} + u_i \, ,$$

and neglecting terms of third or higher order in u_i, we obtain:

$$\mathcal{L} = \sum_\alpha \frac{m_\alpha}{2} \dot{u}_\alpha^2 - \frac{1}{2} \sum_{\alpha,\beta} V_{\alpha,\beta} u_\alpha u_\beta - V(0) \, .$$

Here all masses m_α can be replaced by 1 in the original equation, if one adds a symbol $\sim$, i.e., by the substitution $\tilde{u}_\alpha := \sqrt{m_\alpha} u_\alpha$.

Thus we have

$$\mathcal{L} = \sum_{\alpha=1}^{3N} \frac{\dot{\tilde{u}}_\alpha^2}{2} - \frac{1}{2} \sum_{\alpha,\beta=1}^{3N} \tilde{V}_{\alpha,\beta} \tilde{u}_\alpha \tilde{u}_\beta - V(0) \, .$$

Here

$$\tilde{V}_{\alpha,\beta} := \frac{V_{\alpha,\beta}}{\sqrt{m_\alpha m_\beta}}$$

is again a symmetric matrix which can also be diagonalized by a rotation in $\mathcal{R}^{3N}$ (and now the rotation leaves also the kinetic energy invariant). The diagonal values ("eigenvalues") of the matrix are positive (with the above-mentioned exception), so they can be written as ω_α^2, with $\omega_\alpha \geq 0$, for $\alpha = 1, \dots, 3N$, including the six zero-frequencies of the Goldstone modes.

The ω_α are called normal frequencies, and the corresponding eigenvectors are called normal modes (see below).

One should of course use a cartesian basis corresponding to the diagonalized quadratic form, i.e., to the directions of the mutually orthogonal eigenvectors. The related cartesian coordinates, Q_ν, with $\nu = 1, \dots, 3N$, are called *normal coordinates*.

After diagonalization[1], the *Lagrangian* is (apart from the unnecessary additive constant $\mathcal{V}(0)$):

$$\mathcal{L} = \frac{1}{2} \sum_{\nu=1}^{3N} \left(\dot{Q}_\nu^2 - \omega_\nu^2 Q_\nu^2 \right) \, .$$

Previously the oscillations were *coupled*, but by *rotation to diagonal form* in $\mathcal{R}^{3N}$ they have been *decoupled*. The Hamiltonian corresponds exactly to $\mathcal{L}$ (the difference is obvious):

$$\mathcal{H} = \frac{1}{2} \sum_{\nu=1}^{3N} \left(P_\nu^2 + \omega_\nu^2 Q_\nu^2 \right) \, .$$

Here P_ν is the momentum conjugate with the normal coordinate Q_ν.

---

[1] The proof of the diagonalizability by a suitable rotation in $\mathcal{R}^{3N}(\omega)$, including the proof of the mutual orthogonality of the eigenvectors, is essentially self-evident. It is only necessary to know that a positive-definite quadratic form, $\sum_{\alpha,\beta=1}^{3N} V_{\alpha,\beta} \omega_\alpha \omega_\beta$, describes a $3N$-dimensional ellipsoid in this Euclidean space, which can be diagonalised by a rotation to the principal axes of the ellipsoid.

10.2 Diagonalization: Evaluation of the Eigenfrequencies and Normal Modes

In the following we replace $3N$ by f. The equations of motion to be solved are:

$$\ddot{u}_i = -\sum_k \tilde{V}_{i,k} \tilde{u}_k \ .$$

For simplicity, we omit the symbol $\sim$ in $\tilde{u}_i$.

Thus, using the *ansatz* $u_i = u_i^{(0)} \cos(\omega t - \alpha)$, we obtain:

$$\sum_k V_{i,k} u_k^{(0)} - \omega^2 u_i^{(0)} = 0 \ , \quad \text{with} \quad i = 1, \ldots, f \ ;$$

and explicitly:

$$\begin{pmatrix} V_{1,1} - \omega^2 \ , & V_{1,2} & , \ldots, & V_{1,f} \\ V_{2,1} & , V_{2,2} - \omega^2 & , \ldots, & V_{2,f} \\ \ldots & , & \ldots, & , \ldots, & \ldots \\ \ldots & , & \ldots, & , \ldots, & \ldots \\ V_{f,1} & , & V_{f,2} & , \ldots, V_{f,f} - \omega^2 \end{pmatrix} \cdot \begin{pmatrix} u_1^{(0)} \\ u_2^{(0)} \\ \ldots \\ \ldots \\ u_f^{(0)} \end{pmatrix} = 0 \ . \qquad (10.1)$$

These equations have of course the trivial solution

$$\left(u_1^{(0)}, \ldots, u_f^{(0)} \right) \equiv 0 \ ,$$

which is not of interest. Nontrivial solutions exist exactly iff the *determinant* of the matrix of the set of equations vanishes. This yields f (not necessarily different) eigenfrequencies ω.

As mentioned, the squares of these eigenfrequencies are all non-negative. The corresponding eigenvectors (normal modes), which are only determinate up to an arbitrary factor, can be typically obtained by inserting the previously determined eigenfrequency into the first $(f-1)$ equations of the system (10.1), from which the

$$\left(u_1^{(0)}, u_2^{(0)}, \ldots, u_{f-1}^{(0)}, 1 \right)$$

can be calculated. Usually, this is straightforward, but cumbersome.

Often, however, one can considerably simplify this procedure, since *for reasons of symmetry*, one already knows the eigenvectors in advance, either completely or at least partially, before one has evaluated the eigenfrequencies, as we shall see in the following example.

10.3 A Typical Example: Three Coupled Pendulums with Symmetry

In the following example, consider a horizontal rod, e.g., a curtain rod, from which three pendulums are hanging from separate threads, not necessarily

of different length l_i, $i = 1, 2, 3$, with point masses m_i at the lower ends.
The pendulums are assumed to move in an (x,z)-plane, and additionally they
are supposed to be coupled by two horizontal springs. The two springs are
fastened to the respective threads at a distance L from the uppermost point
of the pendulum considered; spring one joins the threads 1 and 2, spring two
joins the threads 2 and 3[2]. The corresponding spring constants are $k_{1,2}$ and
$k_{2,3}$. (Instead of the three threads one can also use three rigid bars made
from an extremely light material.)

The kinetic and potential energies of the system are thus given by:

$$T = \sum_{i=1}^{3} \frac{1}{2} m_i l_i^2 \dot{\varphi}^2 \,, \tag{10.2}$$

$$V = \sum_{i=1}^{3} m_i g l_i \cdot (1 - \cos \varphi_j) + \frac{L^2}{2} \cdot \left\{ k_{1,2}(\varphi_2 - \varphi_1)^2 + k_{2,3}(\varphi_3 - \varphi_2)^2 \right\} \,. \tag{10.3}$$

As usual,

$$\mathcal{L} = T - V \,;$$

g is the acceleration due to gravity.

In the following we shall replace

$$1 - \cos \varphi_j \quad \text{by} \quad \frac{\varphi_j^2}{2} \,;$$

i.e., we consider the approximation of small oscillations around the equilibrium position $\varphi_j = 0$. The three Lagrangian equations of the 2nd kind are:

$$\frac{\mathrm{d}}{\mathrm{d}t} \frac{\partial \mathcal{L}}{\partial \dot{\varphi}_j} - \frac{\partial \mathcal{L}}{\partial \varphi_j} = 0 \,,$$

with $j = 1, 2, 3$. They lead to

$$\begin{aligned}
m_1 l_1 \ddot{\varphi}_1 + m_1 g l_1 \varphi_1 + k_{1,2} L^2 \cdot (\varphi_1 - \varphi_2) &= 0 \\
m_2 l_2 \ddot{\varphi}_2 + m_2 g l_2 \varphi_2 + k_{1,2} L^2 \cdot (\varphi_2 - \varphi_1) + k_{2,3} L^2 \cdot (\varphi_2 - \varphi_3) &= 0 \\
m_3 l_3 \ddot{\varphi}_3 + m_3 g l_3 \varphi_3 + k_{2,3} L^2 \cdot (\varphi_3 - \varphi_2) &= 0 \,.
\end{aligned} \tag{10.4}$$

With the *ansatz*

$$\varphi_j(t) = \varphi_j^{(0)} \cdot \cos(\omega t - \alpha_j)$$

and by dividing the line j by $m_j l_j^2$ we obtain the following linear algebraic
3×3-equation:

[2] It is again suggested that the reader should make his/her own sketch.

$$\left(\begin{array}{ccc} \frac{g}{l_1} + \frac{k_{1,2}L^2}{m_1 l_1^2} - \lambda\,, & -\frac{k_{1,2}L^2}{m_1 l_1^2}\,, & 0 \\ -\frac{k_{1,2}L^2}{m_2 l_2^2}\,, & \frac{g}{l_2} - \frac{(k_{1,2}+k_{2,3})L^2}{m_2 l_2^2} - \lambda\,, & -\frac{k_{2,3}L^2}{m_2 l_2^2} \\ 0\,, & -\frac{k_{2,3}L^2}{m_3 l_3^2}\,, & \frac{g}{l} + \frac{k_{2,3}L^2}{m_3 l_3^2} - \lambda \end{array}\right)$$

$$\cdot \begin{pmatrix} \varphi_1^{(0)} \\ \varphi_2^{(0)} \\ \varphi_3^{(0)} \end{pmatrix} = 0\,, \quad \text{where} \quad \lambda = \omega^2\,. \tag{10.5}$$

The eigenfrequencies ω_j^2 are routinely obtained by searching for the zeroes of the determinant of this set of equations; subsequently, also routinely, one can determine the eigenvectors, as described above. In the present case of three coupled equations this task is still feasible, although tedious and rather dull. However, in the case of "left-right symmetry", $1 \Leftrightarrow 3$ (see below), one can simplify the calculation considerably, as follows:

Assume that all parameters reflect this left-right symmetry, i.e., the *Lagrangian* $\mathcal{L}$ shall be invariant against permutation of the indices $j = 1$ and $j = 3$, such that the system possesses mirror symmetry with respect to the central pendulum $j = 2$.

We then find (without proof)[3] that the eigenvectors correspond to two different classes, which can be treated separately, *viz*

a) to the class I of *odd normal modes*:

$$\varphi_1(t) \equiv -\varphi_3(t)\,, \quad \varphi_2(t) \equiv 0$$

i.e., the external pendulums, 1 and 3, oscillate against each other, while the central pendulum, 2, is at rest, and
b) to the class II of *even normal modes*,

$$\varphi_1(t) \equiv +\varphi_3(t)(\neq \varphi_2(t))\,.$$

One also speaks of *odd* or *even* parity (see Part III).
For class I there is only one eigenfrequency,

$$\omega_1^2 = \frac{g}{l_1} + \frac{k_{1,2}L^2}{m_1 l_1^2}\,.$$

Here the first term corresponds to oscillations of pendulum 1, i.e., with length l_1, in a gravitational field of acceleration g; the second term represents the additional stress induced by the horizontal spring, which is $\propto k_{1,2}$.

---

[3] For generalizations, one can refer to the script by one of the authors (U.K.) on "*Gruppentheorie und Quantenmechanik*".

In contrast, for the second class one obtains two equations with two unknowns:

$$\begin{pmatrix} \frac{g}{l_1} + \frac{k_{1,2}L^2}{m_1 l_1^2} - \omega^2 \, , & -\frac{k_{1,2}L^2}{m_1 l^2} \\ -\frac{2k_{1,2}L^2}{m_2 l_2^2} \, , & \frac{g}{l_2} + 2\frac{k_{1,2}L^2}{m_2 l_2^2} - \omega^2 \end{pmatrix} \cdot \begin{pmatrix} \varphi_1^{(0)} \\ \varphi_2^{(0)} \end{pmatrix} = 0 \, . \tag{10.6}$$

With the abbreviations

$$\Omega_{1,2} := \frac{k_{1,2}L^2}{m_1 l_1^2} \, ,$$

$$\Omega_{2,1} := \frac{k_{1,2}L^2}{m_2 l_2^2} \, ,$$

$$\Omega_{1,1} := \frac{g}{l_1} + \frac{k_{1,2}L^2}{m_1 l_1^2} \, , \quad \text{and}$$

$$\Omega_{2,2} := \frac{g}{l_2} + \frac{k_{1,2}L^2}{m_2 l_2^2}$$

we have:

$$\omega_{2,3}^2 = \omega_{\pm}^2 = \frac{\Omega_{1,1}^2 + \Omega_{2,2}^2}{2} \pm \sqrt{\frac{(\Omega_{1,1}^2 - \Omega_{2,2}^2)^2}{4} + 2\Omega_{1,2}^2 \Omega_{2,1}^2} \, . \tag{10.7}$$

Thus, of the two eigenfrequencies, one is lower, $\omega = \omega_-$, the other one higher, $\omega = \omega_+$. For the lower eigenfrequency, all three pendulums oscillate almost in phase, i.e., the springs are almost unstressed. In contrast, for the higher eigenfrequency, $\omega = \omega_+$, only pendulum 1 and pendulum 3 oscillate almost in phase, whereas pendulum 2 moves in anti-phase ("push-pull scenario"), so that the horizontal springs are strongly stressed. (We advise the reader to make a sketch of the normal modes).

10.4 Parametric Resonance: Child on a Swing

For a single pendulum of length l, the eigenfrequency of the oscillation,

$$\omega_0 = \sqrt{\frac{g}{l}} \, ,$$

does not depend on the amplitude φ_0.

This is true as long as $\varphi_0^2 \ll 1$. If this condition is violated, then ω_0 decreases and depends on φ_0, i.e., for

$$0 < \varphi_0 < \pi \quad \text{one has:} \quad \omega_0 = \frac{2\pi}{\tau_0} \, .$$

Here τ_0 is the time period. Using the principle of conservation of energy as in section 3.2 one obtains the result:

$$\frac{\tau_0}{4} = \sqrt{\frac{2l}{g}} \cdot \int\limits_0^{\varphi_0} \mathrm{d}\varphi \frac{1}{\sqrt{\cos\varphi - \cos\varphi_0}} \ .$$

Therefore, although it contains an amplitude-dependent factor, the oscillation frequency is still inversely proportional to the square-root of l/g. If the parameter l/g is periodically changed, then one obtains the phenomenon of so-called *parametric resonance*. In the resonance case, the frequency ω_P of this parameter variation is a *non-trivial integral multiple of ω_0*. If, for example, the length of the pendulum is *shortened*, whenever it reaches one of the two *points of return* (i.e., the points where the total energy is identical to the potential energy; there the pendulum is in effect "pulled upwards"), and then the pendulum length is *increased* when the next *zero-crossing* is reached (i.e., there the "child on the swing" stretches out, such that the maximum value of the kinetic energy, which in this case is identical to the total energy, is enhanced; there the pendulum is in effect "pushed down"), then, by this periodic parameter variation,

$$l \rightarrow l \mp \delta L \ ,$$

one can increase the mechanical energy of the motion of the pendulum. In this way the oscillation amplitude can be *pumped*, until it becomes stationary due to frictional losses.

The period of the parameter variation leading to the building up of the oscillation amplitude is

$$\omega_P = 2 \cdot \omega_0$$

for the above example. This mechanical example is actually most instructive. In fact, the equations are more complex, and besides paramagnetic resonance one should also consider the usual driving-force resonance, extended to non-linearity. However, we shall omit the mathematical details, which are far from being straightforward.

11 Rigid Bodies

11.1 Translational and Rotational Parts of the Kinetic Energy

Consider a rigid body consisting of N ($\approx 10^{23}$) atoms, with fixed distances $|r_i - r_k|$ between the atoms. The velocity of an arbitrary atom is written as

$$v_i = \dot{r}_0 + \omega \times (r_i - r_0) \, .$$

Here r_0 corresponds to an arbitrary point of reference, and ω is the vector of the angular velocity, i.e., we have

$$\omega = n_\omega \cdot \omega \, ,$$

where the unit vector n_ω describes the axis of the rotation and

$$\omega := |\omega|$$

the magnitude of the angular velocity. (We assume $\omega > 0$; this is no restriction, because ω is a so-called *axial vector*[1]: For example, a right-handed rotation around the unit vector n with positive angular velocity $\dot{\varphi}$ is identical to a rotation, also right-handed, around $(-n)$ with negative angular velocity $(-\dot{\varphi})$; for left-handed rotations one has similar statements.)

Changing the reference point r_0 does not change ω. This is an essential statement. We return to this freedom of choice of the reference point in connection with the so-called Steiner theorem below.

In spite of this fact it is convenient to proceed as usual by choosing the center of mass as reference point, even if this changes with time, which often happens:

$$r_0(t) \overset{!}{=} R_{\mathrm{s}}(t) \, , \quad \text{with} \quad R_{\mathrm{s}}(t) := M^{-1} \sum_{i=1}^{N} m_i r_i(t) \, .$$

This choice has the decisive advantage that the kinetic energy $\mathcal{T}$ of the rigid body can be separated into two parts corresponding to the translational motion of the center of mass, and to a rotational energy, whereas with other reference points it can be shown that mixed terms would also appear.

---

[1] e.g., the vector product $v_1 \times v_2$ of two ordinary (i.e., *polar*) vectors v_i is an *axial* vector.

With v_s as the velocity of the center of mass we have:

$$T = \sum_i \frac{m_i}{2} v_i^2 = \frac{M}{2} v_\mathrm{s}^2 + \mathrm{T_{rot}} \ ,$$

including the *rotational energy*, which is $\propto \omega^2$:

$$\mathrm{T_{rot}} = \sum_{i=1}^{N} \frac{m_i}{2} [\boldsymbol{\omega} \times (\boldsymbol{r} - \boldsymbol{R}_\mathrm{s}(t))]^2 \ . \tag{11.1}$$

11.2 Moment of Inertia and Inertia Tensor; Rotational Energy and Angular Momentum

The *moment of inertia* $\Theta(\boldsymbol{n}_\omega)$ of a system $\mathcal{K}$ with respect to rotations around a given axis $\boldsymbol{n}_\omega$ through the center of mass can thus be shown from (11.1) to satisfy the equation

$$\mathcal{T}_\mathrm{rot} = \frac{1}{2} \Theta(\boldsymbol{n}_\omega) \omega^2 \ ,$$

and can be calulated as follows:

$$\Theta(\boldsymbol{n}_\omega) = \sum_{i=1}^{N} m_i r_{i,\perp}^2 \ , \quad \text{or as an integral:}$$

$$\Theta(\boldsymbol{n}_\omega) = \int_{\mathcal{K}} \varrho(\boldsymbol{r}) \mathrm{d}V r_\perp^2 \ .$$

Here $\varrho(\boldsymbol{r})$ is the mass density and $r_\perp$ is the perpendicular distance from the axis of rotation. At first glance it appears as if we would be forced, for extremely asymmetric systems, to re-calculate this integral for every new rotation direction. Fortunately the situation is much simpler. For a given system a maximum of six integrals suffices. This is at the expense of defining a mathematical entity $\overset{\leftrightarrow}{\boldsymbol{\theta}}$ with two indices and with transformation properties similar to a product of two components of the same vector (i.e., similar to $v_i v_k$), the components $\theta_{i,k} = \theta_{k,i}$ of the so-called *inertia tensor*, as derived in the following:

Firstly we apply the so-called Laplace identity

$$[\boldsymbol{a} \times \boldsymbol{b}]^2 \equiv a^2 b^2 - (\boldsymbol{a} \cdot \boldsymbol{b})^2 \ ,$$

and obtain explicitly, by using the distance vector $\boldsymbol{r}_i - \boldsymbol{R}_\mathrm{s}$ with components (x_i, y_i, z_i):

$$\mathrm{T_{rot}} = \frac{1}{2} \sum_{i=1}^{N} m_i \left[\omega^2 (\boldsymbol{r}_i - \boldsymbol{R}_\mathrm{s})^2 - (\boldsymbol{\omega} \cdot (\boldsymbol{r}_i - \boldsymbol{R}_\mathrm{s}))^2 \right]$$

$$= \frac{1}{2} \sum_{i=1}^{N} m_i \cdot \left[\omega_x^2 \cdot (y_i + z_i)^2 + \omega_y^2 \cdot (z_i + x_i)^2 + \omega_z^2 \cdot (x_i + y_i)^2 \right.$$

$$\left. - 2\omega_x \omega_y \cdot x_i y_i - 2\omega_y \omega_z \cdot y_i z_i - 2\omega_z \omega_x \cdot z_i y_i \right] \ . \tag{11.2}$$

As a consequence, the kinetic energy is

$$T \equiv \frac{1}{2} M v_{\mathrm{s}}^2 + \frac{1}{2} \Theta(\boldsymbol{n}_\omega) \omega^2 \ ,$$

with a moment of inertia $\Theta(\boldsymbol{n}_\omega)$, which depends on the direction $\boldsymbol{n}_\omega$ of the rotation vector according to the following statement:

For

$$\hat{\boldsymbol{\omega}} = \boldsymbol{n}_\omega = \frac{\boldsymbol{\omega}}{|\boldsymbol{\omega}|}$$

we have

$$\Theta(\boldsymbol{n}_\omega) \equiv \sum_{j,k=1}^{3} \theta_{j,k} \hat{\omega}_j \hat{\omega}_k \ ,$$

with the so-called inertia tensor $\theta_{j,k}$, $j, k = 1, 2, 3$, which is a real symmetric 3×3-matrix, where the diagonal elements and off-diagonal elements are defined by

$$\theta_{x,x} := \int_\mathcal{K} \mathrm{d}V \varrho \cdot \left(y^2 + z^2 \right), \quad \theta_{x,y} = \theta_{y,x} := - \int_\mathcal{K} \mathrm{d}V \varrho \cdot xy, \quad \text{etc.} \quad (11.3)$$

These formulae can be unified (with $x =: x_1$, $y =: x_2$, $z =: x_3$) to the following expression:

$$\theta_{j,k} = \int_\mathcal{K} \mathrm{d}V \varrho \cdot \left\{ (x_1^2 + x_2^2 + x_3^2) \cdot \delta_{j,k} - x_j x_k \right\} \ .$$

Here $\delta_{j,k}$ is the Kronecker delta, defined as $\delta_{j,k} = 1$ for $j = k$, and $\delta_{j,k} = 0$ for $j \neq k$.

At this point we shall summarize the results. For the *rotational energy* one obtains

$$\mathrm{T}_{\mathrm{rot}} = \frac{1}{2} \Theta(\hat{\omega}) \omega^2 \equiv \frac{1}{2} \sum_{j,k=1}^{3} \omega_j \theta_{j,k} \omega_k \ . \qquad (11.4)$$

Here the double sum can be shortened to

$$\mathrm{T}_{\mathrm{rot}} \equiv \frac{1}{2} \boldsymbol{\omega} \cdot \overset{\leftrightarrow}{\boldsymbol{\theta}} \cdot \boldsymbol{\omega} \ .$$

Similarly one can show that the *angular momentum*, $\boldsymbol{L}$, typically with respect to an axis of rotation through the center of mass, can also be expressed by the inertia tensor, e.g., for $j = 1, 2, 3$:

$$L_j = \sum_{k=1}^{3} \theta_{j,k} \omega_k \ . \qquad (11.5)$$

This corresponds to the short version $\boldsymbol{L} \equiv \overset{\leftrightarrow}{\boldsymbol{\theta}} \cdot \boldsymbol{\omega}$.

11.3 Steiner's Theorem; Heavy Cylinder on an Inclined Plane; Physical Pendulum

Steiner's theorem states that for a rotation about an axis $\boldsymbol{n}_\omega$ not passing through the center of mass, but with a (perpendicular) distance $\boldsymbol{l}_\perp$ from this point, the moment of inertia is related to the "central moment" $\Theta^{(cm)}(\boldsymbol{n}_\omega)$ (i.e., for a parallel rotation axis through the center of mass) by the following simple expression:

$$\Theta(\boldsymbol{l}_\perp, \boldsymbol{n}_\omega) \equiv M \cdot l_\perp^2 + \Theta^{(\mathrm{cm})}(\boldsymbol{n}_\omega) \ .$$

The proof of Steiner's theorem will be omitted here, since it is very elementary. We only note that the reference point $\boldsymbol{r}_0$ is not $\boldsymbol{R}_\mathrm{s}$, but $\boldsymbol{R}_\mathrm{s} + \boldsymbol{l}_\perp$.

As an example consider a *heavy roller* on an inclined plane. We assume that the mass distribution of the system has cylindrical symmetry. Let M be the total mass of the cylinder, while the radius is R. We assume further that the moment of inertia w.r.t. a longitudinal axis through the center of mass of the cylinder is $\Theta^{(s)}$.

Let the slope of the inclined plane, which is assumed to be parallel to the cylinder axis, be characterized by the angle α. We then have the *Lagrangian*

$$\mathcal{L} = \frac{1}{2}\left(M\boldsymbol{v}_\mathrm{s}^2 + \Theta^{(s)}\dot{\varphi}^2\right) + Mg_\mathrm{eff} \cdot s \ .$$

Here s stands for the distance rolled, while $g_\mathrm{eff} = g \cdot \tan\alpha$.

Furthermore, one can set

$$\boldsymbol{v}_\mathrm{s}^2 = \dot{s}^2 \quad \text{and} \quad \varphi = \frac{s}{R} \ ,$$

thus obtaining a *Lagrangian* that depends only on s and $\dot{s}$. The rest of the analysis is elementary[2].

Up till now, in our description of the motion of the roller, we have concentrated on the cylinder axis and the center of mass of the cylinder, which lies on this axis; and so we have explicitly obtained the translational part of the kinetic energy. But equivalently, we can also concentrate on the *rolling motion* of the *tangential* point on the surface of the cylinder, i.e., where the surface of the roller touches the inclined plane. Then the translational part does not enter explicitly; instead, it is concealed in the new moment of inertia. This moment must be calculated according to Steiner's theorem with respect to the *axis of contact*. Therefore, we have

$$\Theta_R := \Theta^{(s)} + MR^2 \ , \quad \text{i.e., we obtain} \quad \mathcal{L} \equiv \mathcal{T} - \mathcal{V} = \frac{1}{2}\Theta_R\dot{\varphi}^2 + Mg_\mathrm{eff}s \ .$$

Both descriptions are of course equivalent.

---

[2] Below, we shall return to this seemingly elementary problem, which (on extension) is more complicated than one might think at first sight.

We should mention in this context that modifications of this problem frequently appear in written examination questions, in particular the problem of the *transition* from rolling motion down an inclined segment to a subsequent horizontal motion, i.e., on a horizontal plane segment, and we also mention the difference between *gliding* and *rolling* motions (see below).

As an additional problem we should also mention the *physical pendulum*, as opposed to the mathematical pendulum (i.e., a point mass suspended from a weightless thread or rod), on which we concentrated previously. The *physical pendulum* (or compound pendulum) is represented by a rigid body, which is suspended from an axis around which it can rotate. The center of mass is at a distance s from the axis of rotation, such that

$$\mathcal{L} \equiv \mathcal{T} - \mathcal{V} \equiv \frac{1}{2}\Theta\dot{\vartheta}^2 - Mgs \cdot (1 - \cos\vartheta) \,.$$

Here M is the mass of the body, and

$$\Theta := \Theta^{(s)} + Ms^2$$

is the moment of inertia of the rigid body w.r.t. the axis of rotation.

The oscillation eigenfrequency for small-amplitude oscillations is therefore

$$\omega_0 = \sqrt{\frac{g}{l_{\text{red}}}} \,,$$

where the so-called *reduced length* of the pendulum, l_{red}, is obtained from the following identity:

$$l_{\text{red}} = \frac{\Theta}{Ms} \,.$$

At this point we shall mention a remarkable physical "non-event" occurring at the famous Gothic cathedral in Cologne. The towers of the medieval cathedral were completed in the nineteenth century and political representatives planned to highlight the completion by ringing the so-called Emperor's Bell, which was a very large and powerful example of the best technology of the time. The gigantic bell, suspended from the interior of one of the towers, was supposed to provide acoustical proof of the inauguration ceremony for the towers. But at the ceremony, the bell failed to produce a sound, since the clapper, in spite of heavy activity of the ringing machinery, never actually struck the outer mantle of the bell[3]. This disastrous misconstruction can of course be attributed to the fact that the angle of deflection $\vartheta_1(t)$ of the clapper as a function of time was always practically identical with the angle of deflection of the outer mantle of the bell, i.e., $\vartheta_2(t) \approx \vartheta_1(t)$ (both angles measured from the vertical direction). This is again supported by a diagram (Fig. 11.1).

---

[3] An exercise can be found on the internet, [2], winter 1992, file 8.

Fig. 11.1. Schematic diagram for a bell. The diagram shows the mantle of the bell (the outer double lines, instantaneously swinging to the right) and a clapper (instantaneously swinging to the left). It is obvious from this sketch that there are two characteristic angles for the bell, the first one characterizing the deviation of the center of mass (without clapper) from the vertical, the second one for the clapper itself.

The reason for the failure, which is often used as a typical examination *question on the so-called double compound pendulum*, is that the geometrical and material parameters of the bell were such that

$$l_{\text{red}|\{\text{bell with fastened clapper}\}} \approx l_{\text{red}|\{\text{clapper alone}\}} .$$

Another remarkable example is the *reversion pendulum*. If we start from the above-mentioned equation for the reduced length of the pendulum, insertion of the relation

$$\Theta \overset{!}{=} \Theta^{(s)} + Ms^2 \quad \text{gives} \quad l_{\text{red}}(s) \overset{!}{=} \frac{\Theta^{(s)}}{Ms} + s .$$

This yields a quadratic equation with two solutions satisfying

$$s_1 + s_2 = l_{\text{red}} .$$

For a given system, i.e., with given center of mass, we thus obtain two solutions, with

$$s_2 = l_{\text{red}} - s_1 .$$

If we let the pendulum oscillate about the first axis, i.e., corresponding to the first solution s_1, then, for small oscillation amplitudes we obtain the oscillation eigenfrequency

$$\omega_0 = \sqrt{\frac{g}{l_{\text{red}}}} .$$

If, subsequently, after *reversion*, we let the pendulum oscillate about the parallel second axis, i.e., corresponding to s_2, then we obtain the same frequency.

11.4 Inertia Ellipsoids; Poinsot Construction

The positive-definite quadratic form

$$2\mathrm{T_{rot}} = \sum_{i,k=1}^{3} \theta_{i,k}\omega_i\omega_k$$

describes an ellipsoid within the space $\mathcal{R}^3(\boldsymbol{\omega})$, the so-called *inertia ellipsoid*.

This ellipsoid can be diagonalized by a suitable rotation, and has the *normal form*

$$2\mathrm{T_{rot}} = \sum_{\alpha=1}^{3} \Theta_\alpha \cdot \omega_\alpha^2 \, ,$$

with so-called *principal moments of inertia* Θ_α and corresponding *principal axes* of lengths

$$a_\alpha = \sqrt{\frac{2\mathrm{T_{rot}}}{\Theta_\alpha}} \, .$$

(See Fig. 11.2.)

Every rigid body, however arbitrarily complex, possesses such an *inertia ellipsoid*. This is a smooth fictitious body, in general possessing three principal axes of different length, and where the directions of the principal axes are not easily determined. However, if the body possesses a (discrete!) n-fold axis of rotation[4], with $n \geq 3$, then the inertia ellipsoid is an *ellipsoid of* (continuous!) *revolution* around the symmetry axis.

As a practical consequence, e.g., for a three-dimensional system with a square base, this implies the following property. If the system is rotated about an arbitrary axis lying in the square, then the rotational energy of the system does *not* depend on the direction of the rotation vector $\boldsymbol{\omega}$, e.g., whether it is parallel to one of the edges or parallel to one of the diagonals; it depends only on the *magnitude* of the rotational velocity.

The *angular momentum* $\boldsymbol{L}$ is calculated by forming the gradient (w.r.t. $\boldsymbol{\omega}$) of $\mathrm{T_{rot}}(\boldsymbol{\omega})$. Thus

$$\boldsymbol{L}(\boldsymbol{\omega}) = \mathrm{grad}_\omega \mathrm{T_{rot}}(\boldsymbol{\omega}) \, , \quad \text{i.e.,} \quad L_\alpha = \frac{\partial \mathrm{T_{rot}}}{\partial \omega_\alpha} = \Theta_\alpha \omega_\alpha \, .$$

Since the gradient vector is perpendicular to the tangent plane of $\mathrm{T_{rot}}(\boldsymbol{\omega})$, one thus obtains the so-called Poinsot construction:

For given vector of rotation the angular momentum $\boldsymbol{L}(\boldsymbol{\omega})$ is by construction perpendicular to the tangential plane of the inertial ellipsoid $T_{rot}(\boldsymbol{\omega}) =$ constant corresponding to the considered value of $\boldsymbol{\omega}$. This can be seen in Fig. 11.2.

---

[4] This means that the mass distribution of the system is invariant under rotations about the symmetry axis by the angle $2\pi/n$.

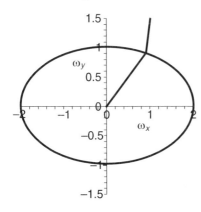

Fig. 11.2. Inertia ellipsoid and Poinsot construction. For the typical values $a := 2$ and $b := 1$, a (ω_x, ω_y)-section of an inertia ellipsoid $\frac{\omega_1^2}{\Theta_1} + \ldots + \frac{\omega_3^2}{\Theta_3} = 2 \cdot T_{\text{rot}}$ (with, e.g., $a^2 \equiv \frac{2 \cdot T_{\text{rot}}}{\Theta_1}$) and an initial straight line (here diagonal) through the origin is plotted as a function of ω_x. The second (external) straight line is *perpendicular to the tangential plane of the first line*, i.e., it has the direction of the angular momentum $\boldsymbol{L}$. This is the so-called Poinsot construction

One can also define another ellipsoid, which is related to the *inertia ellipsoid*, but which is *complementary* or *dual* to it, *viz* its *Legendre transform*. The transformed ellipsoid or so-called *Binet ellipsoid* is not defined in the space $\mathcal{R}^3(\boldsymbol{\omega})$, but in the *dual space* $\mathcal{R}^3(\boldsymbol{L})$ by the relation

$$2T_{\text{rot}} = \sum_{\alpha=1}^{3} \frac{L_\alpha^2}{\Theta_\alpha} .$$

Instead of the *angular velocity* $\boldsymbol{\omega}$, the *angular momentum* $\boldsymbol{L}$ is now central. In fact, $\boldsymbol{L}$ and $\boldsymbol{\omega}$ are not parallel to each other unless $\boldsymbol{\omega}$ has the direction of a principal axis of the inertia ellipsoid.

We shall return to this point in Part II, while discussing crystal optics, when the difference between the directions of the electric vectors $\boldsymbol{E}$ and $\boldsymbol{D}$ is under debate.

11.5 The Spinning Top I: Torque-free Top

A *spinning top* is by definition a rigid body supported at a *canonical reference point* $\boldsymbol{r}_0$ (see above), where in general $\boldsymbol{v}_0 \neq 0$. For example, $\boldsymbol{r}_0$ is the point of rotation of the rigid body on a flat table. In this case, the gravitational forces, virtually concentrated at the center of mass, produce the following torques w.r.t. the point of rotation of the top on the table:

$$\boldsymbol{\mathcal{D}}_{\boldsymbol{r}_0} = (\boldsymbol{R}_{\text{s}} - \boldsymbol{r}_0) \times (-Mg\hat{\boldsymbol{z}}) .$$

A *torque-free top* is by definition supported at its center of mass; therefore no gravitational torque does any work on the top. The total angular momentum $\boldsymbol{L}$ is thus conserved, as long as external torques are not applied, which is assumed. Hence the motion of the *torque-free top* can be described as a "rolling of the inertia ellipsoid along the Poinsot plane", see above. In this way, $\boldsymbol{\omega}$-lines are generated on the surface of the inertia ellipsoid as described in the following section.

11.6 Euler's Equations of Motion and the Stability Problem

We now wish to describe the stability of the orbits of a *torque-free top*. Therefore we consider as perturbations only very weak transient torques. Firstly, we have

$$\boldsymbol{\omega}(t) = \sum_{\alpha=1}^{3} w_{\alpha}(t) \boldsymbol{e}_{\alpha}(t) \quad \text{and} \quad \boldsymbol{L} = \sum_{\alpha=1}^{3} L_{\alpha}(t) \boldsymbol{e}_{\alpha}(t) \,.$$

Here the (time-dependent) unit vectors $\boldsymbol{e}_{\alpha}(t)$ describe the motion of the principal axes of the (moving) inertia ellipsoid; $w_{\alpha}(t)$ and $L_{\alpha}(t)$ are therefore *co-moving* cartesian coefficients of the respective vectors $\boldsymbol{\omega}$ and $\boldsymbol{L}$; i.e., the co-moving cartesian axes $\boldsymbol{e}_{\alpha}(t)$, for $\alpha = 1, 2, 3$, are moving with the rigid body, to which they are *fixed* and they depend on t, whereas the axes $\boldsymbol{e}_{j}$, for $j = x, y, z$, are fixed in space and do *not* depend on t:

$$L_{\alpha}(t) \equiv L_{\alpha}(t)_{|\text{co-moving}} \,,$$

$$w_{\alpha}(t) \equiv w_{\alpha}(t)_{|\text{co-moving}} \,.$$

But

$$\frac{\mathrm{d}}{\mathrm{d}t} \boldsymbol{e}_{\alpha} = \boldsymbol{\omega} \times \boldsymbol{e}_{\alpha} \,;$$

hence

$$\frac{\mathrm{d}}{\mathrm{d}t} \boldsymbol{L} = \sum_{\alpha=1}^{3} \left(\frac{\mathrm{d}L_{\alpha}}{\mathrm{d}t} \right)_{|\text{co-moving}} \boldsymbol{e}_{\alpha}(t) + \boldsymbol{\omega} \times \boldsymbol{L} = \delta \boldsymbol{\mathcal{D}}(t) \,, \tag{11.6}$$

where the perturbation on the right describes the transient torque.

We also have

$$\frac{\mathrm{d}}{\mathrm{d}t} \boldsymbol{\omega} = \sum_{\alpha=1}^{3} \left(\frac{\mathrm{d}w_{\alpha}}{\mathrm{d}t} \right)_{|\text{co-moving}} \boldsymbol{e}_{\alpha}(t) \,.$$

Hence in the co-moving frame of the rigid body:

$$\left(\frac{\mathrm{d}L_{\alpha}}{\mathrm{d}t} \right) + w_{\beta} L_{\gamma} - w_{\gamma} L_{\beta} = \delta \mathcal{D}_{\alpha}$$

etc., and with $L_{\alpha} = \Theta_{\alpha} w_{\alpha}$ we obtain the (non-linear) *Euler equations* for the calculation of the orbits $\boldsymbol{\omega}(t)$ on the surface of the *inertia ellipsoid*:

$$\Theta_{\alpha} \cdot \left(\frac{\mathrm{d}w_{\alpha}}{\mathrm{d}t} \right) + w_{\beta} w_{\gamma} \cdot (\Theta_{\gamma} - \Theta_{\beta}) = \delta \mathcal{D}_{\alpha}(t) \,,$$

$$\Theta_{\beta} \cdot \left(\frac{\mathrm{d}w_{\beta}}{\mathrm{d}t} \right) + w_{\gamma} w_{\alpha} \cdot (\Theta_{\alpha} - \Theta_{\gamma}) = \delta \mathcal{D}_{\beta}(t) \,,$$

$$\Theta_\gamma \cdot \left(\frac{d\omega_\gamma}{dt} \right) + \omega_\alpha \omega_\beta \cdot (\Theta_\beta - \Theta_\alpha) = \delta \mathcal{D}_\gamma(t) \; . \tag{11.7}$$

Let us now consider the case of an ellipsoid with three different principal axes and assume without lack of generality that

$$\Theta_\alpha < \Theta_\beta < \Theta_\gamma \; .$$

In (11.7), of the three prefactors

$$\propto \omega_\beta \omega_\gamma \, , \propto \omega_\gamma \omega_\alpha \quad \text{and} \quad \propto \omega_\alpha \omega_\beta$$

the first and third are positive, while only the second (corresponding to the so-called *middle axis*) is negative. As a consequence, small transient perturbations $\delta \mathcal{D}_\alpha$ and $\delta \mathcal{D}_\gamma$ lead to *elliptical* motions around the respective principal axis, i.e.,

$$\propto \left(\varepsilon_\beta \omega_\beta^2 + \varepsilon_\gamma \omega_\gamma^2 \right) \quad \text{and} \quad \propto \left(\varepsilon_\alpha \omega_\alpha^2 + \varepsilon_\beta \omega_\beta^2 \right) \; ,$$

whereas perturbations corresponding to $\delta \mathcal{D}_\beta$ may have fatal effects, since in the vicinity of the second axis the $\boldsymbol{\omega}$-lines are *hyperbolas*,

$$\propto \left(\varepsilon_\alpha \omega_\alpha^2 - \varepsilon_\gamma \omega_\gamma^2 \right) \; ,$$

i.e., along one diagonal axis they are *attracted*, but along the other diagonal axis there is *repulsion*, and the orbit runs far away from the axis where it started.

The appearance of *hyperbolic fixed points* is typical for the *transition to chaos*, which is dicussed below.

Motion about the middle axis is thus unstable.

This statement can be shown to be plausible by plotting the orbits in the frame of the rotating body for infinitesimally small perturbations, $\delta \boldsymbol{\mathcal{D}} \to 0$, The lines obtained in this way on the surface of the inertia ellipsoid by integration of the three coupled equations (11.7) describe the *rolling motion of the inertia ellipsoid on the Poinsot plane*. Representations of these lines can be found in almost all relevant textbooks.

Another definition for the $\boldsymbol{\omega}$-lines on the inertia ellipsoid is obtained as follows: they are obtained by cutting the *inertia ellipsoid*

$$\sum_{\alpha=1}^{3} \Theta_\alpha \omega_i^2 = 2 \mathrm{T_{rot}} = \text{constant}' \; .$$

with the so-called $\boldsymbol{L}^2$ or *swing ellipsoid*, i.e., the set

$$\sum_{\alpha=1}^{3} \Theta_\alpha^2 \omega_i^2 = \boldsymbol{L}^2 = \text{constant}' \; .$$

We intentionally omit a proof of this property, which can be found in many textbooks.

To calculate the *Hamiltonian* it is not sufficient to know

$$\frac{\mathrm{d}}{\mathrm{d}t}\omega_{\alpha}|_{\mathrm{co-moving}} \; .$$

Therefore we proceed to the next section.

11.7 The Three Euler Angles φ, ϑ and ψ; the Cardani Suspension

Apart from the location of its center of mass, the position of a rigid body is characterized by the location in *space* of two orthogonal axes fixed within the *body*: e_3 and e_1. For example, at $t = 0$ we may assume

$$e_3 \equiv \hat{z} \quad \text{and} \quad e_1 \; (\text{without } \textit{prime}) \equiv \hat{x} \; .$$

For $t > 0$ these vectors are moved to new directions:

$$e_3(0) \to e_3(t)(\equiv e_3) \quad \text{and} \quad e_1 \to e_1' \; (\text{now with } \textit{prime}) \; .$$

This occurs by a *rotation* $\mathcal{D}_R$, which corresponds to the three *Euler angles* φ, χ and ψ, as follows (the description is now *general*):

a) Firstly, the rigid body is rotated about the z-axis (which is fixed in space) by an angle of rotation φ, in such a way that the particular axis of the rigid body corresponding to the $\hat{x}$ direction is moved to a so-called *node direction* e_1.
b) Next, the particular axis, which originally (e.g., for $t = 0$) points in the $\hat{z}$-direction, is tilted around the *nodal axis* e_1 in the e_3 direction by an angle $\vartheta = \arccos(\hat{z} \cdot e_3)$.
c) Finally, a rotation by an angle ψ around the e_3 axis follows in such a way that the *nodal direction*, e_1, is rotated into the final direction e_1'.

As a consequence we obtain

$$\mathcal{D}_R \equiv \mathcal{D}_{e_3}^{3)}(\psi) \cdot \mathcal{D}_{e_1}^{2)}(\vartheta) \cdot \mathcal{D}_{\hat{z}}^{1)}(\varphi) \; ,$$

where one reads from right to left, and the correct order is important, since finite rotations do not commute (only infinitesimal rotations would commute). A technical realization of the *Euler angles* is the well-known *Cardani suspension*, a construction with a sequence of three intertwined rotation axes in independent axle bearings (see the following text). In particular this is a realization of the relation

$$\omega = \dot{\psi}e_3 + \dot{\vartheta}e_1 + \dot{\varphi}\hat{z} \; ,$$

which corresponds to the special case of infinitesimal rotations; this relation will be used below.

In the following we shall add a verbal description of the *Cardani suspension*, which (as will be seen) corresponds perfectly to the *Euler angles*.

The lower end of a dissipationless *vertical* rotation axis, 1 ($\hat{=} \pm \hat{z}$, corresponding to the rotation angle φ), branches into two axle bearings, also dissipationless, supporting a *horizontal* rotation axis, 2 ($= \pm e_1$), with corresponding rotation angle ϑ around this axis.

This horizontal rotation axis corresponds in fact to the above-mentioned *nodal line*; in the middle, this axis branches again (for $\vartheta = 0$: upwards and downwards) into dissipationless axle bearings, which support a third rotation axis, 3, related to the direction e_3 (i.e., the *figure axis*), with rotation angle ψ. For finite ϑ, this third axis is tilted with respect to the vertical plane through $\hat{z}$ and e_1. The actual top is attached to this third axis.

Thus, as mentioned, for the so-called *symmetrical top*, i.e., where one is dealing with an inertia ellipsoid with rotational symmetry, e.g.,

$$\Theta_1 \equiv \Theta_2 \, ,$$

the (innermost) third axis of the Cardani suspension corresponds to the so-called *figure axis* of the top.

The reader is strongly recommended to try to transfer this verbal description into a corresponding sketch! The wit of the Cardani suspension is that the vertical axis 1 branches to form the axle bearings for the horizontal nodal axis, 2, and this axis provides the axle bearings for the figure axis 3.

To aid the individual imagination we present our own sketch in Fig. 11.3.

The importance of the *Euler angles* goes far beyond theoretical mechanics as demonstrated by the technical importance of the *Cardani suspension*.

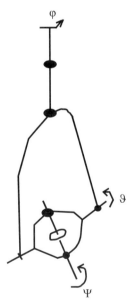

Fig. 11.3. The figure illustrates a so-called Cardani suspension, which is a construction involving three different axes of rotation with axle bearings, corresponding to the Euler angles φ, ϑ and ψ

11.8 The Spinning Top II: Heavy Symmetric Top

The "heavy symmetric top" is called *heavy*, because it is *not* supported at the center of mass, so that gravitational forces now exert a torque, and it is called *symmetric*, because it is assumed, e.g., that

$$\Theta_1 \equiv \Theta_2 (\neq \Theta_3) \ .$$

The axis corresponding to the vector e_3 is called the *figure axis*. Additionally one uses the abbreviations

$$\Theta_\perp := \Theta_1 \equiv \Theta_2 \quad \text{and} \quad \Theta_{||} := \Theta_3 \ .$$

Let us calculate the *Lagrangian*

$$\mathcal{L}(\varphi, \vartheta, \psi, \dot\varphi, \dot\vartheta, \dot\psi) \ .$$

Even before performing the calculation we can expect that ψ and φ are *cyclic coordinates*, i.e., that

$$\frac{\partial\mathcal{L}}{\partial\varphi} = \frac{\partial\mathcal{L}}{\partial\psi} = 0 \ ,$$

such that the corresponding *generalized momenta*

$$p_\varphi := \frac{\partial\mathcal{L}}{\partial\dot\varphi} \quad \text{and} \quad p_\psi := \frac{\partial\mathcal{L}}{\partial\dot\psi}$$

are *conserved* quantities.

Firstly, we express the *kinetic energy*

$$\mathrm{T}_{\mathrm{rot}} := \frac{1}{2} \cdot \left[\Theta_\perp \cdot (\omega_1^2 + \omega_2^2) + \Theta_{||}\omega_3^2 \right]$$

in terms of the Euler angles by

$$\boldsymbol{\omega} = \dot\psi e_3 + \dot\vartheta e_1 + \dot\varphi \hat{z} \ .$$

Scalar multiplication with e_3, e_1 and e_2 yields the result[5]:

$$\omega_3 = \dot\psi + \dot\varphi \cdot \cos\vartheta \ ,$$
$$\omega_1 = \dot\vartheta \quad \text{and}$$
$$\omega_2 = \dot\varphi \cdot \sin\vartheta \ .$$

As a consequence, we have

$$\mathrm{T}_{\mathrm{rot}} = \frac{\Theta_\perp}{2} \cdot \left(\dot\vartheta^2 + \sin^2\vartheta\dot\varphi^2 \right) + \frac{\Theta_{||}}{2} \cdot \left(\dot\psi + \dot\varphi\cos\vartheta \right)^2 \ . \tag{11.8}$$

---

[5] because e_3, $\hat{z}$ and e_2 are co-planar, such that $e_3 \cdot \hat{z} = \cos\vartheta$ and $e_2 \cdot \hat{z} = \sin\vartheta$.

Additionally we have the *potential energy*

$$V = Mgs \cdot \cos \vartheta , \quad \text{where} \quad s \cdot \cos \vartheta$$

is the z-component of the vector $(\boldsymbol{R}_{\mathrm{s}} - \boldsymbol{r}_0)$. The *Lagrangian* is

$$\mathcal{L} = \mathrm{T}_{\mathrm{rot}} - \mathcal{V} .$$

Obviously ψ and φ are cyclic, and the above-mentioned *generalized momenta* are *conserved*, *viz* the following two components of the *angular momentum* $\boldsymbol{L}$:

$$p_\psi = \frac{\partial \mathrm{T}_{\mathrm{rot}}}{\partial \dot\psi} = \Theta_{||} \cdot (\dot\psi + \dot\varphi \cos \vartheta) = \Theta_{||}\omega_3 \equiv L_3 \equiv \boldsymbol{L}(t) \cdot \boldsymbol{e}_3(t) , \quad \text{and}$$

$$\begin{aligned} p_\varphi = \frac{\partial \mathrm{T}_{\mathrm{rot}}}{\partial \dot\varphi} &= \Theta_\perp \cdot \sin^2 \vartheta \dot\varphi + \Theta_{||} \cdot (\dot\psi + \dot\varphi \cos \vartheta) \cos \vartheta \\ &= \Theta_2 \omega_2 \sin \vartheta + \Theta \omega_3 \cos \vartheta \\ &= L_2 \sin \vartheta + L_3 \cos \vartheta \equiv L_z \equiv \boldsymbol{L}(t) \cdot \hat{\boldsymbol{z}} . \end{aligned}$$

The *symmetric heavy top* thus has exactly as many *independent*[6] *conserved quantities* as there are *degrees of freedom*. The number of degrees of freedom is $f = 3$, corresponding to the three *Euler angles*, and the *three independent conserved quantities* are:

a) The conservation law for the (co-moving!) component L_3 of the angular momentum, i.e., the projection onto the (moving!) *figure axis*. (This conservation law is remarkable, in that the *figure axis* itself performs a complicated motion. On the one hand the figure axis *precesses* around the (fixed) z-axis; on the other hand, it simultaneously performs so-called *nutations*, i.e., the polar angle $\vartheta(t)$ does not remain constant during the precession, but performs a more-or-less rapid periodic motion between two extreme values.)
   For the *asymmetric* heavy top this conservation law would *not* be true.
b) The component L_z of the angular momentum (fixed in space!).
c) As a third conservation law we have, of course, *conservation of energy*, since on the one hand the Lagrangian $\mathcal{L}$ does not depend explicitly on time, while on the other hand we have used *skleronomous* coordinates, i.e., the relation between the Euler angles and the cartesian coordinates x, y and z does not explicitly depend on time.

---

[6] Two *dependent* conserved quantities are e.g., the Hamiltonian $\mathcal{H}$ of a conservative system and any function $F(\mathcal{H})$.

12 Remarks on Non-integrable Systems: Chaos

Systems for which the number of independent conserved quantities agrees with the number f of *degrees of freedom* are termed *integrable*. They are quasi *especially simple* and set the standards in many respects.

If one fixes the values of the independent conserved quantities, then in $2f$-dimensional phase-space (with the canonical phase-space variables $p_1, \ldots, p_f$ and $q_1, \ldots, q_f$) one generates an f-dimensional hypersurface, which for $f = 2$) has the topology of a *torus*.

However, it is obvious that most systems are *non-integrable*, since generically the number of degrees of freedom is larger than the number of *conservation theorems*. This applies, e.g., to so-called *three-body problems*, and it would also apply to the *asymmetric heavy top*, as mentioned above, or to the *double pendulum*. It is no coincidence that in the usual textbooks[1] little attention is paid to *non-integrable systems* because they involve complicated relations, which require more mathematics than can be assumed on a more or less elementary level.[2]

Linear systems are, as we have seen above, always simple, at least in principle. In contrast, for *non-linear* non-integrable systems *chaotic behavior* occurs. In most cases this behavior is qualitatively *typical* and can often be understood from simple examples or so-called *scenarios*. One of these *scenarios* concerns the so-called *sensitive dependence on the initial conditions*, e.g. as follows.

Consider a non-linear system of differential equations

$$\frac{\mathrm{d}\boldsymbol{X}(t)}{\mathrm{d}t} = \boldsymbol{f}(t; \boldsymbol{X}_0) \, ,$$

where $\boldsymbol{X}_0$ are the initial values of $\boldsymbol{X}(t)$ for $t = t_0$. We then ask whether the orbits $\boldsymbol{X}(t)$ of this non-linear "dynamical system" depend continuously on the initial values in the limit $t \to \infty$; or we ask how long the orbits remain in an ε-neighborhood of the initial values.

---

[1] This text makes no exception.

[2] It is mostly unknown and symptomatic for the complexity of nonintegrable systems that Sommerfeld, who was one of the greatest mathematical physicists of the time, wrote in the early years of the twentieth century a voluminous book containing three volumes on the "Theory of the Spinning Top".

In particular one might ask for the properties of the derivative

$$\frac{\mathrm{d}\boldsymbol{X}(t;\boldsymbol{X}_0+\varepsilon)}{\mathrm{d}\varepsilon}$$

as a function of t. For "malicious" behavior of the system. e.g., for turbulent flow, weather or stock-exchange or traffic forecasting etc., the limits $\varepsilon \to 0$ and $t \to \infty$ cannot be interchanged. This has the practical consequence that very small errors in the data accumulation or in the weighting can have irreparable consequences beyond a characteristic time for the forecasting or turbulent flow (the so-called *butterfly effect*).

However, these and other topics lead far beyond the scope of this text and shall therefore not be discussed in detail, especially since a very readable book on chaotic behavior already exists [9].

Some of these aspects can be explained by the above example of the *spherical pendulum*, see section 8.2. As long as the length of the thread of the pendulum is constant, the system is *conservative* in the *mechanical* sense; or in a mathematical sense it is describable by an *autonomous* system of two coupled ordinary differential equations for the two variables $\varphi(t)$ and $\vartheta(t)$. Thus the number of degrees of freedom is $f = 2$, which corresponds perfectly to the number of independent conserved quantities, these being the vertical component of the angular momentum plus the sum of kinetic and potential energies: the system is *integrable*.

However, the integrability is lost if the length of the pendulum depends explicitly on t and/or the rotational invariance w.r.t. the azimuthal angle φ is destroyed[3]. In the first case the system of coupled differential equations for the variables ϑ and φ becomes *non-autonomous*; i.e., the time variable t must be explicitly considered as a third relevant variable.

In the following we shall consider autonomous systems. In a $2f$-dimensional *phase space* $\varPhi$ of (generalized) coordinates and momenta we consider a two-dimensional sub-manifold, e.g., a plane, and mark on the plane the points, one after the other, where for given initial values the orbit intersects the plane. In this way one obtains a so-called *Poincaré section*, which gives a condensed impression of the trajectory, which may be very "chaotic" and may repeatedly intersect the plane. For example, for periodic motion one obtains a deterministic sequence of a finite number of discrete intersections with the plane; for nonperiodic motion one typically has a more or less *chaotic* or random sequence, from which, however, on detailed inspection one can sometimes still derive certain nontrivial quantitative laws for large classes of similar systems.

---

[3] This applies, e.g., to the so-called *Henon-Heiles potential* $V(x_1,x_2) := x_1^2 + x_2^2 + \varepsilon \cdot (x_1^2 x_2 - \frac{x_2^3}{3})$, which serves as a typical example of *non-integrability* and "chaos" in a simple two-dimensional system; for $\varepsilon \neq 0$ the potential is no longer rotationally invariant, but has only *discrete triangular symmetry*.

A typical nonintegrable system, which has already been mentioned, is the *double pendulum*, i.e., with $\mathcal{L}(\vartheta_1, \vartheta_2, \dot{\vartheta}_1, \dot{\vartheta}_2)$, which was discussed in Sect. 11.2. If here the second pendulum sometimes "flips over", the sequence of these times is of course deterministic, but practically "random", i.e., non-predictable, such that one speaks of *deterministic chaos*; this can easily be demonstrated experimentally.

13 Lagrange Formalism II: Constraints

13.1 D'Alembert's Principle

Consider a system moving in n-dimensional space with (generalized) coordinates $q_1, .., q_n$, but now under the influence of *constraints*. The *constraints* can be either (i) *holonomous* or (ii) *anholonomous*. In the first case *a single* constraint can be formulated, as follows: $f(q_1, q_2, \ldots, q_n, t) \overset{!}{=} 0$, i.e., the constraint defines a time-dependent $(n-1)$-dimensional hypersurface in $\mathcal{R}^n(\boldsymbol{q})$. If one has *two* holonomous constraints, another condition of this kind is added, and the hypersurface becomes $(n-2)$-dimensional, etc..

In the second case we may have:

$$\sum_{\alpha=1}^{n} a_\alpha(q_1, \ldots, q_n, t)\, \mathrm{d}q_\alpha + a_0(q_1, \ldots, q_n, t)\, \mathrm{d}t \overset{!}{=} 0 \ , \qquad (13.1)$$

where (in contrast to holonomous constraints, for which necessarily $a_\alpha \equiv \frac{\partial f}{\partial q_\alpha}$, and generally but not necessarily, $a_0 \equiv \frac{\partial f}{\partial t}$) the conditions of integrability,

$$\frac{\partial a_\alpha}{\partial q_\beta} - \frac{\partial a_\beta}{\partial q_\alpha} \equiv 0, \forall \alpha, \beta = 1, \ldots, n \ ,$$

are *not all* satisfied; thus in this case one has only local hypersurface elements, which do not fit together.

If the constraints depend (explicitly) on the time t, they are called *rheonomous*, otherwise *skleronomous*.

In the following we shall define the term *virtual displacement*: in contrast to *real displacements*, for which the full equation (13.1) applies and which we describe by exact differentials $\mathrm{d}q_\alpha$, the *virtual dispacements* δq_α are written with the variational sign δ, and instead of using the full equation (13.1), for the δq_α the following *shortened condition* is used:

$$\sum_{\alpha=1}^{n} a_\alpha(q_1, \ldots, q_2, t)\delta q_\alpha \overset{!}{=} 0 \ . \qquad (13.2)$$

In the transition from equation (13.1) to (13.2) we thus always put $\delta t \equiv 0$ (although dt may be $\neq 0$), which corresponds to the special role of time in a Galilean transformation[1], i.e., to the formal limit $c \to \infty$.

To satisfy a *constraint*, the system must exert an n-dimensional *force of constraint* $\mathbf{Z}$.[2] (For more than one constraint, $\mu = 1, \ldots, \lambda$, of course, a corresponding set of forces $\mathbf{Z}^{(\mu)}$ would be necessary, but for simplicity in the following we shall only consider the case $n = 1$ explicitly, where the index μ can be omitted.)

D'Alembert's principle applies to the constraining forces:

a) *For all virtual displacements a force of constraint does no work*

$$\sum_{\alpha=1}^{n} Z_\alpha \delta q_\alpha \equiv 0 \ . \tag{13.3}$$

The following two statements are equivalent:

b) A *force of constraint* is always *perpendicular* to the instantaneous hypersurface or to the local virtual hypersurface element, which corresponds to the constraint; e.g., we have $\delta f(q_1, \ldots, q_n, t) \equiv 0$ or equation (13.2).

c) A so-called *Lagrange multiplier* λ exists, such that for all

$$\alpha = 1, \ldots, n : \quad Z_\alpha = \lambda \cdot a_\alpha(q_1, \ldots, q_n, t) \ .$$

These three equivalent statements have been originally formulated in cartesian coordinates; but they also apply to generalized coordinates, if the term *force of constraint* is replaced by a *generalized force of constraint*.

Now, since without constraints

$$\mathcal{L} = \mathcal{T} - \mathcal{V}$$

while for cartesian coordinates the forces are

$$F_\alpha = -\frac{\partial \mathcal{V}}{\partial x_\alpha}$$

it is natural to modify the *Hamilton principle of least action* in the presence of a single holonomous constraint, as follows:

$$\frac{dS[q_1 + \varepsilon\delta q_1, \ldots, q_n + \varepsilon\delta q_n]}{d\varepsilon}\bigg|_{\varepsilon \equiv 0} =$$

$$= \int_{t=t_1}^{t_2} dt\{\mathcal{L}(q_1, \ldots, q_n, \dot{q}_1, \ldots, \dot{q}_n, t) + \lambda \cdot f(q_1, \ldots, q_n, t)\} \overset{!}{=} 0 \ . \tag{13.4}$$

---

[1] This is also the reason for using a special term in front of dt in the definition (13.1) of anholonomous constraints.

[2] As stated below, here one should add a slight generalization: *force* $\to$ *generalized force*; i.e., $\sum_{\beta=1}^{n} \tilde{Z}_\beta \delta x_\beta \equiv \sum_{\alpha=1}^{n} Z_\alpha \delta q_\alpha$, where the $\tilde{Z}_\beta$ are the (cartesian) components of the constraining force and the Z_α the components of the related generalized force.

Analogous relations apply for more than one holonomous or anholonomous constraint. Instead of the *Lagrange equations of the second kind* we now have:

$$-\frac{\mathrm{d}}{\mathrm{d}t}\frac{\partial \mathcal{L}}{\partial \dot{q}_\alpha} + \frac{\partial \mathcal{L}}{\partial q_\alpha} + \lambda \cdot \frac{\partial f}{\partial q_\alpha} = 0 \ . \tag{13.5}$$

For anholonomous constraints the term

$$\lambda \cdot \frac{\partial f}{\partial q_\alpha}$$

is replaced by $\lambda \cdot a_\alpha$, and for additional constraints one has a sum of similar terms:

$$\sum_\mu \lambda^{(\mu)} \cdot a_\alpha^{(\mu)} \ .$$

These are the *Lagrange equations of the 2nd kind with constraints.* Originally they were formulated only in cartesian coordinates as so-called Lagrange equations of the 1st kind, which were based on the principle of *d' Alembert* and from which the Lagrange equations of the 2nd kind were derived. Many authors prefer this historical sequence.

Textbook examples of *anholonomous* constraints are not very common. We briefly mention here the example of a skater. In this example the constraint is such that the gliding direction is given by the angular position of the skates. The constraints would be similar for skiing.

13.2 Exercise: Forces of Constraint for Heavy Rollers on an Inclined Plane

For the above-mentioned problem of *constraints* additional insight can again be gained from the seemingly simple problem of a *roller on an inclined plane*. Firstly we shall consider the *Euler angles*:

- φ is the azimuthal rotation angle of the symmetry axis $\pm e_3$ of the roller about the fixed vertical axis ($\pm\hat{z}$-axis); this would be a "dummy" value, if the plane were not inclined. We choose the value $\varphi \equiv 0$ to correspond to the condition that the roller is just moving down the plane, always in the direction of *steepest descent*.
- ϑ is the tilt angle between the vector $\hat{e}_3$ and the vector $\hat{z}$; usually $\vartheta = \pi/2$.
- Finally, ψ is the azimuthal rotation angle corresponding to the distance

$$\Delta s = R \cdot \Delta \psi$$

moved by the perimeter of the roller; R is the radius of the roller.

In the following we consider the standard assumptions[3]

$$\varphi \equiv 0 \quad \text{and} \quad \vartheta \equiv \frac{\pi}{2} ,$$

i.e., we assume that the axis of the circular cylinder lies horizontally on the plane, which may be not always true.

In each case we assume that the Lagrangian may be written

$$\mathcal{L} = \mathcal{T} - \mathcal{V} = \frac{M}{2} v_s^2 + \frac{\Theta_{||}}{2} \dot{\psi}^2 + Mgs \cdot \sin\alpha ,$$

where α corresponds to the slope of the plane, and s is the distance corresponding to the motion of the center of mass, i.e., of the axis of the roller.

We now consider *three cases*, with $v_s = \dot{s}$:

a) Let the plane be perfectly *frictionless*, i.e., the roller slides down the inclined plane. The number of degrees of freedom is, therefore, $f = 2$; they correspond to the generalized coordinates s and ψ. As a consequence, the equation

$$\frac{d}{dt}\frac{\partial \mathcal{L}}{\partial \dot{s}} - \frac{\partial \mathcal{L}}{\partial s} = 0$$

results in

$$M\ddot{s} = Mg\sin\alpha .$$

In contrast the angle ψ is *cyclic*, because

$$\frac{\partial \mathcal{L}}{\partial \psi} = 0 ;$$

therefore one has

$$v_s = v_{s|0} + g_{\text{eff}} \cdot (t - t_0) \quad \text{and} \quad \dot{\psi} = \text{constant} ,$$

i.e., the heavy roller slides with constant angular velocity and with effective gravitational acceleration

$$g_{\text{eff}} := g\sin\alpha ,$$

which is given by the slope $\tan\alpha$ of the inclined plane.

b) In contrast, let the plane be perfectly *rough*, i.e., the cylinder rolls down the plane. Now we have $f \equiv 1$; and since

$$\dot{\psi} = \frac{v_s}{R} : \mathcal{L} \equiv \frac{M}{2} v_s^2 + \frac{\Theta_{||}}{2} \frac{v_s^2}{R^2} + Mg_{\text{eff}} s .$$

---

[3] One guesses that the problem can be made much more complex, if the roller does not simply move down the plane in the direction of steepest descent, but if φ and/or ϑ were allowed to vary; however, even to *formulate* these more complex problems would take some effort.

Therefore

$$v_s \equiv R\dot\psi = v_{s|0} + \frac{M \cdot g_{\text{eff}}}{M_{\text{eff}}} \cdot (t - t_0) , \quad \text{with} \quad M_{\text{eff}} := M \cdot \left(1 + \frac{\Theta_{||}}{MR^2}\right) .$$

The holonomous (and skleronomous) constraint $s - R\psi = 0$ has been explicitly eliminated, such that for the remaining degree of freedom the simple Lagrange equation of the second kind without any constraint could be used directly.
But how do the constraining forces originate? (These are frictional forces responsible for the transition from sliding to rolling.) The answer is obtained by detailed consideration, as follows:

c) The plane is rough, but initially $0 \le R_s\dot\psi < v_s$ (e.g., the rolling is slow or zero). Now consider the *transition* from $f = 2$ to $f \equiv 1$.
We have,

$$(1), \quad M\dot v_s = F_g - F_{fr} \quad \text{and} ,$$
$$(2), \quad \Theta_{||}\ddot\psi = R \cdot F_g .$$

(Here $F_g = Mg\sin\alpha$ is the constant *gravitational force* applied to the axis, directed downwards, while $(-F_{fr})$ is the *frictional force*, $\propto \dot\psi$, applied to the tangential point of rolling, and with upward direction.)
According to (b), the angular velocity $\dot\psi$ *increases* (e.g., from zero) $\propto t$ as the cylinder rolls downwards; at the same time the gravitational force is constant; thus, after a certain time τ_c the frictional force counteracts the gravitational one and $v_s = R\dot\psi$. As a consequence, a weighted sum of the two above equations yields after this time:

$$\left(M + \frac{\Theta_{||}}{R^2}\right)\dot v_s = F_g .$$

Here one sees explicity how d'Alembert's principle (that the forces of constraint *do no work*, one considers virtual displacements) becomes satisfied after τ_c: Then we have

$$\delta A = \boldsymbol{F}_{fr} \cdot \delta\boldsymbol{r} + \mathcal{D}_\psi\delta\psi = -F_g ds + RF_{fr}\, d\psi = 0 \quad \text{for} \quad ds = R d\psi .$$

Another consequence of the above facts is that a soccer player should avoid letting the football roll on the grass, because the ball is slowed down due to rolling:

$$M \to \left(M + \frac{\Theta_{||}}{R^2}\right) .$$

14 Accelerated Reference Frames

14.1 Newton's Equation
in an Accelerated Reference Frame

Thus far we have considered reference frames moving at a *constant* velocity. In the following, however, we shall consider the transition to *accelerated coordinate systems*, where the accelerated coordinates and basis vectors are again denoted by a prime, while $r_0(t)$ is the radius vector of the accelerated origin.

Keeping to the limit $v^2 \ll c^2$, i.e., to the Newtonian theory, the basic equation is

$$m_{\mathrm{t}} \frac{\mathrm{d}^2}{\mathrm{d}t^2} \left\{ r_0(t) + \sum_{i=1}^{3} x_i'(t) e_i'(t) \right\} \equiv F .$$

The force F on the r.h.s. will be called the *true force*[1], in contrast to *fictitious forces*, which are also called *inertial forces*, appearing below in equation (14.1) on the r.h.s., all multiplied by the *inertial mass* m_{t}.

By systematic application of the product rule and the relation

$$\frac{\mathrm{d}e_i'}{\mathrm{d}t} = \omega \times e_i' , \quad \text{with} \quad v'(t) = \sum_{i=1}^{3} \dot{x}_i'(t) e_i'(t) ,$$

the following result for the velocity is obtained:

$$v(t) = v_0(t) + v'(t) + \omega \times r' .$$

In the same way one obtains for the acceleration:

$$a(t) = a_0(t) + a'(t) + 2\omega \times v' + \omega \times [\omega \times r'] + \dot{\omega} \times r' .$$

Newton's equation of motion $m_{\mathrm{t}} a = F$, transformed to the primed (accelerated) system, is therefore given by:

$$m_{\mathrm{t}} a' \equiv F$$
$$-m_{\mathrm{t}} \ddot{r}_0(t) - 2m_{\mathrm{t}} \omega \times r' - m_{\mathrm{t}} \omega \times [\omega \times r'] - m_{\mathrm{t}} \dot{\omega} \times r' . \quad (14.1)$$

---

[1] Here we remind ourselves that in General Relativity, [7] and [8], all inertial forces and the gravitational part of the "true" forces are transformed into geometrical properties of a curved Minkowski spacetime.

This is the equation of motion in the *accelerated* (i.e., linearly accelerated and/or rotating) reference frame. On the r.h.s. of (14.1) the first term represents the *true force* (involving, e.g., the *gravitational mass*), from which the following terms are *subtracted*: these terms are the *fictitous* forces, recognizable by the factor m_t, the *inertial mass* of the corresponding point, i.e.

- the so-called *elevator force*, $m_t \cdot \ddot{\boldsymbol{r}}_0(t)$;
- the *Coriolis force* $m_t \cdot 2\boldsymbol{\omega} \times \boldsymbol{v}$;
- the *centrifugal force* $m_t \cdot \boldsymbol{\omega} \times [\boldsymbol{\omega} \times \boldsymbol{r}']$,
- and finally, a force $m_t \cdot \dot{\boldsymbol{\omega}} \times \boldsymbol{r}'$, which has no specific name.

Since a major part of the *true force*, the *gravitational force*, is proportional to the *gravitational mass* m_s, this can be compensated by the *inertial forces*, because of the equality (according to the pre-Einstein viewpoint: *not* identity![2]) $m_s = m_t$.

In particular, for a linearly accelerated system, which corresponds to an elevator falling with acceleration $-g\hat{\boldsymbol{z}}$ downwards in "free fall", the difference between the gravitational "pull" $-m_s \cdot g\hat{\boldsymbol{z}}$ and the inertial elevator *"push"* is exactly zero. From this *thought experiment*, Einstein, some years after he had formulated his *special theory of relativity*, was led to the postulate that

a) no principal difference exists between gravitational and inertial forces (*Einstein's equivalence principle*); moreover,

b) no *global* inertial frames as demanded by Mach exist, but only "free falling" local inertial frames, or more accurately: relative to the given gravitating bodies *freely moving* local inertial frames exist, where for small *trial masses* the gravitational forces are exactly *compensated* by the *inertial forces* corresponding to the "free motion" in the gravitational field; in particular

c) the general motion of a small *trial point* of infinitesimal mass $m_t = m_s$ takes place along *extremal paths* in a curved Minkowski space, where the *proper time* does *not* obey, as in a "flat" Minkowski space, the formula

$$-\,\mathrm{d}s^2 \left(\equiv c^2 \mathrm{d}\tau^2\right) = c^2\,\mathrm{d}t^2 - \mathrm{d}x^2 - \mathrm{d}y^2 - \mathrm{d}z^2\;,$$

but a more general formula corresponding to a nontrivial *differential geometry* in a *curved Minkowski manifold*, i.e.,

$$-\,\mathrm{d}s^2 \left(\equiv c^2 \mathrm{d}\tau^2\right) = \sum_{i=1}^{4} g_{i,k}\left(\tilde{x}\right)\,\mathrm{d}x^i\,\mathrm{d}x^k\;,$$

with a so-called *metric fundamental tensor*[3] $g_{i,k}(\tilde{x})$, which depends in a nontrivial manner on the distribution of the gravitating masses, and

---

[2] It was again Einstein, who postulated in 1910 that the equality should actually be replaced by an identity.

[3] As to the sign and formulation of $\mathrm{d}s^2$ there are, unfortunately, different equivalent conventions.

which has, as for "flat" Minkowski spaces, one eigenvalue of one sign and three eigenvalues of the opposite sign.

The space of *Minkowski four-vectors*

$$\tilde{x} := (x, y, z, \mathrm{i}ct)$$

thus becomes a *curved Minkowski manifold,* where local coordinates of space and time may be defined by means of radar signals.

Further details which would lead to Einstein's *general theory of relativity,* will not be treated here, although the effects of curvature of space and time not only play a part in present-day astrophysics, e.g., in the neighborhood of neutron stars and *black holes,* but have also entered our daily lives through the *Global Positioning System (GPS),* a satellite navigation system, which is presently used for many purposes.

14.2 Coriolis Force and Weather Pattern

Apart from *elevator forces,* Coriolis forces are perhaps the most important *inertial force.* In contrast to *centrifugal forces* they are proportional to ω and *not* $\propto \omega^2$ (i.e., not of second-order):

$$\boldsymbol{F}'_{\text{Coriolis}} = -2m_{\text{t}} \cdot \boldsymbol{\omega} \times \boldsymbol{v}' \ .$$

Particularly important are the consequences of this force in the weather pattern, where the Coriolis force governs the deflection of wind currents from regions of high atmospheric pressure to those of low pressure. For example, if there were a high pressure region at the equator, with coordinates

$$x' = -H, y' = 0 \ ,$$

and if the nearest low-pressure region were at

$$x' = +T, y' = 0 \ ,$$

then without rotation of the earth, i.e., for $\omega = 0$, the wind would only have a velocity component v_x, i.e., it would directly move from west to east on the shortest path from high pressure to low pressure.

However, due to the rotation of the earth,

$$\boldsymbol{\omega} = \frac{2\pi}{24\text{h}} \hat{z} \ ,$$

where $\hat{z}$ denotes the axis of rotation, we have

$$\boldsymbol{F}'_{\text{Coriolis}} = -2m_{\text{t}} \cdot \omega v_x \boldsymbol{e}_y \ ,$$

i.e., a force directed from north to south. As a consequence, in the northern hemisphere the wind flows out of the high-pressure region with a deflection to the right and into the low-pressure region with a deflection to the left.

In addition, for the so-called *trade winds* (Passat wind) over the oceans, which without rotation of the earth would blow directly towards the equator, i.e., southwards (in the northern hemisphere), the rotation of the earth leads to a shift of the direction. As a consequence the *trade winds* are directed from north-east to south-west.

A further subject for which the rotation of the earth plays an essential part concerns the *Foucault pendulum*, a very long pendulum swinging from the ceiling of a large building, such that one can directly infer the rotation of the earth from the varying plane of oscillations of the swings, c.f. subsection 14.4.

14.3 Newton's "Bucket Experiment" and the Problem of Inertial Frames

We return to the problem of ascertaining whether a frame of reference is rotating, $\omega \neq 0$, or whether, perhaps, it is an *inertial frame*. For this purpose, we can return to the proposal already known by Newton, of observing the surface of water in a bucket rotating with the reference frame.

Due to the internal friction of the liquid, its surface shows a profile given by

$$z(r) = \frac{r^2 \omega^2}{2g} \ ,$$

where $z(r)$ is the height of the liquid surface measured from the central (lowest) point. In addition, the local slope of the surface of the liquid in the bucket is given by

$$\tan \alpha = \frac{\mathrm{d}z}{\mathrm{d}r} = \frac{r\omega^2}{g} \ .$$

This relation is obtained by equating the centrifugal force (per unit mass of liquid)

$$(F'_x)_{\text{centrifugal}} = m_{\text{t}} \omega^2 r$$

with the gravitational force (per unit mass)

$$F_z = -m_{\text{s}} g \ .$$

In this way one can therefore ascertain whether one is in a rotating reference frame, and here the third *inertial force*, the *centrifugal force*, explicitly comes into play.

In view of the difficulties in defining an *inertial reference frame*, towards the end of the nineteenth century, the physicist and important Viennese philosopher, Ernst Mach, postulated that a *global inertial system* can only be defined in a constructive way by the totality of all stars. But as mentioned above it was only Einstein, who, some years later, cut the "Gordian knot" by stating that only *local* inertial frames exist, thus solving in an elegant way an essential problem left over from Newtonian mechanics. More details can be found above in Sect. 14.1.

14.4 Application: Free Falling Bodies with Earth Rotation; the Foucault Pendulum

In a cartesian coordinate system fixed to the moving surface of the earth, let e_1 correspond to the direction "east", e_2 to "north" and e_3 to the *geometrical* vertical direction ($e_3 \equiv e_1 \times e_2$).

Let the geographical latitude correspond to the angle ψ such that $\psi = 0$ at the equator and $\psi = \frac{\pi}{2}$ at the North Pole. Hence for the angular velocity of the rotation of the earth we have

$$\boldsymbol{\omega} = \omega \cos \psi \, \boldsymbol{e}_2 + \omega \sin \psi \, \boldsymbol{e}_3 \ .$$

The three equations of motion are, therefore:

$$m_t \ddot{x}_1 = m_s G_1 - 2m_t \omega \cdot (\cos \psi v_3 - \sin \psi v_2) \ ,$$
$$m_t \ddot{x}_2 = m_s G_2 - 2m_t \omega \sin \psi v_1 + m_t \omega^2 R \cos \psi \sin \psi \ ,$$
$$m_t \ddot{x}_3 = m_s G_3 + 2m_t \omega \cos \psi v_1 + m_t \omega^2 R \cos^2 \psi. \qquad (14.2)$$

If the earth had the form of an exact sphere, and if the mass distribution were exactly spherical, then G_1 and G_2 would vanish, and $G_3 \equiv -g$. The deviations correspond to *gravitational anomalies*, which are measured and mapped by geophysicists, e.g., in mineral prospecting. As usual, for didactical reasons we distinguish between the *inertial* mass m_t and the *gravitational* mass m_s, although

$$m_s = m_t \ ; \quad R(= R(\psi))$$

is the radius of the earth at the latitude considered (where we average over mountains and depressions at this value of ψ).

We shall now discuss the terms which are (a) $\propto \omega^2$, i.e., the centrifugal force, and (b) $\propto |\boldsymbol{\omega}\boldsymbol{v}|$, i.e., the Coriolis force.

– (ai) The very last term in (14.2) leads to a weak, but significant flattening of the sphere ($\rightarrow$ *"geoid" model of the earth = oblate spheroid*), since

$$-m_s g + m_t \omega^2 R \cos^2 \psi =: -m_t g_{\text{eff}}(\psi)$$

depends on the geographical latitude. The gravitational weight of a *kilogram* increases towards the poles.

– (aii) The terms $\propto \omega^2$ in the second equation and the *north anomaly* G_2[4] of the gravitational force lead to a deviation in the direction of the gravitational force both from the direction of e_3 and also from the G direction. The deviations are described by the angles δ and δ', respectively:

$$\tan \delta = \frac{G_2 - \omega^2 R \cos \psi \sin \psi}{G_3}$$

and

$$\tan \delta' = \frac{-\omega^2 R \cos \psi \sin \psi}{G_3} .$$

For realistic values

$$G_3 \approx 981 \ \mathrm{cm/s}^2 \quad \text{and} \quad \omega^2 R \approx 3.4 \ \mathrm{cm/s}^2$$

these effects are roughly of the relative order of magnitude $3 \cdot 10^{-3}$.

With regard to the terms which are linear in ω (case b), we think of the Foucault pendulum and assume therefore that $v_3 \equiv 0$, whereas v_1 and v_2 are *not* $\equiv 0$. The equations of motion are

$$m_t \ddot{x}_1 = F_1 + 2m_t \omega v_2 \sin \psi \quad \text{and}$$
$$m_t \ddot{x}_2 = F_2 - 2m_t \omega v_1 \sin \psi , \quad \text{i.e.,}$$
$$\dot{v} = \frac{F}{m_t} + \omega_{\mathrm{eff}}(\psi) \times v .$$

This can be formulated with an effective rotational velocity that depends on the geographical latitude ψ, *viz*

$$\omega_{\mathrm{eff}}(\psi) = -2\omega \sin \psi e_3 , \quad \text{with} \quad \omega = \frac{2\pi}{24\mathrm{h}} .$$

A superb example of a Foucault pendulum can be seen in the Science Museum in London.

---

[4] The averaged *east/west anomaly* G_1 vanishes for reasons of symmetry, or with perturbations of the east-west symmetry it is usually much smaller in magnitude than $|G_2|$.

15 Relativity II: E=mc²

The Lorentz transformations,

$$x = \frac{x' + \frac{v}{c}ct'}{\sqrt{1 - \frac{v^2}{c^2}}} \,, \qquad ct = \frac{ct' + \frac{v}{c}x'}{\sqrt{1 - \frac{v^2}{c^2}}} \,,$$

have already been treated in Sect. 9.1, together with the difference between the invariant *proper time (eigenzeit)* $d\tau$ – this is the time in the *co-moving system* – and the noninvariant time dt in the so-called *laboratory system*.

Let us recall the invariant quantity

$$-(d\tilde{x})^2 := c^2\,dt^2 - dx^2 - dy^2 - dz^2 \,,$$

which can be either positive, negative or zero. *Positive* results apply to *time-like Minkowski four-vectors*

$$d\tilde{x} := (\,dx,\,dy,\,dz, ic\,dt)$$

in all inertial frames. For *space-like* four-vectors the invariant expression on the r.h.s. of the above equation would be *negative* and for *light-like* Minkowski four-vectors it would be zero. These inequalities and this equality characterize the above terms.

In fact, *time-like, space-like* and *light-like* Minkowski four-vectors exhaust and distinguish all possible cases, and for time-like four-vectors one obtains (with the proper time τ):

$$c^2\,d\tau^2 \equiv -d\tilde{x}^2 \,,$$

i.e., apart from the sign this is just the square of the invariant pseudo-length in a pseudo-Euclidean Minkowski space. As mentioned above, Lorentz tranformations can be considered as pseudo-rotations in this space.

Since $d\tilde{x}$ is an (infinitesimal) Minkowski four-vector and $d\tau$ an (infinitesimal) four-scalar[1], i.e., Lorentz invariant, the above-mentioned four-velocity,

$$\tilde{v} := \frac{d\tilde{x}}{d\tau} = \frac{(v_x, v_y, v_z, ic)}{\sqrt{1 - \frac{v^2}{c^2}}}$$

---

[1] In contrast to $d\tau$ the differential dt is not a Minkowski scalar, i.e., not invariant against Lorentz transforms.

is a Minkowski four-vector. The factor

$$\frac{1}{\sqrt{1 - \frac{v^2}{c^2}}}$$

is thus *compulsory*! In addition, if this result is multiplied by the *rest mass* m_0, one obtains a Minkowski four-vector, i.e.

$$\tilde{p} := \frac{m_0}{\sqrt{1 - \frac{v^2}{c^2}}} (v_x, v_y, v_z, \mathrm{i}c) \; .$$

This four-vector is the relativistic generalization of the momentum, and the dependence of the first factor on the velocity, i.e., the velocity dependence of the mass,

$$m = \frac{m_0}{\sqrt{1 - \frac{v^2}{c^2}}} \; ,$$

is also as *compulsory* as previously that of $\tilde{v}$.

The imaginary component of the four-momentum is thus

$$\mathrm{i} \cdot \frac{m_0 c}{\sqrt{1 - \frac{v^2}{c^2}}} \; .$$

On the other hand, in the co-moving system ($\boldsymbol{v} = 0$), a force vector can be uniquely defined by the enforced velocity change, i.e.,

$$\frac{\mathrm{d}(m_0 \boldsymbol{v})}{\mathrm{d}\tau} =: \boldsymbol{F} \; ,$$

analogously to Newton's 2nd law.

The work done by this force, $\delta \hat{A}$, serves to enhance the kinetic energy

$$E_{\mathrm{kin}} : \delta\hat{A} - \dot{E}_{\mathrm{kin}} \, \mathrm{d}t = \boldsymbol{F} \cdot \mathrm{d}\boldsymbol{r} - \frac{l_F}{c} c \, \mathrm{d}t \equiv 0 \; ,$$

with the so-called *power* $l_F := \dot{E}_{\mathrm{kin}}$.

Formally this means the following:

The force-power four-vector

$$\tilde{F} := \left(F_x, F_y, F_z, \mathrm{i}\frac{l_F}{c} \right)$$

is pseudo-perpendicular to

$$\mathrm{d}\tilde{x} := (\mathrm{d}x, \mathrm{d}y, \mathrm{d}z, \mathrm{i}c\mathrm{d}t) : \tilde{F} \cdot \mathrm{d}\tilde{x} \equiv 0 \; .$$

This is valid in all inertial frames (Lorentz frames).

We now define
$$E := E_{\text{kin}} + m_0 c^2 \ .$$

It is then natural to supplement the relation
$$l_F = \dot{E} \quad \text{by} \quad \boldsymbol{F} = \dot{\boldsymbol{p}}$$

and in this way to define the *energy-momentum four-vector*
$$\tilde{p} = \frac{m_0}{\sqrt{1 - \frac{v^2}{c^2}}}(v_x, v_y, v_z, \text{i}c) \ ,$$

i.e.,
$$\tilde{p} = \left(p_x, p_y, p_z, \text{i}\frac{E}{c} \right)$$

with the following two relations, which belong together, equivalent to
$$\tilde{F} \cdot \text{d}\tilde{x} \equiv 0 :$$

$$\boldsymbol{p} = \frac{m_0 \boldsymbol{v}}{\sqrt{1 - \frac{v^2}{c^2}}} \quad \text{and} \quad E = \frac{m_0 c^2}{\sqrt{1 - \frac{v^2}{c^2}}} \ . \tag{15.1}$$

In fact, one can explicitly evaluate that in this way the equation
$$\tilde{F} \cdot \text{d}\tilde{x} \equiv 0$$

is satisfied.

The invariant pseudo-length of the *energy-momentum four-vector* is thus
$$\tilde{p}^2 = -m_0^2 c^2 \ ; \quad \text{the energy} \quad E = mc^2$$

is the sum of the *rest energy* $m_0 c^2$ and the *kinetic energy*
$$\mathcal{T} := \frac{m_0 c^2}{\sqrt{1 - \frac{v^2}{c^2}}} - m_0 c^2 \ ,$$

which in a first approximation leads to the usual nonrelativistic expression,
$$\mathcal{T} \approx \frac{m_0 v^2}{2} \ .$$

In addition we have the relativistic form of the equation of motion, which is that in every inertial system:
$$\frac{\text{d}\boldsymbol{p}}{\text{d}t} = \boldsymbol{F} \ , \tag{15.2}$$

with

$$p = \frac{m_0 \boldsymbol{v}}{\sqrt{1 - \frac{v2}{c^2}}} \ .$$

By integrating this differential equation for the velocity at constant force and with the initial values

$$v(t = 0) \equiv 0 \ ,$$

one obtains

$$p^2 = F^2 t^2 \ ,$$

or

$$\frac{v^2}{c^2} = \frac{\hat{F}^2 t^2}{1 + \hat{F}^2 t^2} \ ,$$

with

$$\hat{F} := \frac{F}{m_0 c} \ .$$

For $t \to \infty$ the r.h.s. of $\frac{v^2}{c^2}$ converges to 1 (monotonically from below); c is thus the *upper limit of the particle velocity v.*

Our derivation of the above relations was admittedly rather formal, but Einstein was bold enough, and gave reasons (see below) for proposing the even more important interpretation of his famous relation,

$$E = mc^2 \ ,$$

which not only implies the above decomposition of the energy into a *rest energy* and a *kinetic part,* but that the formula should additionally be interpreted as the *equivalence* of mass and energy; i.e., he suggested that *mass differences* ($\times c^2$) can be transformed quantitatively into *energy differences*[2]. This is the essential basis, e.g., of *nuclear energetics*; nuclei such as He, which have a strong binding energy δE_B, also have a measurable *mass defect*

$$\delta m_B = \delta E_B / c^2 \ .$$

One could give many more examples.

The above reasons can, in fact, be based on considerations of the *impact* between different particles. For example, viewed from the *rest frame* of a very heavy point mass, which forms the target of other particles moving towards it with very high velocity (almost c) the kinetic energy of any of the moving particles is essentially identical to its (velocity dependent) mass times c^2. Thus in this case the *rest mass* is negligible.

All these considerations, which in our presentation have been partly stringent and partly more or less heuristic, have proved valid over decades, without

---

[2] Here one should not forget the fact that $m_0 c^2$ (in contrast to E) is relativistically invariant.

any restriction, not only by thought experiments and mathematical derivations, but also by a wealth of nontrivial empirical results. As a consequence, for decades they have belonged to the canonical wisdom not only of university physics, but also of school physics and gradually also in daily life, although there will always be some people who doubt the above reasoning and experience. We shall therefore conclude Part I of the book by restating Einstein's relation

$$E = mc^2 \ , \quad where \quad m = m_0 / \sqrt{1 - \frac{v^2}{c^2}} \ ,$$

almost exactly one century after it was originally conceived in the "miraculous year" of 1905.

Electrodynamics and Aspects of Optics

16 Introduction and Mathematical Preliminaries to Part II

The theory of electrodynamics provides the foundations for most of our present-day way of life, e.g., electricity, radio and television, computers, radar, mobile phone, etc.; Maxwell's theory lies at the heart of these technologies, and his equations at the heart of the theory.

Exercises relating to this part of the book (originally in German, but now with English translations) can be found on the internet, [2].

Several introductory textbooks on theoretical electrodynamics and aspects of optics can be recommended, in particular "Theoretical Electrodynamics" by Thorsten Fliessbach (taken from a comprehensive series of textbooks on theoretical physics). However, this author, like many others, exclusively uses the Gaussian (or cgs) system of units (see below). Another book of lasting value is the text by Bleaney and Bleaney, [10], using mksA units, and containing an appendix on how to convert from one system to the other. Of similar value is also the 3rd edition of the book by Jackson, [11].

16.1 Different Systems of Units in Electromagnetism

In our treatment of electrodynamics we shall adopt the international system (SI) throughout. SI units are essentially the same as in the older mksA system, in that length is measured in metres, mass in kilograms, time in seconds and electric current in ampères. However other systems of units, in particular the centimetre-gram-second (cgs) (or Gaussian) system, are also in common use. Of course, an mks system (without the "A") would be essentially the same as the cgs system, since 1 m = 100 cm and 1 kg = 1000 g. In SI, the fourth base quantity, the unit of current, ampère (A), is now defined via the force between two wires carrying a current. The unit of charge, coulomb (C), is a derived quantity, related to the ampère by the identity 1 C = 1 A.s. (The coulomb was originally defined via the amount of charge collected in 1 s by an electrode of a certain electrolyte system.) For theoretical purposes it might have been more appropriate to adopt the elementary charge $|e| = 1.602\ldots \cdot 10^{-19}$C as the basic unit of charge; however, this choice would be too inconvenient for most practical purposes.

The word "electricity" originates from the ancient Greek word $\varepsilon\lambda\varepsilon\kappa\tau\rho o\nu$, meaning "amber", and refers to the phenomenon of *frictional or static electricity*. It was well known in ancient times that amber can be given an electric charge by rubbing it, but it was not until the eighteenth century that the law of force between two electric charges was derived quantitatively: Coulomb's law[1] states that two infinitesimally small charged bodies exert a force on each other (along the line joining them), which is proportional to the product of the charges and inversely proportional to the square of the separation. In SI Coulomb's law is written

$$F_{1\leftarrow 2} = \frac{q_1 q_2}{4\pi\varepsilon_0 r_{1,2}^2} e_{1\leftarrow 2} , \tag{16.1}$$

where the unit vector $e_{1\leftarrow 2}$ describes the direction of the line joining the charges,

$$e_{1\leftarrow 2} := (r_1 - r_2)/|r_1 - r_2| , \quad \text{while} \quad r_{1,2} := |r_1 - r_2|$$

is the distance between the charges.

According to (16.1), ε_0, the *permittivity of free space*, has the physical dimensions of $charge^2/(length^2 \cdot force)$.

In the cgs system, which was in general use before the mksA system had been introduced[2], the quantity $4\pi\varepsilon_0$ in Coulomb's law (16.1) does not appear at all, and one simply writes

$$F_{1\leftarrow 2} = \frac{q_1' q_2'}{r_{1,2}^2} e_{1\leftarrow 2} . \tag{16.2}$$

Hence, the following expression shows how charges in cgs units (primed) appear in equivalent equations in mksA or SI (unprimed) (1a):

$$q' \Leftrightarrow \frac{q}{\sqrt{4\pi\varepsilon_0}} ,$$

i.e., both charges differ only by a factor, which has, however, a physical dimension.

(Also electric currents, dipoles, etc., are transformed in a similar way to electric charges.) For other electrical quantities appearing below, the relations are different, e.g., (2a):

$$E' \Leftrightarrow E \cdot \sqrt{4\pi\varepsilon_0}$$

(i.e., $q' \cdot E' = qE(\equiv F)$); and, (3a):

$$D' \Leftrightarrow D \cdot \sqrt{\frac{4\pi}{\varepsilon_0}}$$

---

[1] Charles Augustin de Coulomb, 1785.

[2] Gauß was in fact an astronomer at the university of Göttingen, although perhaps most famous as a mathematician.

(where *in vacuo* $\boldsymbol{D} := \varepsilon_0\boldsymbol{E}$, but $\boldsymbol{D}' := \boldsymbol{E}'$). For the magnetic quantities we have analogously, (1b):

$$\boldsymbol{m}' \Leftrightarrow \frac{\boldsymbol{m}}{\sqrt{4\pi\mu_0}} \; ;$$

(2b):

$$\boldsymbol{H}' \Leftrightarrow \boldsymbol{H} \cdot \sqrt{4\pi\mu_0} \; ;$$

(3b):

$$\boldsymbol{B}' \Leftrightarrow \boldsymbol{B} \cdot \sqrt{\frac{4\pi}{\mu_0}} \; , \quad \text{where} \quad \mu_0 := \frac{1}{\varepsilon_0 c^2} \, ,$$

with the light velocity (*in vacuo*) c; $\boldsymbol{m}$ is the magnetic moment, while $\boldsymbol{H}$ and $\boldsymbol{B}$ are magnetic field and magnetic induction, respectively.

Equations (1a) to (3b) give the complete set of transformations between (unprimed) mksA quantities and the corresponding (primed) cgs quantities. These transformations are also valid in polarized matter.

Since ε_0 does not appear in the cgs system – although this contains a complete description of electrodynamics – it should *not* have a fundamental significance *per se* in SI. However, Maxwell's theory (see later) shows that ε_0 is unambiguously related to μ_0, the equivalent quantity for magnetic behavior, via c, the velocity of light (i.e., electromagnetic waves) in *vacuo*, *viz*

$$\varepsilon_0\mu_0 \equiv \frac{1}{c^2} \; .$$

Now since c has been measured to have the (approximate) value $2.998\ldots\cdot 10^8$ m/s, while μ_0, the *vacuum permeability*, has been defined by international convention[3] to have the exact value

$$\mu_0 \equiv 4\pi \cdot 10^{-7} \, \text{Vs/(Am)} \, , \text{[4]}$$

then ε_0 is essentially determined in terms of the velocity of light; ε_0 ($\equiv 1/(\mu_0 \cdot c^2)$) has the (approximate) value $8.854\ldots\cdot 10^{-12}$ As/(Vm).

In SI, Maxwell's equations (*in vacuo*) are:

$$\mathrm{div}\boldsymbol{E} = \varrho/\varepsilon_0, \mathrm{div}\boldsymbol{B} = 0, \mathrm{curl}\boldsymbol{E} = -\frac{\partial \boldsymbol{B}}{\partial t}, \mathrm{curl}\boldsymbol{B} = \mu_0\boldsymbol{j} + \frac{1}{c^2}\frac{\partial \boldsymbol{E}}{\partial t} \; . \qquad (16.3)$$

These equations will be discussed in detail in the following chapters.

The equivalent equations in a Gaussian system are (where we write the quantities with a prime):

$$\mathrm{div}\boldsymbol{E}' = 4\pi\varrho', \mathrm{div}\boldsymbol{B}' = 0, \mathrm{curl}\boldsymbol{E}' = -\frac{\partial \boldsymbol{B}'}{\partial ct}, \mathrm{curl}\boldsymbol{B}' = \frac{4\pi}{c}\boldsymbol{j}' + \frac{\partial \boldsymbol{E}'}{\partial ct} \; . \qquad (16.4)$$

---

[3] essentially a political rather than a fundamentally scientific decision!
[4] This implies that $B' = 10^4$ Gauss corresponds exactly to $B = 1$ Tesla.

One should note the different powers of c in the last terms of (16.3) and (16.4).

Another example illustrating the differences between the two equivalent measuring systems is the equation for the force exerted on a charged particle moving with velocity v in an electromagnetic field, the so-called *Lorentz force*: In SI this is given by

$$F = q \cdot (E + v \times B) , \qquad (16.5)$$

whereas in a Gaussian system the expression is written

$$F = q' \cdot \left(E' + \frac{v}{c} \times B' \right)$$

We note that all relationships between electromagnetic quantities can be expressed equally logically by the above transformations in either an SI system ("SI units") or a Gaussian system ("Gaussian units"). However, one should never naively mix equations from different systems.

16.2 Mathematical Preliminaries I: Point Charges and Dirac's δ Function

A *slightly smeared* point charge of strength q at a position r' can be described by a *charge density*

$$\varrho_\varepsilon(r) = q \cdot \delta_\varepsilon(r - r') , \quad \text{where} \quad \delta_\varepsilon$$

is a suitable rounded function (see Fig. 16.1) and ε is a corresponding *smearing parameter*. For all positive values of ε the integral

$$\int d^3 r \delta_\varepsilon(r - r') \overset{!}{=} 1 ,$$

although at the same time the limit $\lim_{\varepsilon \to 0} \delta_\varepsilon(r)$ shall be zero for all $r \neq 0$. It can be shown that this is possible, e.g., with a *Gaussian function*

$$\delta_\varepsilon(r - r') = \frac{\exp\left(-\frac{|r-r'|}{2\varepsilon^2} \right)}{(2\pi\varepsilon^2)^{3/2}}$$

(in three dimensions, $d = 3$).

The functions $\delta_\varepsilon(\boldsymbol{r} - \boldsymbol{r}')$ are well-behaved for $\varepsilon \neq 0$ and only become very "spiky" for $\varepsilon \to 0$ [5], i.e.,

$$\delta(\boldsymbol{r} - \boldsymbol{r}') := \lim_{\varepsilon \to 0} \delta_\varepsilon(\boldsymbol{r} - \boldsymbol{r}') ,$$

where the limit should always be performed *in front of* the integral, as in the following equations (16.6) and (16.7).

For example, let $f(\boldsymbol{r})$ be a test function, which is differentiable infinitely often and nonzero only on a compact set $\in \mathcal{R}^3$; for all these f we have

$$\int d^3 r' \delta(\boldsymbol{r}' - \boldsymbol{r}) \cdot f(\boldsymbol{r}') := \lim_{\varepsilon \to 0} \int d^3 r' \delta_\varepsilon(\boldsymbol{r}' - \boldsymbol{r}) \cdot f(\boldsymbol{r}') \equiv f(\boldsymbol{r}) . \qquad (16.6)$$

In the "weak topology" case considered here, the δ-function can be differentiated arbitrarily often[6]: Since the test function $f(\boldsymbol{r})$ is (continuously) differentiable with respect to the variable x, one obtains for all f by a partial

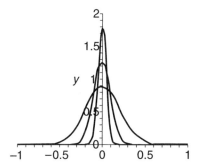

Fig. 16.1. Gaussian curves as an approximation for Dirac's δ-function. Here we consider one-dimensional Gaussian curves (for y between 0 and 2) $\delta_\varepsilon(x) := (2\pi\varepsilon^2)^{-0.5} \exp(-x^2/(2\varepsilon^2))$ plotted versus x (between -1 and 1) with increasing sharpness, $\varepsilon = 0.2$, 0.1 and 0.05, which serve as approximations of Dirac's δ-function for $d = 1$; this function is obtained in the limit $\varepsilon \to 0$, where the limit is performed in front of an integral of the kind mentioned in the text

---

[5] Dirac's δ-function is sometimes referred to as a "spike function"; in particular, the function $\delta_\varepsilon(x)$ (for d=1) is often defined via the limit $\varepsilon \to 0$ of $\delta_\varepsilon(x) = 1/\varepsilon$ for $|x| \leq \varepsilon/2$, otherwise $= 0$. This is convenient in view of the normalization; however, because of (16.7) we recommend modifying these functions slightly by smearing them somewhat, i.e., at the sharp edges they should be made differentiable infinitely often.

[6] This may be somewhat surprising, if one abides by the simple proposition of an infinitely high spike function corresponding to the popular definition $\delta(x) \equiv 0$ for $x \neq 0$, but $\delta(0)$ is so large that $\int dx \delta(x) = 1$, i.e., if one does not keep in mind the *smooth* definition of the approximants $\delta_\varepsilon(x)$.

integration:

$$\int d^3r' \frac{\partial}{\partial x'} \delta(r' - r) \cdot f(r') := \lim_{\varepsilon \to 0} \int d^3r' \frac{\partial}{\partial x'} \delta_\varepsilon(r' - r) \cdot f(r')$$

$$\equiv (-1) \frac{\partial}{\partial x} f(r) . \tag{16.7}$$

To summarize: a point charge of strength q, located at r', can be formally described by a charge density given by:

$$\varrho(r) = q \cdot \delta(r - r') . \tag{16.8}$$

16.3 Mathematical preliminaries II: Vector Analysis and the Integral Theorems of Gauss and Stokes

In this section we shall remind ourselves of a number of important mathematical operators and theorems used in electrodynamics. Firstly, the vector operator *nabla*

$$\nabla := \left(\frac{\partial}{\partial x}, \frac{\partial}{\partial y}, \frac{\partial}{\partial z} \right) \tag{16.9}$$

is used to express certain important operations, as follows:

a) The **gradient** of a scalar function $f(r)$ is a vector field defined by

$$\mathrm{grad} f(r) := \nabla f(r) = (\partial_x f(r), \partial_y f(r), \partial_z f(r)) , \tag{16.10}$$

where we have used the short-hand notation

$$\partial_j := \frac{\partial}{\partial x_j} .$$

This vector, $\mathrm{grad} f(r)$, is perpendicular to the surfaces of constant value of the function $f(r)$ and corresponds to the steepest increase of $f(r)$. In electrodynamics it appears in the law relating the electrostatic field $E(r)$ to its potential $\phi(r)$ ($E = -\mathrm{grad}\phi$). The minus sign is mainly a matter of convention, as with the force F and the *potential energy* V in mechanics.)

b) The **divergence** (or *source density*) of $v(r)$ is a scalar quantity defined as

$$\mathrm{div} v(r) := \nabla \cdot v(r) = \partial_x v_x + \partial_y v_y + \partial_z v_z . \tag{16.11}$$

It is thus formed from $v(r)$ by differentiation plus summation[7]. The fact that this quantity has the meaning of a *source density* follows from the integral theorem of Gauss (see below).

---

[7] One should not forget to write the "dot" for the scalar multiplication (also called "dot product") of the vectors v_1 and v_2, since for example $c := \nabla v$ (i.e., without $\cdot$), would not be a scalar quantity, but could be a second-order tensor, with the nine components $c_{i,k} = \frac{\partial v_k}{\partial x_i}$. However, one should be aware that one can encounter different conventions concerning the presence or absence of the dot in a scalar product.

c) The **curl** (or *circulation density*) of a vector field $\boldsymbol{v}(\boldsymbol{r})$ is a vector quantity, defined as

$$\operatorname{curl}\boldsymbol{v}(\boldsymbol{r}) := \nabla \times \boldsymbol{v}(\boldsymbol{r}) = [\partial_y v_z - \partial_z v_y, \partial_z v_x - \partial_x v_z, \partial_x v_y - \partial_y v_x] . \quad (16.12)$$

(The term $\operatorname{rot}\boldsymbol{v}$ is also used instead of $\operatorname{curl}\boldsymbol{v}$ and the $\wedge$ symbol instead of $\times$.) Additionally, curl can be conveniently expressed by means of the antisymmetric unit tensor and Einstein's summation convention[8], $(\operatorname{curl}\boldsymbol{v})_i = e_{ijk}\partial_j v_k$).

The fact that this quantity has the meaning of a *circulation density* follows from the integral theorem of Stokes, which we shall describe below.

Having introduced the operators grad, div and curl, we are now in a position to describe the integral theorems of Gauss and Stokes.

a) *Gauss's theorem*

Let V (volume) be a sufficiently smooth oriented[9] 3d-manifold with closed boundary ∂V and (outer) *normal* $\boldsymbol{n}(\boldsymbol{r})$. Moreover, let a vector field $\boldsymbol{v}(\boldsymbol{r})$ be continuous on ∂V and continuously differentiable in the interior of V[10]. Then Gauss's theorem states that

$$\oiint_{\partial V} \boldsymbol{v}(\boldsymbol{r}) \cdot \boldsymbol{n}(\boldsymbol{r}) \mathrm{d}^2 A = \iiint_V \operatorname{div}\boldsymbol{v}(\boldsymbol{r}) \mathrm{d}^3 V . \quad (16.13)$$

Here $\mathrm{d}^2 A$ is the area of an infinitesimal surface element of ∂V and $\mathrm{d}^3 V$ is the volume of an infinitesimal volume element of V. The integral on the l.h.s. of equation (16.13) is the "flux" of $\boldsymbol{v}$ out of V through the surface ∂V. The integrand $\operatorname{div}\boldsymbol{v}(\boldsymbol{r})$ of the volume integral on the r.h.s. is the "source density" of the vector field $\boldsymbol{v}(\boldsymbol{r})$.

In order to make it more obvious that the l.h.s. of (16.13) represents a flux integral, consider the following: If the two-dimensional surface ∂V is defined by the parametrization

$$\partial V := \{\boldsymbol{r}|\boldsymbol{r} = (x(u,v), y(u,v), z(u,v))\} , \quad (16.14)$$

with $(u,v) \in G(u,v)$, one then has on the r.h.s. of the following equation an explicit double-integral, *viz*

$$\oiint_{\partial V} \boldsymbol{v}(\boldsymbol{r}) \cdot \boldsymbol{n}(\boldsymbol{r}) \mathrm{d}^2 A \equiv \iint_{G(u,v)} \boldsymbol{v}(\boldsymbol{r}(u,v)) \cdot \left[\frac{\partial \boldsymbol{r}}{\partial u} \times \frac{\partial \boldsymbol{r}}{\partial v}\right] \mathrm{d}u\mathrm{d}v . \quad (16.15)$$

---

[8] Indices appearing two times (here j and k) are summed over.

[9] "Inner" and "outer" normal direction can be globally distinguished.

[10] These prerequisites for Gauss's integral theorem (and similar prerequites for Stokes's theorem) can of course be weakened.

b) *Stokes's theorem*

Let Γ be a *closed curve* with orientation, and let F be a sufficiently smooth 2d-manifold with (outer) normal $\boldsymbol{n}(\boldsymbol{r})$ *inserted*[11] *into* Γ, i.e., $\Gamma \equiv \partial F$. Moreover, let $\boldsymbol{v}(\boldsymbol{r})$ be a vector field, which is continuous on ∂F and continuously differentiable in the interior of F. The theorem of Stokes then states that

$$\oint_{\partial F} \boldsymbol{v}(\boldsymbol{r}) \cdot \mathrm{d}\boldsymbol{r} = \iint_F \mathrm{curl}\boldsymbol{v}(\boldsymbol{r}) \cdot \boldsymbol{n}(\boldsymbol{r})\mathrm{d}^2 A. \qquad (16.16)$$

The line integral on the l.h.s. of this equation defines the "circulation" of $\boldsymbol{v}$ along the closed curve $\Gamma = \partial F$. According to Stokes's theorem this quantity is expressed by a 2d-surface integral over the quantity $\mathrm{curl}\boldsymbol{v} \cdot \boldsymbol{n}$. Hence it is natural to define $\mathrm{curl}\boldsymbol{v}$ as the (vectorial) *circulation density* of $\boldsymbol{v}(\boldsymbol{r})$.

Detailed proofs of these integral theorems can be found in many mathematical textbooks. Here, as an example, we only consider Stokes's theorem, stressing that the surface F is not necessarily planar. But we assume that it can be *triangulated*, e.g., paved with infinitesimal triangles or rectangles, in each of which "virtual circulating currents" flow, corresponding to the chosen orientation, such that *in the interior* of F the edges of the triangles or rectangles *run pairwise in opposite directions*, see Fig. 16.2. It is thus sufficient to prove the theorem for infinitesimal rectangles R. Without lack of generality, R is assumed to be located in the (x,y)-plane; the four vertices are assumed to be

$$\begin{aligned} P_1 &= (-\Delta x/2, -\Delta y/2) \quad \text{(i.e., lower left)}, \\ P_2 &= (+\Delta x/2, -\Delta y/2) \quad \text{(i.e., lower right)}, \\ P_3 &= (+\Delta x/2, +\Delta y/2) \quad \text{(i.e., upper right)} \quad \text{and} \\ P_4 &= (-\Delta x/2, +\Delta y/2) \quad \text{(i.e., upper left)}. \end{aligned}$$

Therefore we have

$$\begin{aligned} \oint_{\partial R} \boldsymbol{v} \cdot \mathrm{d}\boldsymbol{r} &\cong [v_x(0, -\Delta y/2) - v_x(0, +\Delta y/2)]\, \Delta x \\ &\quad + [v_y(+\Delta x/2, 0) - v_y(-\Delta x/2), 0)]\, \Delta y \\ &\cong \left(-\frac{\partial v_x}{\partial y} + \frac{\partial v_y}{\partial x} \right) \Delta x \Delta y \\ &= \iint_R \mathrm{curl}\boldsymbol{v} \cdot \boldsymbol{n}\mathrm{d}^2 A. \end{aligned} \qquad (16.17)$$

---

[11] To aid understanding, it is recommended that the reader produces his or her own diagram of the situation. Here we are somewhat unconventional: usually one starts with a fixed F, setting $\partial F = \Gamma$; however we prefer to start with Γ, since the freedom to choose F, for fixed $\Gamma (\to \partial F)$, is the geometrical reason for the freedom of gauge transformations, see below.

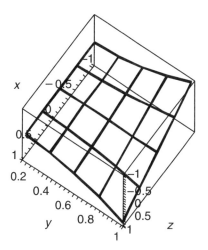

Fig. 16.2. Stokes's theorem: triangulating a 2d-surface in space. As an example, the surface $z = xy^2$ is triangulated (i.e., paved) by a grid of small squares. If all the squares are oriented in the same way as the outermost boundary line, then in the interior of the surface the line integrals are performed in a pairwise manner in the opposite sense, such that they compensate each other, leaving only the outermost boundary lines. As a consequence, if the *integral theorem of Stokes* applies for the small squares, then it also applies for a general surface

With $n = (0, 0, 1)$ and $d^2A = dxdy$ this provides the statement of the theorem.[12]

---

[12] Because of the greater suggestive power, in equations (16.17) we have used $\oint_{\partial R}$ instead of $\int_{\partial R}$.

17 Electrostatics and Magnetostatics

17.1 Electrostatic Fields in Vacuo

17.1.1 Coulomb's Law and the Principle of Superposition

The force on an electric point charge of strength q at the position $\boldsymbol{r}'$ in an electric field $\boldsymbol{E}(\boldsymbol{r}')$ (passive charge, local action) is given by

$$\boldsymbol{F}(\boldsymbol{r}') = q\boldsymbol{E}(\boldsymbol{r}') . \tag{17.1}$$

This expression defines the electric field $\boldsymbol{E}$ via the force acting *locally* on the passive charge q. However, the charge itself *actively* generates a field $\boldsymbol{E}(\boldsymbol{r})$ at a distant position $\boldsymbol{r}$ (active charge, action at a distance) given by:

$$\boldsymbol{E}(\boldsymbol{r}) = \frac{q \cdot (\boldsymbol{r} - \boldsymbol{r}')}{4\pi\varepsilon_0 |\boldsymbol{r} - \boldsymbol{r}'|^3} . \tag{17.2}$$

The fact that the *active* and *passive* charges are the same (apart from a multiplicative constant of nature such as c, which can be simply replaced by 1 without restriction) is not self-evident, but is essentially equivalent to Newton's third axiom. One consequence of this is the absence of torques in connection with Coulomb's law for the force between two point charges:

$$\boldsymbol{F}_{1,2} = \frac{q_1 q_2 \cdot (\boldsymbol{r}_1 - \boldsymbol{r}_2)}{4\pi\varepsilon_0 |\boldsymbol{r}_1 - \boldsymbol{r}_2|^3} (= -\boldsymbol{F}_{2,1}) . \tag{17.3}$$

Similar considerations apply to gravitational forces; in fact, "gravitational mass" could also be termed "gravitational charge", although an essential difference from electric charges is that *inter alia* electric charges can either have a *positive* or a *negative* sign, so that the force between two electric charges q_1 and q_2 can be *repulsive* or *attractive* according to the sign of $q_1 q_2$, whereas the gravitational force is always attractive, since "gravitational charges" always have the same sign, while the gravitational constant leads to attraction (see (17.4)).

Certainly, however, Newton's law of gravitation,

$$(\boldsymbol{F}_{\text{gravitation}})_{1,2} = -\gamma \frac{m_1 m_2 \cdot (\boldsymbol{r}_1 - \boldsymbol{r}_2)}{|\boldsymbol{r}_1 - \boldsymbol{r}_2|^3} , \tag{17.4}$$

where γ is the *gravitational constant*, is otherwise similar to Coulomb's law in electrostatics.

For a *distribution* of point charges, the electric field $\boldsymbol{E}(\boldsymbol{r})$ generated by this ensemble of charges can be calculated according to *the principle of superposition*:

$$\boldsymbol{F}_{1\leftarrow 2,3,\dots} = q_1 \cdot \boldsymbol{E}(\boldsymbol{r}_1) = q_1 \cdot \sum_{k=2,3,\dots} \frac{q_k(\boldsymbol{r}_1 - \boldsymbol{r}_k)}{4\pi\varepsilon_0 |\boldsymbol{r}_1 - \boldsymbol{r}_k|^3} \ . \tag{17.5}$$

This principle, which also applies for Newton's gravitational forces, is often erroneously assumed to be *self-evident*. However for other forces, such as those generated by nuclear interactions, it does not apply at all. Its validity in electrodynamics is attributable to the fact that Maxwell's equations are *linear* w.r.t. the fields $\boldsymbol{E}$ and $\boldsymbol{B}$, and to the charges and currents which generate them. On the other hand: the equations of chromodynamics, a theory which is formally rather similar to Maxwell's electrodynamics but describing nuclear forces, are *non-linear*, so the *principle of superposition* is not valid there.

17.1.2 Integral for Calculating the Electric Field

For a continuous distribution of charges one may "smear" the discrete point-charges, $(q_k \to \varrho(\boldsymbol{r}_k)\Delta V_k)$, and the so-called Riemann sum in equation (17.5) becomes the following integral:

$$\boldsymbol{E}(\boldsymbol{r}) = \iiint \mathrm{d}V' \frac{\varrho(\boldsymbol{r}')(\boldsymbol{r} - \boldsymbol{r}')}{4\pi\varepsilon_0 |\boldsymbol{r} - \boldsymbol{r}'|^3} \ . \tag{17.6}$$

This integral appears to have a singularity $\propto |\boldsymbol{r}-\boldsymbol{r}'|^{-2}$, but this singularity is only an apparent one, since in spherical coordinates near $\boldsymbol{r}' = \boldsymbol{r}$ one has

$$\mathrm{d}V' \propto |\boldsymbol{r}' - \boldsymbol{r}|^2 \mathrm{d}(|\boldsymbol{r}' - \boldsymbol{r}|) \ .$$

The electric field $\boldsymbol{E}(\boldsymbol{r})$ is thus necessarily continuous if $\varrho(\boldsymbol{r})$ is continuous; it can even be shown that under this condition the field $\boldsymbol{E}(\boldsymbol{r})$ is necessarily continuously differentiable if the region of integration is bounded[1]. It can then be shown that

$$\mathrm{div}\,\boldsymbol{E}(\boldsymbol{r}) = \varrho(\boldsymbol{r})/\varepsilon_0 \ .$$

This is the first of Maxwell's equations (16.3), often referred to as Gauss's law. A proof of this law is outlined below. (This law is not only valid under static conditions, but also quite generally, i.e., even if all quantities depend explicitly on time.)

---

[1] This is plausible, since for $d = 1$ the integral of a continuous function is a continuously differentiable function of the upper integration limit.

17.1.3 Gauss's Law

Gauss's law[2] gives the relation between the electric flux flowing out of a closed surface and the electric charge enclosed inside the surface. The proof is subtle and we only give a few essential hints. Consider two complementary regions of integration: For the first integration region, "1", we consider a small sphere of radius ε (an "ε-sphere"), whose center is situated exactly at the singularity $r' = r$. The second integration region, "2", is the complement of our "ε-sphere", i.e., the outer part. (A sketch is recommended.) In this outer region, all differentiations with respect to the components of the vector r [3] can be performed under the integral, and one obtains the exact result 0, because we have for

$$r \neq r' : \frac{(r - r')}{|r - r'|} = 0$$

(since for example $\partial_x \left(x/r^3 \right) = 1/r^3 - 3x^2/r^5$, such that $\nabla \cdot \left(r/r^3 \right) = 0$). As a consequence there remains

$$\operatorname{div} \boldsymbol{E}(\boldsymbol{r}) \equiv \operatorname{div} \boldsymbol{E}_1(\boldsymbol{r}) \,, \quad \text{where} \quad \boldsymbol{E}_1(\boldsymbol{r}) \cong \varrho(\boldsymbol{r}) \Delta V \frac{(\boldsymbol{r} - \boldsymbol{r}')}{4\pi\varepsilon_0 \, |\boldsymbol{r} - \boldsymbol{r}'|^3} \,,$$

where we assume that the volume ΔV of our ε-sphere,

$$\Delta V = \frac{4\pi\varepsilon^3}{3} \,,$$

is small enough that throughout the sphere $\varrho(\boldsymbol{r}')$ can be considered as constant.

We now use the statement, equivalent to the integral theorem of Gauss, that $\operatorname{div} \boldsymbol{E}_1$ can be interpreted as a *source density*, i.e.,

$$\operatorname{div} \boldsymbol{E}_1(\boldsymbol{r}) = \lim_{\Delta V \to 0} (\Delta V)^{-1} \oiint_{\partial \Delta V} \boldsymbol{E}_1(\boldsymbol{r}) \cdot \boldsymbol{n}(\boldsymbol{r}) \mathrm{d}^2 A \,.$$

As a consequence, after a short elementary calculation with

$$R = \varepsilon \quad \text{and} \quad \boldsymbol{n} = \frac{\boldsymbol{r} - \boldsymbol{r}'}{|\boldsymbol{r} - \boldsymbol{r}'|}$$

---

[2] Here we should be aware of the different expression: "Gauss's theorem" is the divergence or *integral theorem*, while "Gauss's law" means the first Maxwell equation.

[3] We must differentiate with respect to r, not r'.

we obtain the result

$$\operatorname{div} \boldsymbol{E}(\boldsymbol{r}) = \frac{\varrho(\boldsymbol{r})(\Delta V)4\pi R^2}{(\Delta V)4\pi\varepsilon_0 R^2} \equiv \varrho(\boldsymbol{r})/\varepsilon_0 \ .$$

This equation,

$$\operatorname{div} \boldsymbol{E}(\boldsymbol{r}) = \varrho(\boldsymbol{r})/\varepsilon_0 \ ,$$

constitutes the *differential form* of Gauss's law. The proof can be simplified using the δ-function:

$$\operatorname{div} \frac{(\boldsymbol{r}-\boldsymbol{r}')}{|\boldsymbol{r}-\boldsymbol{r}'|^3} \equiv 4\pi\delta(\boldsymbol{r}-\boldsymbol{r}') \ , \quad \forall \boldsymbol{r} \ , \quad \text{including} \quad \boldsymbol{r}=\boldsymbol{r}' \ . \tag{17.7}$$

One can now formally differentiate infinitely often under the integral, and one immediately obtains

$$\operatorname{div} \boldsymbol{E}(\boldsymbol{r}) \equiv \iiint \mathrm{d}V' \varrho(\boldsymbol{r}')\delta(\boldsymbol{r}-\boldsymbol{r}')/\varepsilon_0 = \varrho(\boldsymbol{r})/\varepsilon_0 \ . \tag{17.8}$$

Inserting (17.8) into Gauss's integral theorem, we then arrive at the *integral form* of Gauss's law:

$$\oint_{\partial V} \boldsymbol{E}(\boldsymbol{r}) \cdot n(\boldsymbol{r}) \mathrm{d}^2 A \equiv \frac{Q(V)}{\varepsilon_0} \ , \tag{17.9}$$

where $Q(V)$ is the total amount of charge contained in the enclosed volume[4]. The theorem can be proved most visually in this *integral version*, since, due to the superposition principle, one can assume that one is dealing only with a single point charge placed at the origin $\boldsymbol{r}' = 0$. The field strength is then $\propto R^{-2}$, or more exactly:

$$\boldsymbol{E} \cdot \boldsymbol{n} = \cos\vartheta \cdot \left(4\pi\varepsilon_0 R^2\right)^{-1} \ , \quad \text{but} \quad \mathrm{d}^2 A = R^2 \mathrm{d}\Omega/\cos\vartheta \ .$$

Here, ϑ is the angle between the field direction (radial direction) and the surface normal $\boldsymbol{n}$ (not necessarily radially directed); $\mathrm{d}\Omega$ is the solid-angle corresponding to a surface element, i.e.,

$$\oint \mathrm{d}\Omega = 4\pi \ .$$

These arguments are supported by Fig. 17.1.

---

[4] However, surface charges, i.e., those located on ∂V, are only counted with a factor $1/2$, whereas charges in the interior of V are counted with the full factor 1.

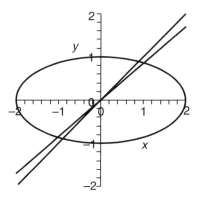

Fig. 17.1. On the proof of Gauss's law (= Maxwell I). At the center of an ellipsoid (here with the axes $a = 2$ and $b = 1$) there is a point charge Q, which generates the electric field $\boldsymbol{E} = Q/(4\pi\varepsilon_0 r^2)\boldsymbol{e}_r$, where $\boldsymbol{e}_r$ is the radial unit vector. Also plotted are two long lines $y \equiv x \cdot 1$ and $y \equiv x \cdot 0.85$, which are meant to illustrate a cone, through which the field flows outwards. The corresponding surface element $\mathrm{d}^{(2)}A$ on the ellipsoid is $r^2\mathrm{d}\Omega/\cos\vartheta$, where r is the distance of the surface from the charge, ϑ the angle between the directions of the surface normal $\boldsymbol{n}$ and the field direction, and $\mathrm{d}\Omega$ the solid-angle element. On the other hand, the scalar product $\boldsymbol{n} \cdot \boldsymbol{E}$ is $Q/(4\pi\varepsilon_0 r^2) \cdot \cos\vartheta$. Therefore, according to obvious geometrical reasons, the result of an integration over the surface of the ellipsoid is Q/ε_0, i.e., Gauss's law, as described above. In contrast, charges *outside* the ellipsoid give two *compensating* contributions, i.e., then the result is zero

17.1.4 Applications of Gauss's Law: Calculating the Electric Fields for Cases of Spherical or Cylindrical Symmetry

The *integral formulation of Gauss's law* is useful for simplifying the calculation of fields in the cases of spherical or cylindrical symmetry (see problems 1 and 3 of the exercises in summer 2002, [2]).

Firstly, we shall consider spherical symmetry:

$$\varrho(\boldsymbol{r}) \equiv \varrho(r) , \quad \text{with} \quad r := \sqrt{x^2 + y^2 + z^2} .$$

As a consequence, the electric field is also spherically symmetrical, i.e.,

$$\boldsymbol{E}(\boldsymbol{r}) = E(r) \cdot \boldsymbol{e}_r , \quad \text{where} \quad \boldsymbol{e}_r := \frac{\boldsymbol{r}}{r}$$

is the radial unit vector.

Applying the integral version of Gauss's law for a sphere of arbitrary radius r, we easily obtain the following result for the amplitude of the field, $E(r)$, i.e. for all $0 \le r < \infty$:

$$E(r) = \frac{1}{\varepsilon_0 r^2} \int_0^r \tilde{r}^2 \mathrm{d}\tilde{r}\varrho(\tilde{r}) . \tag{17.10}$$

A similar result is obtained for cylindrical symmetry, i.e., with

$$\varrho(\boldsymbol{r}) \to \varrho(r_\perp) \,,\,^5$$

where

$$r_\perp := \sqrt{x^2 + y^2} \,.$$

In this case we obtain

$$\boldsymbol{E}(\boldsymbol{r}) = E_\perp(r_\perp)\boldsymbol{e}_\perp \,, \quad \text{with} \quad \boldsymbol{r} = (x, y, z)\,, \quad \text{but} \quad \boldsymbol{e}_\perp := (x, y, 0)/r_\perp \,,$$

and

$$E_\perp(r_\perp) = \frac{1}{\varepsilon_0 r_\perp} \int_0^{r_\perp} \tilde{r}_\perp \mathrm{d}\tilde{r}_\perp \varrho(\tilde{r}_\perp) \,. \tag{17.11}$$

Using these formulae, calculation of the fields in the *interior* and *exterior* of many hollow or massive bodies with spherical or cylindrical symmetry can be considerably simplified. Externally the field appears as if the total charge of the sphere or cylinder is at the center of the sphere or on the axis of the cylinder.

17.1.5 The Curl of an Electrostatic Field; The Electrostatic Potential

One knows "by experience" that the *curl* of an electrostatic field always vanishes (i.e., the electrostatic field is *irrotational*). This is in agreement with the third of Maxwell's equations or Faraday's law of induction

$$\mathrm{curl}\boldsymbol{E} = -\frac{\partial \boldsymbol{B}}{\partial t} \quad \text{(see below)}.$$

One should keep in mind here that although Maxwell's equations are based on experimental observations. Therefore they should not really be regarded as universal laws or axioms, which are – as it were – self-evident or valid "before all experience", although one could conceivably think of them in that way.[6]

The *experience* mentioned above is that by calculating the work done along a closed path in 3-space,

$$\oint \mathrm{d}\boldsymbol{r} \cdot \boldsymbol{F}(\boldsymbol{r}) = q \oint \mathrm{d}\boldsymbol{r} \cdot \boldsymbol{E}(\boldsymbol{r}) \,,$$

---

[5] We do not consider any dependence on z.

[6] The excellent textbook by Sommerfeld *derives* the whole of electrodynamics from Maxwell's equations, from which it starts without any reasoning or justification. This is permissible in this case, but it should not be regarded naively as an acceptable modus operandi for theroretical scientists in general.

no energy can be gained, i.e., one always has

$$\oint_\Gamma \boldsymbol{E}(\boldsymbol{r}) \cdot \mathrm{d}\boldsymbol{r} = 0 \,, \quad \text{for any closed path} \quad \Gamma \in \mathcal{R}^3 \,.$$

In the above we are assuming that the boundary of the region considered is "connected" (i.e., one considers "one-fold connected regions" without any "holes"), such that any closed line Γ in the considered region G can be written as the boundary line, $\Gamma = \partial F$, of a 2d-manifold F which is completely contained in G. With these conditions one obtains from the integral theorem of Stokes:

$$\oint_\Gamma \boldsymbol{E}(\boldsymbol{r}) \cdot \mathrm{d}\boldsymbol{r} = \iint_F \boldsymbol{E}(\boldsymbol{r}) \cdot \boldsymbol{n}(\boldsymbol{r}) \mathrm{d}^2 A \,.$$

As a consequence one has

$$\boldsymbol{E}(\boldsymbol{r}) \equiv 0 \,.$$

However, according to Poincaré's lemma, any vector field $\boldsymbol{v}(\boldsymbol{r})$ which is *irrotational* ($\mathrm{curl}\boldsymbol{v} \equiv 0$) in a "one-fold connected open region" G, possesses in G a potential $\phi(\boldsymbol{r})$ such that

$$\boldsymbol{v}(\boldsymbol{r}) = -\mathrm{grad}\phi(\boldsymbol{r}) \,, \quad \text{i.e.,} \quad v_i(\boldsymbol{r}) = -\partial_i \phi(\boldsymbol{r}) \quad \text{for} \quad i = 1, 2, 3 \,.$$

The potential $\phi(\boldsymbol{r})$ is only determined up to an arbitrary additive constant, similar to the potential energy in mechanics, and the minus sign is mainly a matter of convention (although there are good reasons for it). For any irrotational vector field $\boldsymbol{E}(\boldsymbol{r})$, and in particular the electric field, this potential can be calculated using

$$\phi(\boldsymbol{r}) = -\int_{\boldsymbol{r}_1}^{\boldsymbol{r}} \boldsymbol{E}(\tilde{\boldsymbol{r}}) \cdot \mathrm{d}\tilde{\boldsymbol{r}} \,, \tag{17.12}$$

where both $\boldsymbol{r}_1$ and the integration path from $\boldsymbol{r}_1$ to $\boldsymbol{r}$ can be arbitrarily chosen.

A potential $\phi(\boldsymbol{r})$ which leads to the standard expression occurring in electrostatics,

$$\frac{\boldsymbol{r} - \boldsymbol{r}'}{|\boldsymbol{r} - \boldsymbol{r}'|^3} \,, \quad \text{is} \quad \phi(\boldsymbol{r}) = \frac{1}{|\boldsymbol{r} - \boldsymbol{r}'|} \,.$$

In fact it can easily be shown, with

$$|\boldsymbol{r} - \boldsymbol{r}'| = [(x - x')^2 + (y - y')^2 + (z - z')^2]^{\frac{1}{2}} \,, \quad \text{that} \quad -\partial_x \frac{1}{|\boldsymbol{r} - \boldsymbol{r}'|} = \frac{x - x'}{|\boldsymbol{r} - \boldsymbol{r}'|^3} \,.$$

For a continuous charge distribution, in accordance with the *principle of superposition*, one thus obtains the following potential:

$$\phi(\boldsymbol{r}) = \iiint \mathrm{d}^3 r' \frac{\varrho(\boldsymbol{r}')}{4\pi\varepsilon_0 |\boldsymbol{r} - \boldsymbol{r}'|} \,, \tag{17.13}$$

which is a result that can easily be memorized.

17.1.6 General Curvilinear, Spherical and Cylindrical Coordinates

At this point it makes sense to include a short mathematical section on polar coordinates. Hitherto we have always written vectors, e.g., $d\mathbf{r}$, as *triples* of three orthogonal Cartesian coordinates. We shall now adopt general *curvilinear* coordinates u, v, w such that

$$d\mathbf{r} = \frac{\partial \mathbf{r}}{\partial u_1} du_1 + \frac{\partial \mathbf{r}}{\partial u_2} du_2 + \frac{\partial \mathbf{r}}{\partial u_3} du_3 . \qquad (17.14)$$

By introducing the unit vectors

$$\mathbf{e}_i := \frac{\partial \mathbf{r}}{\partial u_i} / |\frac{\partial \mathbf{r}}{\partial u_i}| \quad (i = 1, 2, 3)$$

one obtains with well-defined functions $a_i(u_1, u_2, u_3)$ (for $i = 1, 2, 3$):

$$d\mathbf{r} = \sum_{i=1}^{3} \{a_i(u_1, u_2, u_3) du_i\} \, \mathbf{e}_i(u_1, u_2, u_3). \qquad (17.15)$$

Note that the products $a_i du_i$ have the physical dimension of "length". By introducing the so-called *dual triplet* of bi-orthogonal vectors $\mathbf{e}_j^*$ (dual to the triplet $\mathbf{e}_i$, i.e., for $i, j = 1, 2$ and 3: $\mathbf{e}_j^* \cdot \mathbf{e}_i \overset{!}{=} \delta_{j,i}$, i.e., $\overset{!}{=} 1$ for $i = j$, but $\overset{!}{=} 0$ for $i \neq j$, implemented by $\mathbf{e}_1^* := \frac{\mathbf{e}_2 \times \mathbf{e}_3}{\mathbf{e}_1 \cdot [\mathbf{e}_2 \times \mathbf{e}_3]}$ etc.), one obtains the short-hand expression

$$\mathrm{grad}\phi = \sum_{i=1}^{3} \frac{\partial \phi}{a_i \partial u_i} \mathbf{e}_i^* . \qquad (17.16)$$

Therefore one always obtains the following relation for the total differential $d\phi$:

$$d\phi = \mathrm{grad}\phi \cdot d\mathbf{r} = \sum_{i=1}^{3} \frac{\partial \phi}{\partial u_i} du_i .$$

This formulation of the vectors $d\mathbf{r}$ and $\mathrm{grad}\phi$ does not depend on the particular choice of *curvilinear coordinates*. In particular, for *spherical* and *cylindrical coordinates* we have $\mathbf{e}_j^* \equiv \mathbf{e}_j$, i.e., one is dealing with the special case of *locally orthogonal curvilinear coordinates*, and the *-symbols can be deleted.

a) For *spherical polar coordinates* we have $\theta \in [0, \pi]$ (latitude), $\varphi \in [0, 2\pi)$ (longitude, meridian, azimuth) and

$$x = r \sin\theta \cos\varphi , \quad y = r \sin\theta \cos\varphi , \quad z = r \cos\theta : \qquad (17.17)$$
$$d\mathbf{r} = dr\mathbf{e}_r + r \cdot d\theta \mathbf{e}_\theta + r \cdot \sin\theta d\varphi \mathbf{e}_\varphi , \qquad (17.18)$$
$$dV = r^2 dr \sin\theta d\theta d\varphi , \qquad (17.19)$$

$$\text{grad}\phi = \frac{\partial\phi}{\partial r}\boldsymbol{e}_r + \frac{\partial\phi}{r\cdot\partial\theta}\boldsymbol{e}_\theta + \frac{\partial\phi}{r\sin\theta\cdot\partial\varphi}\boldsymbol{e}_\varphi, \quad \text{and (see below)} \quad (17.20)$$

$$\nabla^2\phi = \frac{1}{r}\frac{\partial^2}{\partial r^2}(r\phi) + \frac{\partial}{r^2\sin\theta\partial\theta}\left(\sin\theta\frac{\partial\phi}{\partial\theta}\right) + \frac{\partial^2\phi}{r^2\sin^2\theta\partial\varphi^2}. \quad (17.21)$$

(The first term on the r.h.s. of equation (17.21) can also be written as $\frac{\partial}{r^2\partial r}\left(r^2\frac{\partial\phi}{\partial r}\right)$, which is easier generalizable, if one considers dimensionalities $d \geq 2$, i.e., $\frac{\partial}{r^{d-1}\partial r}\left(r^{d-1}\frac{\partial\phi}{\partial r}\right)$, but is less useful for the comparison with $d = 1$, see Part III.)

b) In *cylindrical coordinates* (or planar polar coordinates)

$$x = r_\perp\cos\varphi, \quad y = r_\perp\sin\varphi, \quad z = z \quad \text{(i.e., unchanged)}: \quad (17.22)$$

$$\mathrm{d}\boldsymbol{r} = \mathrm{d}r_\perp\boldsymbol{e}_{r_\perp} + r_\perp\cdot\mathrm{d}\varphi\boldsymbol{e}_\varphi + \mathrm{d}z\boldsymbol{e}_z, \quad (17.23)$$

$$\mathrm{d}V = r_\perp\mathrm{d}r_\perp\mathrm{d}\varphi\mathrm{d}z, \quad (17.24)$$

$$\text{grad}\phi = \frac{\partial\phi}{\partial r_\perp}\boldsymbol{e}_{r_\perp} + \frac{\partial\phi}{r_\perp\cdot\partial\varphi}\boldsymbol{e}_\varphi + \frac{\partial\phi}{\partial z}\boldsymbol{e}_z, \quad (17.25)$$

$$\nabla^2\phi = \frac{\partial}{r_\perp\partial r_\perp}\left(r_\perp\frac{\partial\phi}{\partial r_\perp}\right) + \frac{\partial^2\phi}{r_\perp^2\partial\varphi^2} + \frac{\partial^2\phi}{\partial z^2}. \quad (17.26)$$

In (17.21) and (17.26) we have added the so-called Laplace operator

$$\nabla^2(= \text{div grad}).$$

These results are particularly important, since they occur in many applications and examples.
In fact, with

$$\nabla^2\phi \equiv \text{div grad}\phi \quad \text{and} \quad \text{grad}\phi = -\boldsymbol{E}$$

they result from the following very general formula for *orthogonal curvilinear coordinates*, which can be proved by elementary considerations on source densities:

$$\text{div}\boldsymbol{E} = \frac{1}{a_1a_2a_3}\left\{a_2a_3\frac{\partial E_1}{\partial u_1} + a_3a_1\frac{\partial E_2}{\partial u_2} + a_1a_2\frac{\partial E_3}{\partial u_3}\right\}. \quad (17.27)$$

Capacitors; Capacity; Harmonic Functions; The Dirichlet Problem

Consider two arbitrary metal plates connected to the opposite poles of a cell or battery. The approximate profiles of the equipotential surfaces $\phi(\boldsymbol{r}) = $ constant and the corresponding electrostatic fields

$$\boldsymbol{E} = -\text{grad}\phi$$

between the capacitor plates can be readily sketched; but how does the field vary *quantitatively* in the space between the two capacitor plates?

This will now be discussed in detail.

We recall that the boundary of a conductor is always an equipotential surface (electrons can flow until the potential is equal everywhere) and that the relation

$$\boldsymbol{E} = -\text{grad}\phi$$

implies that the field is perpendicular to the equipotential surfaces

$$\phi(\boldsymbol{r}) = \text{constant} .$$

Furthermore in the interior of the conductor all three components of the electric field are zero.

In the space S outside or between the conductors there are no charges, so the equation

$$\nabla^2 \phi(\boldsymbol{r}) \equiv 0$$

holds. But on the surfaces ∂V_i of the two metal plates 1 and 2 the electrostatic potential ϕ must have different, but constant values, say

$$\phi_{|\partial V_1} \equiv c + U ; \quad \phi_{|\partial V_2} \equiv c ,$$

where U is the potential difference (or "voltage") provided by the battery.[7]

A function $\phi(\boldsymbol{r})$ which satisfies *Laplace's equation*,

$$\nabla^2 \phi = 0 ,$$

in an open region G (here $G = S$) is called *harmonic*.

As a consequence, in the space S between V_1 and V_2 we must find a *harmonic function* $\phi(\boldsymbol{r})$ which assumes the values

$$\phi_{|\partial V_1} \equiv U , \quad \text{but} \quad \phi_{|\partial V_2} \equiv 0$$

(and vanishes, of course, sufficiently quickly as $|\boldsymbol{r}| \to \infty$).

This is a simplification of the somewhat more general *Dirichlet problem*: for given charges $\varrho(\boldsymbol{r})$ one has the task of finding a function $\varPhi(\boldsymbol{r})$, which (i) *in the interior of S* satisfies the Poisson equation

$$\nabla^2 \varPhi \overset{!}{=} -\varrho(\boldsymbol{r})/\varepsilon_0 ,$$

and which (ii) *at the boundary of S* takes prescribed values:

$$\varPhi_{|\partial S} \overset{!}{=} f(\boldsymbol{r}) , \quad \text{with fixed} \quad \varrho(\boldsymbol{r}) \quad \text{and} \quad f(\boldsymbol{r}) .$$

---

[7] We remind the reader that the potential is only determined relative to some arbitrary reference value. Therefore it is always advisable to "ground" the second metal plate, since otherwise the field can be changed by uncontrolled effects, e.g., electrically charged dust particles.

The *Neumann problem*, where one prescribes boundary values not for the function $\phi(\boldsymbol{r})$ itself, but for the normal derivative

$$\partial_{\boldsymbol{n}}\phi := \mathrm{grad}\phi \cdot \boldsymbol{n}$$

is only slightly different.

One can show that a solution of the *Dirichlet problem*, if it exists, is always *unique*. The proof is based on the *linearity* of the problem and on another important statement, which is that *a harmonic function cannot have a local maximum or local minimum in the interior of an open region.*

For example, in the case of a local maximum in three dimensions the equipotential surfaces culminate in a "peak", and the field vectors, i.e., the negative gradients of the potential, are oriented away from the peak. According to Gauss's law there would thus be a positive charge on the peak. This would contradict

$$\varrho_{|S} \equiv 0 \, ,$$

i.e., the *harmonicity* of the potential.

Let us now assume that in S a given Dirichlet problem has two different solutions, ϕ_1 and ϕ_2; then (because of the *linearity*) the *difference*

$$w := \phi_2 - \phi_1$$

would satisfy in S the differential equation

$$\nabla^2 w(\boldsymbol{r}) = 0 \, ,$$

with boundary values

$$w_{|\partial S} = 0 \, .$$

Because of the non-existence of a local maximum of a harmonic function in the interior of S the function $w(\boldsymbol{r})$ must vanish everywhere in S, i.e.,

$$\phi_2 \equiv \phi_1 \, ,$$

in contradiction with the assumption.

By applying voltages U and 0 to the capacitor plate V_1 and V_2, respectively, one induces charges Q_1 (on ∂V_1) and Q_2 (on ∂V_2). We shall now show that

$$Q_2 = -Q_1 \, ,$$

even if only V_1 is connected to pole 1 of the battery, while V_2 is *grounded*, but not directly connected to pole 2;

$$Q_2 \left(= \oiint_{\partial V_2} \varepsilon_0 \cdot \boldsymbol{E} \cdot \boldsymbol{n} \mathrm{d}^2 A \right)$$

is called the *induced charge* on ∂V_2 (see below).

Proof of the above statements is again based on Gauss's integral theorem, according to which (with $E_n := \boldsymbol{E} \cdot \boldsymbol{n}$)

$$Q_1 + Q_2 = \oiint_{\partial V_1 + \partial V_2} \varepsilon_0 E_n \mathrm{d}^2 A . \tag{17.28}$$

However

$$\partial S = -\partial V_1 - \partial V_2 ,$$

where the change of sign reflects that, e.g., the *outer* normal of V_1 is an *inner* normal of S. But according to Gauss's integral theorem:

$$\oiint_{\partial S} E_n \mathrm{d}^2 A = \iiint_S \mathrm{div}\,\boldsymbol{E} \mathrm{d}V. \tag{17.29}$$

Since

$$\mathrm{div}\,\boldsymbol{E} = 0 , \quad \text{we have} \quad Q_1 + Q_2 = 0 ,$$

as stated.

The solution of the corresponding *Neumann problem* (i.e., not U, but Q is given) is also essentially unique for the capacitor. Here the proof of uniqueness is somewhat more subtle (N.B. we have used the term *essentially* above).

For a given Q_j two different solutions ϕ_k $(k = 1, 2)$ must both satisfy

$$-\varepsilon_0 \cdot \oiint_{\partial V_j} (\nabla\phi_k \cdot \boldsymbol{n}) \, \mathrm{d}^2 A = Q_j \quad (\text{for } j = 1, 2) ;$$

thus their difference $w := \phi_2 - \phi_1$ satisfies

$$\oiint_{\partial V_j} \nabla w \cdot \boldsymbol{n} \mathrm{d}^2 A = 0 ,$$

and since w is constant on ∂V_j, one even has

$$\oiint_{\partial(I_s)} (w \nabla w) \cdot \boldsymbol{n} \mathrm{d}^2 A = 0 .$$

Moreover, according to Gauss's integral theorem it follows that

$$\iiint_{I_s} \nabla \cdot (w \nabla w) \mathrm{d}V = 0 , \quad \text{i.e.,}$$

$$\iiint_{I_s} \left[(\nabla w)^2 + w \nabla^2 w \right] \mathrm{d}V = 0 .$$

Therefore, since

$$\nabla^2 w = 0 \quad \text{we have} \quad \iiint_{Is} (\nabla w)^2 \mathrm{d}V = 0 ;$$

as a consequence
$$\nabla w \equiv 0$$
everywhere (apart from a set of zero measure). Therefore, it is not now the potential that is unique but its *essence*, the gradient, i.e., the electric field itself.

The *capacity* of a condenser is defined as the ratio of charge to voltage across the plates, or $C := \frac{Q}{U}$. Elementary calculations contained in most textbooks give the following results: For a *plate capacitor* (with plate area F and plate separation d), a *spherical capacitor* (with inner radius R and outer radius $R + \Delta R$), and for a *cylindrical capacitor* (with length L and inner radius $R_\perp$) one obtains, respectively:

$$C = \varepsilon_0 \frac{F}{d} \; , \quad C = 4\pi\varepsilon_0 \frac{R \cdot (R + \Delta R)}{\Delta R} \; , \quad C = 2\pi\varepsilon_0 \frac{L}{\ln \frac{R_\perp + \Delta R_\perp}{R_\perp}} \; . \quad (17.30)$$

In the limit $\Delta R \ll R$ ($\Delta R_\perp \ll R_\perp$) these results are all identical.

17.1.7 Numerical Calculation of Electric Fields

The *uniqueness theorem* for the Dirichlet problem is useful, amongst other reasons, because it allows one immediately to accept *a solution* (found or guessed by any means available) as *the only* solution. One should not underestimate the practical importance of this possibility!

If all analytical methods fail, numerical methods always remain. These methods are not necessarily as complicated as one might think, and they can often be used in the context of school physics. If one considers, for example, a simple-cubic grid of edge length a between lattice points, then one has (up to a small discretization error $\propto a^2$):

$$\left(\nabla^2 \phi\right)_{\boldsymbol{n}} \equiv \sum_{j=1}^{6} (\phi_{\boldsymbol{n}+\boldsymbol{\Delta}_j} - \phi_{\boldsymbol{n}})/a^2. \quad (17.31)$$

Here the six vectors $\boldsymbol{n} + \boldsymbol{\Delta}_j$ denote the nearest neighbours (right – left, backwards – forwards, up – down) of the lattice point considered, and

$$\phi_{\boldsymbol{n}} := \phi(\boldsymbol{r}_{\boldsymbol{n}}) \; .$$

To obtain a *harmonic function*, it is thus only necessary to iterate (to the desired accuracy) until at any lattice point one has obtained

$$\phi_{\boldsymbol{n}} \equiv \frac{1}{6} \sum_{j=1}^{6} \phi_{\boldsymbol{n}+\boldsymbol{\Delta}_j} \; ,$$

i.e., the l.h.s. must agree with the r.h.s.

17.2 Electrostatic and Magnetostatic Fields in Polarizable Matter

17.2.1 Dielectric Behavior

For a polarizable medium, such as water, the *permittivity of free space* ε_0 must be replaced by the product $\varepsilon_0 \cdot \varepsilon$, where ε is the *relative dielectric constant* of the medium, which is assumed to be isotropic here (this applies to gases, liquids, polycrystalline solids and crystals with cubic symmetry[8]; ε is a dimensionless constant of the order of magnitude $\mathcal{O}(\varepsilon) = 10$ to 100 (e.g., for water $\varepsilon \cong 81)$[9].

When a dielectric (or polarizable material) is placed between the plates of a condenser, the capacity is increased by a factor ε. For a parallel plate condenser, for example, the capacity becomes:

$$C = \varepsilon \varepsilon_0 \frac{F}{d} \ .$$

This enhancement is often considerable, and in practice a dielectric material, such as an insulating plastic, is inserted between condenser plates in order to increase the amount of charge that can be stored for a given U.

In addition, the electric displacement vector $\boldsymbol{D}$ (see later)[10] is defined for such materials as $\boldsymbol{D} := \varepsilon_0 \varepsilon \boldsymbol{E}$. Then, from Gauss's law we obtain the first of Maxwell's equations (i) in *integral form*

$$\oint_{\partial V} \boldsymbol{D} \cdot \boldsymbol{n} \mathrm{d}^2 A = Q(V) \quad \text{(Maxwell I)} \ , \tag{17.32}$$

and (ii) in *differential form*

$$\operatorname{div} \boldsymbol{D} = \varrho \ . \tag{17.33}$$

What are the atomistic reasons for *dielectric behavior*? The answer to this short question requires several sections (see below).

17.2.2 Dipole Fields; Quadrupoles

An *electric dipole* at a position $\boldsymbol{r}'$ *in vacuo* can be generated by the following elementary *dumbbell approach*.

---

[8] Generalizations to non-cubic crystals will be treated below in the context of crystal optics.

[9] The product $\varepsilon_0 \cdot \varepsilon$ is sometimes called the *absolute dielectric constant of the material* and also written as ε; however this convention is not followed below, i.e., by ε we always understand the *relative* dielectric constant, especially since ε_0 does not appear in the cgs system whereas the *relative* dielectric constant is directly taken over into that system. In addition it should be mentioned that the dielectric constant is not always *constant* but can be frequency dependent.

[10] So-called for historical reasons.

Firstly consider two exactly opposite point charges each of strength $\pm q$, placed at the ends of a small dumbbell, $\boldsymbol{r}' + (\boldsymbol{a}/2)$ and $\boldsymbol{r}' - (\boldsymbol{a}/2)$. Then take the limit (i) $a \to 0$ while (ii) $q \to \infty$, in such a way that $q\boldsymbol{a} \to \boldsymbol{p}(\neq 0)$, whereas the limit $qa^2 \to 0$. The result of this procedure (the so-called "dipole limit") is the vector $\boldsymbol{p}$, called the *dipole moment* of the charge array. Similarly to Dirac's δ-function the final result is largely independent of intermediate configurations.

The electrostatic potential ϕ (before performing the limit) is given by

$$\phi(\boldsymbol{r}) = \frac{q}{4\pi\varepsilon_0} \left(\frac{1}{|\boldsymbol{r} - \boldsymbol{r}' - \frac{a}{2}|} - \frac{1}{|\boldsymbol{r} - \boldsymbol{r}' + \frac{a}{2}|} \right) . \tag{17.34}$$

Using a Taylor expansion w.r.t. $\boldsymbol{r}$ and neglecting quadratic terms in $|\boldsymbol{a}|$, one then obtains:

$$\phi(\boldsymbol{r}) \cong -\frac{q\boldsymbol{a}}{4\pi\varepsilon_0} \cdot \mathrm{grad}_{\boldsymbol{r}} \frac{1}{|\boldsymbol{r} - \boldsymbol{r}'|} = \frac{q\boldsymbol{a}}{4\pi\varepsilon_0} \cdot \frac{\boldsymbol{r} - \boldsymbol{r}'}{|\boldsymbol{r} - \boldsymbol{r}'|^3} . \tag{17.35}$$

The electrostatic potential ϕ_{Dp} due to an electric dipole with dipole vector $\boldsymbol{p}$ at position $\boldsymbol{r}'$ is thus given by

$$\phi_{Dp}(\boldsymbol{r}) = \frac{\boldsymbol{p}}{4\pi\varepsilon_0} \cdot \frac{\boldsymbol{r} - \boldsymbol{r}'}{|\boldsymbol{r} - \boldsymbol{r}'|^3} , \tag{17.36}$$

and the corresponding electric field, $\boldsymbol{E}_{Dp} = -\mathrm{grad}\phi_{Dp}$, is ($\to$ exercises):

$$\boldsymbol{E}(\boldsymbol{r})_{Dp} = \frac{(3(\boldsymbol{r} - \boldsymbol{r}') \cdot \boldsymbol{p})\,(\boldsymbol{r} - \boldsymbol{r}') - |\boldsymbol{r} - \boldsymbol{r}'|^2 \boldsymbol{p}}{4\pi\varepsilon_0 |\boldsymbol{r} - \boldsymbol{r}'|^5} . \tag{17.37}$$

In particular one should keep in mind the characteristic *kidney-shaped* appearance of the field lines (once more a sketch by the reader is recommended).

A similar calculation can be performed for an *electric quadrupole*. The corresponding array of charges consists of two opposite dipoles shifted by a vector $\boldsymbol{b}$.[11] The Taylor expansion must now be performed up to second order. Monopole, dipole and quadrupole potentials thus decay $\sim r^{-1}$, $\sim r^{-2}$ and $\sim r^{-3}$, whereas the fields decay $\propto r^{-2}$, $\propto r^{-3}$ and $\propto r^{-4}$, respectively.

17.2.3 Electric Polarization

In a fluid (i.e., a gas or a liquid) of dielectric molecules, an external electric field $\boldsymbol{E}$ *polarizes* these molecules, which means that the *charge center* of the (negatively charged) electron shells of the molecules shifts relative to the (positively charged) nuclei of the molecule. As a consequence, a molecular electric dipole moment is induced, given by

---

[11] Another configuration corresponds to a sequence of equal charges of alternating sign at the four vertices of a parallelogram.

$$p = \alpha \varepsilon_0 E \; ,$$

where α is the so-called *molecular polarizability*, which is calculated by quantum mechanics.

Now, a volume element ΔV of the fluid contains

$$\Delta N := n_V \Delta V$$

molecules of the type considered. Moreover, the electric moment of this volume element is simply the sum of the electric moments of the molecules contained in ΔV, i.e.,

$$\Delta p \equiv P \Delta V \; .$$

This is the definition of the so-called *electric polarization*:

$$P \equiv \frac{\Delta p}{\Delta V} \; , \quad \text{where} \quad \Delta p := \sum_{r_i \in \Delta V} p(r_i) \; ;$$

i.e., *the polarization P is the (vector) density of the electric moment (dipole density).* Under the present conditions this definition leads to the result

$$P = n_V \varepsilon_0 \alpha E \; .$$

Furthermore, the displacement D, is generally defined via the dipole density as

$$D := \varepsilon_0 E + P \; . \tag{17.38}$$

For dielectric material we obtain

$$D = \varepsilon_0 \cdot (1 + \chi) \cdot E \quad \text{where} \quad \chi = n_V \alpha$$

is the *electric susceptibility*, i.e., $\varepsilon = 1 + \chi$.

(N.B.: In a cgs system, instead of (17.38) we have: $D' := E' + 4\pi P'$, where $P' = n_v \alpha E'$, for unchanged α. Hence $\varepsilon = 1 + 4\pi \chi' = 1 + \chi$, i.e., $\chi = 4\pi \chi'$. Unfortunately the prime is usually omitted from tables of data, i.e., the authors of the table rely on the ability of the reader to recognize whether given data correspond to χ or χ'. Often this can only be decided if one knows which system of units is being used.)

17.2.4 Multipole Moments and Multipole Expansion

The starting point for this rather general subsection is the formula for the potential $\phi(r)$ of a charge distribution, which is concentrated in a region of vacuum G':

$$\phi(\boldsymbol{r}) = \int_{G'} \frac{\varrho(\boldsymbol{r}')\mathrm{d}V'}{4\pi\varepsilon_0|\boldsymbol{r}-\boldsymbol{r}'|} \ .$$

We assume that for all $\boldsymbol{r}' \in G'$ the following inequality holds: $|\boldsymbol{r}| \gg |\boldsymbol{r}'|$, i.e., that the sampling points $\boldsymbol{r}$ are very far away from the sources $\boldsymbol{r}'$.

A Taylor expansion w.r.t. $\boldsymbol{r}'$ can then be performed, leading to

$$\frac{1}{|\boldsymbol{r}-\boldsymbol{r}'|} \cong \frac{1}{r} + \sum_{i=1}^{3}(-x_i')\frac{\partial}{\partial x_i}\frac{1}{r} + \frac{1}{2!}\sum_{i,k=1}^{3}(-x_i')(-x_k')\frac{\partial^2}{\partial x_i \partial x_k}\frac{1}{r} + \dots \quad (17.39)$$

Substitution of (17.39) into the formula for $\phi(\boldsymbol{r})$ finally results in

$$\phi(\boldsymbol{r}) = \frac{1}{4\pi\varepsilon_0}\left\{\frac{Q}{r} + \frac{\boldsymbol{p}\cdot\boldsymbol{r}}{r^3} + \frac{1}{2}\sum_{i,k=1}^{3}q_{i,k}\frac{3x_i x_k - r^2\delta_{i,k}}{r^5} + \dots\right\}, \quad (17.40)$$

with

a) the *total charge* of the charge distribution,

$$Q := \int_{G'}\varrho(\boldsymbol{r}')\mathrm{d}V' \ ;$$

b) the *dipole moment* of the charge distribution, a vector with the three components

$$p_i := \int_{G'}\varrho(\boldsymbol{r}')x_i'\mathrm{d}V' \ ;$$

and

c) the *quadrupole moment* of the distribution, a symmetric second-order tensor, with the components

$$q_{i,k} = \int_{G'}\varrho(\boldsymbol{r}')x_i'x_k'\mathrm{d}V' \ .$$

Higher *multipole moments* $q_{i_1,\dots,i_l}$ are calculated analogously, i.e., in terms of order l $(l = 0, 1, 2, \dots)$ one obtains

$$\phi(\boldsymbol{r}) \cong \frac{Q}{4\pi\varepsilon_0 r} + \frac{\boldsymbol{p}\cdot\boldsymbol{r}}{4\pi\varepsilon_0 r^3}$$
$$+ \sum_{l=2}^{\infty}\sum_{i_1,\dots,i_l=1}^{3}\frac{(-1)^l}{l!}\frac{1}{4\pi\varepsilon_0}q_{i_1,\dots,i_l}\cdot\frac{\partial^l}{\partial x_{i_1}\dots\partial x_{i_l}}\left(\frac{1}{r}\right) . \quad (17.41)$$

However, only a minor fraction of the many 2^l-pole moments $q_{i_1,\dots,i_l}$ actually influence the result; in every order l, there are only $2l + 1$ linear independent terms, and one can easily convince oneself that the quadrupole tensor can be changed by the addition of an arbitrary diagonal tensor,

$$q_{i,k} \to q_{i,k} + a \cdot \delta_{i,k} ,$$

without any change in the potential.

As a result, for $l = 2$ there are not six, but only $2l + 1 = 5$ linearly independent quadrupole moments $q_{i,k} = q_{k,i}$. Analogous results also apply for $l > 2$, involving so-called *spherical harmonics* $Y_{lm}(\theta, \varphi)$, which are listed in (almost) all relevant books or collections of formulae, especially those on quantum mechanics. One can write with suitable complex expansion coefficients $c_{l,m}$:

$$\phi(\boldsymbol{r}) = \frac{Q}{4\pi\varepsilon_0 r} + \frac{\boldsymbol{p} \cdot \boldsymbol{r}}{4\pi\varepsilon_0 r^3} + \sum_{l=2}^{\infty} \frac{1}{4\pi\varepsilon_0 r^{l+1}} \cdot \sum_{m=-l}^{+l} c_{l,m} Y_{l,m}(\theta, \varphi) . \qquad (17.42)$$

Due to the orthogonality properties of the spherical harmonics, which are described elsewhere (see Part III), the so-called *spherical multipole moments* $c_{l,m}$, which appear in (17.42), can be calculated using the following integral involving the complex-conjugates $Y_{l,m}^*$ of the spherical harmonics:

$$c_{l,m} = \frac{4\pi}{2l+1} \int_0^\infty dr r^2 \int_0^\pi d\theta \sin\theta \int_0^{2\pi} d\varphi \varrho(\boldsymbol{r}) r^l Y_{l,m}^*(\theta, \varphi) . \qquad (17.43)$$

Dielectric, Paraelectric and Ferroelectric Systems; True and Effective Charges

a) In *dielectric* systems, the molecular dipole moment is only *induced*, $\boldsymbol{p}_{\text{molec.}} = \varepsilon_0 \alpha \boldsymbol{E}$.

b) In contrast, for *paraelectric* systems one has a *permanent* molecular dipole moment, which, however, for $\boldsymbol{E} = 0$ vanishes on average, i.e., by performing an average w.r.t. to space and/or time. (This is the case, e.g., for a dilute gas of HCl molecules.)

c) Finally, for *ferroelectric* systems, e.g., BaTiO$_3$ crystals, below a so-called *critical temperature* T_c a *spontaneous long-range order* of the electric polarization $\boldsymbol{P}$ exists, i.e., even in the case of infinitesimally small external fields (e.g. $\boldsymbol{E} \to 0^+$) everywhere within the crystal a finite expectation value of the vector $\boldsymbol{P}$ exists, i.e., for $T < T_c$ one has $\langle \boldsymbol{P} \rangle \neq 0$.

In each case where $\boldsymbol{P} \neq 0$ Gauss's law does *not* state that

$$\oint_{\partial V} \varepsilon_0 \boldsymbol{E} \cdot \boldsymbol{n} d^2 A = Q(V) ,$$

but, instead (as already mentioned, with $D = \varepsilon_0 E + P$):

$$\oiint_{\partial V} D \cdot n \mathrm{d}^2 A = Q(V) \,. \tag{17.44}$$

Alternatively, in differential form:

$$\operatorname{div} D = \varrho \quad \text{i.e., precisely:} \quad \operatorname{div}(\varepsilon_0 E + P) = \varrho \,, \tag{17.45}$$

and *not* simply

$$\operatorname{div} \varepsilon_0 E = \varrho \,.$$

For a parallel plate condenser filled with a dielectric, for given charge Q, the electric field E between the plates is smaller than *in vacuo*, since part of the charge is *compensated* by the induced polarization. The expression

$$\varrho_E := \operatorname{div} \varepsilon_0 E$$

represents only the remaining non-compensated (i.e. *effective*) charge density,

$$\varrho_E = \varrho_{\text{true}} - \operatorname{div} P \,,$$

(see below).

In the following, ϱ will be systematically called the *true charge density* ($\varrho \equiv \varrho_{\text{true}}$). In contrast, the expression

$$\varrho_E := \varepsilon_0 \operatorname{div} E$$

will be called the *effective charge density*. These names are semantically somewhat arbitrary. In each case the following equation applies:

$$\varrho_E = \varrho - \operatorname{div} P \,.$$

Calculating the Electric Field in the "Polarization Representation" and the "Effective Charge Representation"

a) We shall begin with the representation in terms of *polarization*, i.e., with the existence of *true electric charges* plus *true electric dipoles*, and then apply the superposition principle. In the case of a continous charge and dipole distribution, one has the following sum:

$$\phi(r) = \iiint \frac{\varrho(r')\mathrm{d}V'}{4\pi\varepsilon_0|r - r'|} + \iiint \frac{\mathrm{d}V' P(r') \cdot (r - r')}{4\pi\varepsilon_0|r - r'|^3} \,. \tag{17.46}$$

This *dipole representation* is simply the superposition of the Coulomb potentials of the true charges plus the contribution of the dipole potentials of the true dipoles.

b) Integrating by parts (see below), equation (17.46) can be directly transformed into the equivalent *effective-charge representation*, i.e., with

$$\varrho_E := \varrho - \operatorname{div} \boldsymbol{P}$$

one obtains:

$$\phi(\boldsymbol{r}) = \iiint \frac{\varrho_E(\boldsymbol{r}')\mathrm{d}V'}{4\pi\varepsilon_0|\boldsymbol{r}-\boldsymbol{r}'|} \ . \tag{17.47}$$

The electric polarization does not appear in equation (17.47), but instead of the density ϱ of true charges one now has the density $\varrho_E = \varrho - \operatorname{div}\boldsymbol{P}$ of effective (i.e., not compensated) charges (see above). For simplicity we have assumed that at the boundary ∂K of the integration volume K the charge density and the polarization do not jump discontinuously to zero but that, instead, the transition is smooth. Otherwise one would have to add to equation (17.47) the Coulomb potential of effective surface charges

$$\sigma_E\mathrm{d}^2A := (\sigma + \boldsymbol{P} \cdot \boldsymbol{n})\mathrm{d}^2 A \ ,$$

i.e., one would obtain

$$\phi(\boldsymbol{r}) = \iiint_K \frac{\varrho_E(\boldsymbol{r}')\mathrm{d}V'}{4\pi\varepsilon_0|\boldsymbol{r}-\boldsymbol{r}'|} + \oiint_{\partial K} \frac{\sigma_E(\boldsymbol{r}')\mathrm{d}^2 A'}{4\pi\varepsilon_0|\boldsymbol{r}-\boldsymbol{r}'|} \ . \tag{17.48}$$

Later we shall deal separately and in more detail with such *boundary divergences* and similar *boundary rotations*. Essentially they are related to the (more or less elementary) fact that the formal derivative of the unit step function (Heaviside function)

$$\Theta(x)(= 1 \text{ for } x > 0 \ ; \quad = 0 \quad \text{for} \quad x < 0 \ ; \quad = \frac{1}{2} \quad \text{for} \quad x = 0)$$

is Dirac's δ-function:

$$\frac{\mathrm{d}\Theta(x)}{\mathrm{d}x} = \delta(x) \ .$$

Proof of the equivalence of (17.47) and (17.46) would be simple: the equivalence follows by partial integration of the relation

$$\frac{\boldsymbol{r}-\boldsymbol{r}'}{|\boldsymbol{r}-\boldsymbol{r}'|^3} = +\nabla_{\boldsymbol{r}'}\frac{1}{|\boldsymbol{r}-\boldsymbol{r}'|} \ .$$

One shifts the differentiation to the left and obtains from the second term in (17.46):

$$-\iiint \frac{\nabla_{\boldsymbol{r}'} \cdot \boldsymbol{P}(\boldsymbol{r}')}{4\pi\varepsilon_0|\boldsymbol{r}-\boldsymbol{r}'|}\mathrm{d}V' \ .$$

Together with the first term this yields the required result.

17.2.5 Magnetostatics

Even the ancient Chinese were acquainted with *magnetic fields* such as that due to the earth, and *magnetic dipoles* (e.g., *magnet needles*) were used by mariners as compasses for navigation purposes. In particular it was known that magnetic dipoles exert forces and torques on each other, which are analogous to those of electric dipoles.

For example, the *torque* $\mathcal{D}$, which a magnetic field $\boldsymbol{H}$ exerts on a magnetic dipole, is given by

$$\mathcal{D} = \boldsymbol{m} \times \boldsymbol{H} \ ,$$

and the force $\boldsymbol{F}$ on the same dipole, if the magnetic field is inhomogeneous, is also analogous to the electric case, i.e.,

$$\boldsymbol{F} = (\boldsymbol{m} \cdot \text{grad})\boldsymbol{H}$$

(see below).

On the other hand a magnetic dipole at $\boldsymbol{r}'$ itself generates a magnetic field $\boldsymbol{H}$, according to[12]

$$\boldsymbol{H}(\boldsymbol{r}) = -\text{grad}\phi_m \ , \quad \text{with} \quad \phi_m = \frac{\boldsymbol{m} \cdot (\boldsymbol{r} - \boldsymbol{r}')}{4\pi\mu_0|\boldsymbol{r} - \boldsymbol{r}'|^3} \ .$$

All this is completely analogous to the electric case, i.e., according to this convention one only has to replace ε_0 by μ_0, $\boldsymbol{E}$ by $\boldsymbol{H}$ and $\boldsymbol{p}$ by $\boldsymbol{m}$.

However, apparently there are *no* individual magnetic charges, or magnetic monopoles, although one has searched diligently for them. Thus, if one introduces analogously to the *electric polarization* $\boldsymbol{P}$ a so-called *magnetic polarization* $\boldsymbol{J}$ [13], which is given analogously to the electric case by the relation

$$\Delta\boldsymbol{m} = \boldsymbol{J}\Delta V \ ,$$

then one can define a quantity

$$\boldsymbol{B} := \mu_0\boldsymbol{H} + \boldsymbol{J} \ ,$$

the so-called "magnetic induction", which is analogous to the "dielectric displacement"

$$\boldsymbol{D} := \varepsilon_0\boldsymbol{E} + \boldsymbol{P} \ ;$$

but instead of

$$\text{div}\boldsymbol{D} = \varrho$$

---

[12] Unfortunately there are different, although equivalent, conventions: many authors write $\mathcal{D} = \boldsymbol{m}_B \times \boldsymbol{B}$, where *in vacuo* $\boldsymbol{B} = \mu_0\boldsymbol{H}$, and define $\boldsymbol{m}_B$ as *magnetic moment*, i.e., $\boldsymbol{m}_B = \boldsymbol{m}/\mu_0$, where (unfortunately) the index B is omitted.

[13] All these definitions, including those for $\boldsymbol{P}$ and $\boldsymbol{J}$, are prescribed by international committees and should not be changed.

one obtains

$$\mathrm{div}\,\boldsymbol{B} \equiv 0 \ ,$$

or better

$$\oiint_{\partial V} \mathrm{d}^2 A (\boldsymbol{B} \cdot \boldsymbol{n}) \equiv 0 \ , \quad \forall V \quad \text{(Maxwell II)} \ ; \tag{17.49}$$

i.e., there are no ("true") magnetic charges at all, but only magnetic dipoles.

The above relation is the second Maxwell equation, which is again essentially based on experimental experience.

Using these relations, the magnetic field $\boldsymbol{H}$ in the absence of electric currents, can be derived from a *magnetic potential ϕ_m*, which is calculated from equations analogous to (17.46) and (17.48). (The reader – if a student undergoing examinations – should write down the equations as preparation for a possible question.)

Solution:

$$\boldsymbol{H}(\boldsymbol{r}) = -\mathrm{grad}\phi_m(\boldsymbol{r}) \ ,$$

with the two equivalent formulae

a) *dipole representation for magnetized bodies*

$$\phi_m(\boldsymbol{r}) = \iiint_G \mathrm{d}V' \frac{\boldsymbol{J}(\boldsymbol{r}') \cdot (\boldsymbol{r} - \boldsymbol{r}')}{4\pi\mu_0 |\boldsymbol{r} - \boldsymbol{r}'|^3} \ ; \tag{17.50}$$

b) *representation in terms of effective magnetic charges*

$$\phi_m(\boldsymbol{r}) = \iiint_G \mathrm{d}V' \frac{-(\mathrm{div}\boldsymbol{J})(\boldsymbol{r}')}{4\pi\mu_0 |\boldsymbol{r} - \boldsymbol{r}'|} + \oiint_{\partial G} \mathrm{d}^2 A' \frac{\boldsymbol{J}(\boldsymbol{r}') \cdot \boldsymbol{n}(\boldsymbol{r}')}{4\pi\mu_0 |\boldsymbol{r} - \boldsymbol{r}'|} \ . \tag{17.51}$$

17.2.6 Forces and Torques on Electric and Magnetic Dipoles

The force on an electric dipole in an electric field may be calculated using the dumbbell approximation of two opposite charges at slightly different positions:

$$\begin{aligned} \boldsymbol{F} &= q \cdot \left(\boldsymbol{E}\left(\boldsymbol{r} + \frac{\boldsymbol{a}}{2}\right) - \boldsymbol{E}\left(\boldsymbol{r} - \frac{\boldsymbol{a}}{2}\right) \right) \cong q \sum_{i=1}^{3} \left(a_i \frac{\partial}{\partial x_i} \right) \boldsymbol{E}(\boldsymbol{r}) \\ &\to (\boldsymbol{p} \cdot \nabla)\,\boldsymbol{E}(\boldsymbol{r}) \ . \end{aligned} \tag{17.52}$$

In the *dipole limit* ($q \to \infty$, $\boldsymbol{a} \to 0$, but $q \cdot \boldsymbol{a} \to \boldsymbol{p}(\neq 0)$, whereas $q a_i a_j \to 0$) one obtains the result

$$\boldsymbol{F} = (\boldsymbol{p} \cdot \mathrm{grad})\boldsymbol{E}(\boldsymbol{r}) \ .$$

A similar calculation yields the formula for the torque $\mathcal{D}$:

$$\begin{aligned} \mathcal{D} &= q\left[\left(\boldsymbol{r} + \frac{\boldsymbol{a}}{2}\right) \times \boldsymbol{E}\left(\boldsymbol{r} + \frac{\boldsymbol{a}}{2}\right) - \left(\boldsymbol{r} - \frac{\boldsymbol{a}}{2}\right) \times \boldsymbol{E}\left(\boldsymbol{r} - \frac{\boldsymbol{a}}{2}\right) \right] \\ &\to \boldsymbol{p} \times \boldsymbol{E}(\boldsymbol{r}) \ . \end{aligned} \tag{17.53}$$

For the magnetic case one only needs to replace $\boldsymbol{p}$ by $\boldsymbol{m}$ and $\boldsymbol{E}$ by $\boldsymbol{H}$.

17.2.7 The Field Energy

We shall now introduce a number of different, but equivalent, expressions for the energy associated with an electric field. The expressions are fundamental and will be used later. Firstly, we shall start with a capacitor with dielectric material of dielectric constant ε between the plates.

Let us transport an infinitesimal amount of charge[14], δQ, from the metal plate at lower potential to that with the higher potential, where a *capacitor voltage*

$$U(Q) = \frac{Q}{C}$$

has already been built up. The (infinitesimal) energy or work done in transporting the charge is

$$\delta E = \delta Q \cdot U(Q) \,.$$

By transporting the total charge in this way we finally obtain

$$E = \int\limits_0^Q \mathrm{d}\tilde{Q} \cdot U(\tilde{Q}) \,, \quad \text{with} \quad U(\tilde{Q}) = \frac{\tilde{Q}}{C} \,,$$

i.e.,

$$1.) \qquad E = \frac{1}{2} U \cdot Q = \frac{Q^2}{2C} = \frac{C}{2} U^2. \tag{17.54}$$

This is our first expression for the electric field energy.

The following (seemingly more general) expression is equivalent:

$$2.) \qquad E = \frac{1}{2} \iiint \phi(\boldsymbol{r}) \varrho(\boldsymbol{r}) \mathrm{d}V \,. \tag{17.55}$$

Since the charge density vanishes in the space between the metal plates, $\varrho \equiv 0$, whereas at the lower plate the potential is $\phi \equiv 0$ and at the upper plate $\phi \equiv U$, and since $\varrho \mathrm{d}V$ can be replaced by $\sigma \mathrm{d}^2 A$ (with $\sigma = \boldsymbol{D} \cdot \boldsymbol{n}$), we obtain the same result as before.

Now, substituting the result

$$\phi(\boldsymbol{r}) = \iiint \mathrm{d}V' \frac{\varrho(\boldsymbol{r}')}{4\pi\varepsilon\varepsilon_0 |\boldsymbol{r} - \boldsymbol{r}'|}$$

into equation (17.55), we obtain a third expression for the field energy, which is the energy of mutual repulsion of the charges at $\boldsymbol{r}'$ and $\boldsymbol{r}$:

$$3.) \qquad E = \frac{1}{2} \iiint\limits_{\boldsymbol{r}} \iiint\limits_{\boldsymbol{r}'} \mathrm{d}V \mathrm{d}V' \frac{\varrho(\boldsymbol{r}')\varrho(\boldsymbol{r})}{4\pi\varepsilon\varepsilon_0 |\boldsymbol{r} - \boldsymbol{r}'|} \,. \tag{17.56}$$

---

[14] "true charge"

(The factor "$\frac{1}{2}$" in all these expressions is most obvious here, since when calculating the mutual repulsion each pair of charges should only be counted once.)

Finally, perhaps the most important expression is the following result, which is obtained from (17.55) by integrating by parts: In equation (17.55) we substitute

$$\varrho = \nabla \cdot \boldsymbol{D}$$

and again shift the ∇-differentiation to the left. The result, with

$$-\nabla\phi = \boldsymbol{E} \ ,$$

is:

$$4.) \qquad E = \frac{1}{2}\iiint_S \mathrm{d}V\,\boldsymbol{E}\cdot\boldsymbol{D}. \qquad (17.57)$$

(It is obvious that one only has to integrate over the space S between the metal plates 1 and 2, since in the interior of the plates the electric field $\boldsymbol{E}$ vanishes.)

The *energy density of the electric field*, w_e, e.g., to build up the field between the plates of a capacitor, is thus given by

$$w_\mathrm{e}(\boldsymbol{r}) := \frac{\varepsilon\cdot\varepsilon_0}{2}\boldsymbol{E}(\boldsymbol{r})^2 \ .$$

Similarly (without proof) in the *magnetic case* we have the following result for the energy density of the field, w_m, e.g., to build up the magnetic field in the interior of a *solenoid* filled with material of *relative permeability* μ:

$$w_m(\boldsymbol{r}) = \frac{\mu\cdot\mu_0}{2}\boldsymbol{H}(\boldsymbol{r})^2 \ .$$

17.2.8 The Demagnetization Tensor

With the aid of the above equations (17.46) and (17.48) one can always calculate the electrostatic field $\boldsymbol{E}(\boldsymbol{r})$ and the magnetostatic field $\boldsymbol{H}(\boldsymbol{r})$ of an electrically or magnetically polarized system, at every sampling point $\boldsymbol{r}$; however, in general the calculation is difficult and the results are complicated, except for the case of an *ellipsoid*.

It can be shown that the field outside an ellipsoid has exactly the same form as the field due to a *dipole* at the center, with dipole moment

$$\boldsymbol{p} = \boldsymbol{P}\Delta V \ ,$$

where ΔV is the volume of the ellipsoid and $\boldsymbol{P}$ is the electric polarization. This field is of course inhomogeneous, but rather simple. In the interior of the ellipsoid one has an even simpler result, a *homogeneous field* with the three components

$$E_i = -\frac{1}{\varepsilon_0} \sum_{k=1}^{3} N_{ik} P_k \ .$$

(The proof is omitted here).

$N_{i,k}(= N_{k,i})$ is a tensor, which is diagonal w.r.t. the principal axes of the ellipsoid and has the property that the eigenvalues depend only on the ratios of these axes. In the magnetic case, this same tensor is called the *demagnetization tensor*. It always satisfies the identity

$$\sum_{i=1}^{3} N_{ii} \equiv 1 \ .$$

The three *eigenvalues* of the demagnetization tensor are called *demagnetization factors*; i.e., for a sphere they have the value $\frac{1}{3}$; for an infinitely-long circular cylinder two of the eigenvalues (the "transverse" ones) are $\frac{1}{2}$, whereas the "longitudinal" one is zero; and finally for a very thin extended plane the two "in-plane" eigenvalues are zero, whereas the "out-of-plane" eigenvalue is 1. For other geometries the eigenvalues can be found in tables.

17.2.9 Discontinuities at Interfaces; "Interface Divergence" and "Interface Curl"

At interfaces between systems with different material properties the fields are usually discontinuous, but the integral formulations of Maxwell's equations are valid, whenever the integrals can be performed, e.g., for piecewise continuous functions, which are non-differentiable. From Gauss's law ("Maxwell I") (17.32), by applying it to a so-called *Gauss interface box* (which is a box running parallel to the 2d-interface that is aligned, e.g., *horizontally*, such that the top of the box is contained in the region *above* the interface and the bottom is *below* the interface, whereas the height of the side surfaces is negligibly small) the following equation can be derived

$$\boldsymbol{n} \cdot \left(\boldsymbol{D}^{(+)} - \boldsymbol{D}^{(-)} \right) = \sigma \ . \tag{17.58}$$

The $\boldsymbol{D}^{(\pm)}$ are the fields at the outer and inner sides of the interface, respectively, and σ is the (2d) interface charge density.

The operation

$$\boldsymbol{n} \cdot (\boldsymbol{v}^+ - \boldsymbol{v}^-)$$

corresponding to the l.h.s. of equation (17.58) is called the *interface divergence* of the vector field $\boldsymbol{v}(\boldsymbol{r})$. This quantity is obtained from div$\boldsymbol{v}$ by formally replacing the (vectorial) *differential operation* $\nabla \cdot \boldsymbol{v}$ by the *difference operation*

$$\boldsymbol{n} \cdot (\boldsymbol{v}^+ - \boldsymbol{v}^-)$$

appearing in (17.58).

One can proceed similarly with the curl operator: Calculating the circulation of the electric field $\boldsymbol{E}(\boldsymbol{r})$ along a *Stokes interface loop* (i.e., a small closed loop running in one direction on the upper side of the horizonal interface and in the opposite direction on the lower side but with negligible vertical height), one obtains from curl$\boldsymbol{E} = \nabla \times \boldsymbol{E} = 0$:

$$\boldsymbol{n} \times \left(\boldsymbol{E}^{(+)} - \boldsymbol{E}^{(-)} \right) = 0 \, . \qquad (17.59)$$

From (17.58) and (17.59) one can derive a *law of refraction the electric field lines* at the interface between two different dielectric materials. This law follows from the fact that the tangential components of $\boldsymbol{E}$ are continuous, whereas the normal components

$$\boldsymbol{n} \cdot \boldsymbol{E}^{(i)} \quad \text{with} \quad i = 1, 2$$

(i.e., corresponding to the two different materials) are inversely proportional to the respective ε_i. It then follows that $\frac{\tan \alpha_2}{\tan \alpha_1} \equiv \frac{\varepsilon_2}{\varepsilon_1}$, with angles α_i to the normal. For $\varepsilon_2/\varepsilon_1 \to \infty$ one obtains conditions such as those for a metal surface *in vacuo*, $\alpha_2 \to 90°$, $\alpha_1 \to 0$ (a sketch is recommended).

18 Magnetic Field of Steady Electric Currents

18.1 Ampère's Law

For centuries it had been assumed that electricity and magnetism were completely separate phenomena. Therefore it was quite a scientific sensation when in 1818 the Danish physicist Hans Christian Ørsted proved experimentally that magnetic fields were not only generated by permanent magnetic dipoles, but also by electric currents, and when slightly later André Marie Ampère showed quantitatively that the *circulation of the magnetic field* $\boldsymbol{H}$ along a closed loop followed the simple relation:

$$\oint_{\partial F} \boldsymbol{H}(\boldsymbol{r}) \cdot \mathrm{d}\boldsymbol{r} = I(F) \quad \text{(Ampere's law)} . \tag{18.1}$$

Here, $I(F)$ is the *flux* of electric current through a surface F inserted into the closed loop $\Gamma = \partial F$[1]

$$I(F) := \iint_F \boldsymbol{j} \cdot \boldsymbol{n} \mathrm{d}^2 A . \tag{18.2}$$

$$\boldsymbol{j} := \varrho(\boldsymbol{r})\boldsymbol{v}(\boldsymbol{r})$$

is the vector of the *current density* (dimensionality: $\mathrm{A/cm}^2 = \mathrm{C/(cm^2 s)}$).

With Stokes's integral theorem it follows that the differential form of Ampère's law (18.1) is given by:

$$\mathrm{curl}\boldsymbol{H} = \boldsymbol{j} . \tag{18.3}$$

For the special case of a thin wire aligned along the z-axis from $(-\infty)$ to $(+\infty)$, in which a steady electric current I flows, using cylindrical coordinates one obtains

$$\boldsymbol{H}_{\text{`z-wire'}} = \boldsymbol{e}_\varphi \frac{I}{2\pi r_\perp} . \tag{18.4}$$

Just as the electrostatic field of a point charge possesses a (three-dimensional) δ-divergence,

---

[1] The surface F is not uniquely defined by Γ, since different surfaces can be inserted into the same closed loop. This is the topological reason underlying *gauge freedom* of the vector potential, which is discussed below.

$$\mathrm{div}\left(\frac{q\boldsymbol{r}}{4\pi\varepsilon_0 r^3}\right) = q\delta(x, y, z) \ ,$$

an analogous relation is also valid for the curl of the magnetic field of the above "z-wire":

$$(\mathrm{curl}\boldsymbol{H}_{\text{``z-wire''}})(x, y, z) = I\delta(x, y)\boldsymbol{e}_z \ .$$

We can formulate these ideas in a general way:

The effective electric charges are the sources of the electrostatic field $\boldsymbol{E}(\boldsymbol{r})$ (whereas the vortices of $\boldsymbol{E}$ vanish); in contrast the vortices of the magneto-static field $\boldsymbol{B}(\boldsymbol{r})$ correspond to effective electric currents (whereas the sources of $\boldsymbol{B}$ vanish).

Generally, a vector field $\boldsymbol{v}(\boldsymbol{r})$ is determined by its sources and vortices.

Note that we have written $\boldsymbol{B}$, not $\boldsymbol{H}$, and "effective" quantities, not "true" ones (see above). In particular, the relations between $\boldsymbol{E}$ and $\boldsymbol{D}$ as well as $\boldsymbol{B}$ and $\boldsymbol{H}$ are not quite simple, and not all magnetic fields are produced by electric currents (Sect. 18.5 → *spin magnetism*).

18.1.1 An Application: 2d Boundary Currents for Superconductors; The Meissner Effect

As already detailed in Sect. 17.2.9, at an interface Ampère's equation

$$\mathrm{curl}\boldsymbol{H} = \boldsymbol{j}$$

must be generalized to

$$\boldsymbol{n} \times (\boldsymbol{H}^+ - \boldsymbol{H}^-) = \boldsymbol{j}_s \ ,$$

where $\boldsymbol{j}_s$ is an *interface-current density* (dimensionality: A/cm, *not* A/cm^2; and we have $\boldsymbol{j}_s \equiv \sigma\boldsymbol{v}$, analogously to $\boldsymbol{j} \equiv \varrho\boldsymbol{v}$).

As we shall see, this formulation yields a simple explanation of the so-called Meissner effect of superconductivity. This effect amounts to "expelling" the magnetic field from the interior of a superconducting material, by loss-free interface (super)currents that flow tangentially at the interface between a superconducting region "1" (e.g., the r.h.s. of a plane) and a normally conducting region "2" (e.g., vacuum on the l.h.s.). For example, if the interface normal (from "1" to "2") is in the $(-x)$-direction and the external magnetic field (in the normal conducting region "2") is (as usual) in the $+z$-direction, then in "1" (at the interface towards "2") supercurrents flow in the y-direction, producing in "1" a field $-B\boldsymbol{e}_z$, which is different from zero only in a very thin layer of typical width

$$\Delta x = \lambda \approx 10\,\mathrm{nm} \ .$$

For energy reasons (the magnetic field energy in region "1" can be saved) the supercurrents flow with such a strength that in the interior of region "1", outside the above-mentioned interface zone of width

$$\Delta x = \lambda \,,$$

the external magnetic field is exactly compensated. Further details cannot be given here.

18.2 The Vector Potential; Gauge Transformations

Since
$$\mathrm{curl}\boldsymbol{H} = \boldsymbol{j}(\neq 0) \,,$$
the magnetic field can no longer be calculated from a scalar potential: With

$$\boldsymbol{H}(\boldsymbol{r}) = -\mathrm{grad}\phi_m(\boldsymbol{r})$$

one would derive
$$\mathrm{curl}\boldsymbol{H} \equiv 0 \,,$$
since
$$\mathrm{curl}\,\mathrm{grad}\phi_m(\boldsymbol{r}) \equiv 0$$
for arbitrary scalar functions $\phi_m(\boldsymbol{r})$. ($\nabla \times (\nabla\phi_m)$ is formally a cross-product of two identical vectors and thus $\equiv 0$.) Fortunately we have

$$\mathrm{div}\boldsymbol{B}(\boldsymbol{r}) \equiv 0 \,,$$

so that one can try:
$$\boldsymbol{B} = \mathrm{curl}\boldsymbol{A}(\boldsymbol{r}) \,,$$
because
$$\mathrm{div}\,\mathrm{curl}\boldsymbol{v}(\boldsymbol{r}) \equiv 0$$
for all vector fields $\boldsymbol{v}(\boldsymbol{r})$, as can easily be shown. (Formally div curl$\boldsymbol{v}$ is a so-called *spate product*, the determinant of a 3×3-matrix, i.e., of the form $\boldsymbol{u} \cdot [\boldsymbol{v} \times \boldsymbol{w}]$, with two identical vectors, $\nabla \cdot [\nabla \times \boldsymbol{v}]$, and therefore it also vanishes identically.)

In fact an important mathematical theorem, *Poincaré's lemma*, states the following: For source-free vector fields $\boldsymbol{B}$, i.e., if

$$\oiint_{\partial G} \mathrm{d}^2 A \boldsymbol{B} \cdot \boldsymbol{n} \equiv 0 \,,$$

in a convex open region G (e.g., in the interior of a sphere) with a sufficiently well-behaved connected boundary ∂G, one can write vector potentials $\boldsymbol{A}$ with

$$\boldsymbol{B} = \mathrm{curl}\boldsymbol{A} \,.$$

One should note that $\boldsymbol{A}$ is not at all unique, i.e., there is an infinity of different vector potentials $\boldsymbol{A}$, but *essentially* they are all identical. If one adds an arbitrary gradient field to $\boldsymbol{A}$, then curl$\boldsymbol{A}$ is *not* changed at all. A so-called *gauge transformation*:

$$\boldsymbol{A} \to \boldsymbol{A}' := \boldsymbol{A} + \mathrm{grad} f(\boldsymbol{r}) \,, \tag{18.5}$$

with arbitrary $f(\boldsymbol{r})$, implies

$$\mathrm{curl}\boldsymbol{A} \equiv \mathrm{curl}\boldsymbol{A}' \,, \quad \text{since} \quad \mathrm{curl}\,\mathrm{grad} f \equiv 0 \,.$$

Therefore, the physical quantity $\boldsymbol{B}$ is unchanged.

18.3 The Biot-Savart Equation

In the following we consider, as usual, $G = \mathcal{R}^3$.

a) Firstly, we shall use a gauge such that $\mathrm{div}\boldsymbol{A}(\boldsymbol{r}) = 0$ (*Landau gauge*).
b) Secondly, from Ampère's law,

$$\mathrm{curl}\boldsymbol{H} = \boldsymbol{j} \,, \quad \text{with} \quad \boldsymbol{B} = \mu_0 \boldsymbol{H} + \boldsymbol{J} \,,$$

we conclude that

$$\mathrm{curl}\boldsymbol{B} = \mu_0\boldsymbol{j} + \mathrm{curl}\boldsymbol{J} =: \mu_0 \boldsymbol{j}_B \,,$$

with the effective current

$$\boldsymbol{j}_B := \boldsymbol{j} + \boldsymbol{M} \,, \quad \text{where} \quad \boldsymbol{M} := \frac{\boldsymbol{J}}{\mu_0}$$

is the *magnetization* and $\boldsymbol{J}$ the *magnetic polarization*[2].
c) Thirdly, we now use the general identity

$$\mathrm{curl}\,\mathrm{curl}\boldsymbol{A} \equiv \mathrm{grad}\,\mathrm{div}\boldsymbol{A} - \nabla^2\boldsymbol{A} \,. \tag{18.6}$$

Hence, due to

$$\mathrm{curl}\,\boldsymbol{B} =: \mu_0 \boldsymbol{j}_B \,,$$

the Cartesian components of $\boldsymbol{A}$ satisfy the Poisson equations

$$-\nabla^2 A_i = \mu_0 \cdot (j_B)_i \,, \quad \text{for} \quad i = x, y, z \,.$$

The solution of these equations is analogous to the electrostatic problem, *viz*

$$\boldsymbol{A}(\boldsymbol{r}) = \iiint \mathrm{d}V' \frac{\mu_0 \boldsymbol{j}_B(\boldsymbol{r}')}{4\pi|\boldsymbol{r} - \boldsymbol{r}'|} \,. \tag{18.7}$$

---

[2] In the cgs system the corresponding quantities are $\boldsymbol{M}' \left(= (\Delta V)^{-1} \sum_{r_i \in \Delta V} \boldsymbol{m}'_i\right)$ and $4\pi\boldsymbol{M}'$.

One can easily show by partial integration that this result also satisfies the equation

$$\operatorname{div}\boldsymbol{A} \equiv 0 \,, \quad \text{since} \quad \operatorname{div}\boldsymbol{j}_B = 0 \,.$$

Later, in the context of the so-called *continuity equation*, this relation will be discussed more generally.

By applying the curl operator, equation (18.7) leads to the formula of *Biot and Savart*:

$$\boldsymbol{B}(\boldsymbol{r}) = \iiint \mathrm{d}V' \frac{\mu_0}{4\pi} \frac{\boldsymbol{j}_B(\boldsymbol{r}') \times (\boldsymbol{r} - \boldsymbol{r}')}{|\boldsymbol{r} - \boldsymbol{r}'|^3} \,. \tag{18.8}$$

*In the integrand one has the same dependence on distance as in Coulomb's law for **E**, but complemented by the well-known right-hand rule connecting the directions of the effective current **j**$_B$ and the magnetic induction **B**, i.e., the product*

$$\frac{1}{\varepsilon_0} \varrho_E(\boldsymbol{r}') \frac{(\boldsymbol{r} - \boldsymbol{r}')}{|\boldsymbol{r} - \boldsymbol{r}'|^3}$$

is replaced by the cross-product

$$\mu_0 \boldsymbol{j}_B(\boldsymbol{r}') \times \frac{(\boldsymbol{r} - \boldsymbol{r}')}{|\boldsymbol{r} - \boldsymbol{r}'|^3} \,.$$

(It is no coincidence that the equation for $\boldsymbol{A}$, (18.7), is easier to remember than its consequence, the *Biot-Savart equation* (18.8).)

18.4 Ampère's Current Loops and their Equivalent Magnetic Dipoles

This section is especially important, since it shows that the relationships between electric currents and magnetic dipoles are very strong indeed. Firstly we state (without proof, but see the next footnote) that the magnetic induction $\boldsymbol{B}(\boldsymbol{r})$ produced by a *current loop* $\Gamma = \partial F$ (current I) is quantitatively identical to the magnetic field that would be produced by an infinitesimal film of magnetic dipoles inserted into the same loop, i.e., for the fictitious 2d-dipole density $\mathrm{d}\boldsymbol{m}$ of that film the following formula would apply:

$$\mathrm{d}\boldsymbol{m} \equiv \mu_0 I \boldsymbol{n} \mathrm{d}^2 A \,.$$

a) For a current loop, one obtains from *Biot and Savart's* equation

$$\boldsymbol{B}(\boldsymbol{r}) = \frac{\mu_0 I}{4\pi} \oint_{\partial F} \mathrm{d}\boldsymbol{r}' \times \frac{\boldsymbol{r} - \boldsymbol{r}'}{|\boldsymbol{r} - \boldsymbol{r}'|^3} \,. \tag{18.9}$$

b) In the dipole case one would obtain outside the (fictitious) dipole film the equivalent result

$$B = \mu_0 H \ ,$$

with

$$H(r) = -\mathrm{grad}\frac{I}{4\pi}\iint_F \mathrm{d}^2 A' \frac{n(r')\cdot(r-r')}{|r-r'|^3} \ . \qquad (18.10)$$

Proof of the equivalence of the two results proceeds analogously to Stokes's theorem, but since it is somewhat difficult in detail, we only give an outline in a footnote[3]. An example is given in Fig. 18.1.

In this context we additionally keep two useful identities in mind:

$$A = m \times \frac{r}{4\pi r^3}$$

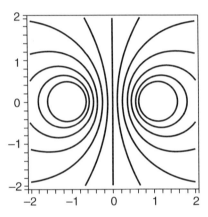

Fig. 18.1. The diagram illustrates a typical section of the magnetic field lines produced by a current loop of two (infinitely) long straight wires. The wires intersect the diagram at the points $(\pm 1, 0)$. The plane of the loop of area A $(\to \infty)$ and carrying a current I (i.e., of opposite signs in the two long wires) is perpendicular to the plane of the diagram. Exactly the same induction $B(= \mu_0 H)$ is also produced by a layer of magnetic dipoles inserted into the current loop, with the quantitative relation, $\mathrm{d}m \equiv \mu_0 I n \mathrm{d}^2 A$, given in the text

---

[3] In the following we use the antisymmetric unit-tensor e_{ijk} and Einstein's summing convention, i.e., all indices which appear twofold are summed over. With these conventions Stokes's theorem becomes: $\oint_{\partial F} E_j \mathrm{d}x_j = \iint_F e_{jlm}\partial_l E_m n_j \mathrm{d}^2 A$. Now the following chain of equations is true: $\oint_{\partial F} e_{ijk}\frac{\mathrm{d}x'_j(x_k-x'_k)}{|r-r'|^3}\left(\equiv \oint_{\partial F} \mathrm{d}x'_j e_{jki}\partial'_k \frac{1}{|r-r'|}\right) = \iint_F e_{jlm}\partial'_l e_{mki}\partial'_k \frac{1}{|r-r'|}n_j \mathrm{d}^2 A' = -\iint_F e_{ikm}e_{jlm}\partial'_{lk}\frac{1}{|r-r'|}n_j \mathrm{d}^2 A'$. With the basic identity $e_{ikm}e_{jlm} = \delta_{ij}\delta_{kl}-\delta_{il}\delta_{kj}$ and the simple relations $\partial'_i \frac{1}{|r-r'|} = -\partial_i \frac{1}{|r-r'|}$ and $\partial_{kk}\frac{1}{|r-r'|} = 0$ (for $r' \neq r$) our statement of equivalence is obtained.

for the vector potential of a magnetic dipole and

$$\operatorname{curl}\left(\boldsymbol{m} \times \frac{\boldsymbol{r}}{r^3}\right) = \boldsymbol{m}\operatorname{div}\frac{\boldsymbol{r}}{r^3} - \operatorname{grad}\frac{\boldsymbol{m} \cdot \boldsymbol{r}}{r^3}$$

(cf. problem 5 of the exercises, summer 2002 [2]).

Because of the above-mentioned equivalence it would be natural to suggest that *all* magnetic dipole moments are generated in this way by Ampèrian current loops. However, this suggestion would be wrong: *There are magnetic moments which cannot be generated in this "classical" way, but which are related to the non-classical concept of "electron spin" (see Part III: Quantum Mechanics).* The following section deals with the difference.

18.5 Gyromagnetic Ratio and Spin Magnetism

An atomic electron orbiting the nucleus on a circular path of radius R with velocity

$$v = \omega R \quad \left(\text{time period} \quad T = \frac{2\pi}{\omega}\right)$$

has an angular momentum of magnitude

$$L = R \cdot m_{\mathrm{e}}v = m_{\mathrm{e}}\omega R^2 \ .$$

According to the Ampèrian "current loop" picture it would be equivalent to a magnetic dipole moment

$$m = \mu_0 \frac{e}{T}\pi R^2 = \frac{\mu_0 e\omega R^2}{2} \ ,$$

where we have used $I = \frac{e}{T}$ (m_{e} is the electron mass).

For a current loop, therefore, the gyromagnetic ratio

$$\gamma := \frac{m}{L}$$

is given by

$$\gamma \equiv \frac{\mu_0 e}{2m_{\mathrm{e}}} \ .$$

However, in the nineteen-twenties due to an experiment by *Einstein and de Haas* it was shown that for the usual magnetic materials, e.g., alloys of Fe, Co and Ni, the gyromagnetic ratio is *twice* as large as the above ratio. For these materials the magnetism is due almost entirely to pure *spin magnetism.* For the angular momentum of these alloys the "classical" orbital contribution (see Part I) is almost negligible; the (dominant!) contribution is essentially "non-classical", i.e., due to *spin magnetism*, which is only understandable in a quantum mechanical context. (In fact, a profound analysis is not even possible in non-relativistic quantum mechanics, but only in Dirac's relativistic version.)

In elementary texts one often reads that the spin angular momentum of a (charged) particle is some kind of "proper angular momentum", this being acceptable if one does not consider the particle to be rotating like a "spinning top", since a spinning charge would have the classical value,

$$\gamma = \frac{\mu_0}{2m_\mathrm{e}} ,$$

for the gyromagnetic ratio and *not* twice this value. One has to admit that these relations are complicated and not understandable at an elementary level.

19 Maxwell's Equations I: Faraday's Law of Induction; the Continuity Equation; Maxwell's Displacement Current

Maxwell's first and second equations, $\operatorname{div}\boldsymbol{D} = \varrho$ and $\operatorname{div}\boldsymbol{B} = 0$ (i.e., Gauss's law for the electric and the (non-existent) magnetic charges, respectively) also apply without change for time-dependent electrodynamic fields. This is different with respect to the third and fourth Maxwell equations:

a) *Faraday's law of induction* (Faraday 1832)

$$\operatorname{curl}\boldsymbol{E} = -\frac{\partial \boldsymbol{B}}{\partial t} \; , \tag{19.1}$$

    and

b) *Ampère's law including Maxwell's displacement current*:

$$\operatorname{curl}\boldsymbol{H} = \boldsymbol{j} + \frac{\partial \boldsymbol{D}}{\partial t} \; . \tag{19.2}$$

These two equations, (19.1) and (19.2), will be discussed in the following subsections. To aid our understanding of the last term in (19.2), known as Maxwell's displacement current, we shall include a subsection on the *continuity equation*. This general equation contains an important conservation law within it, the conservation of total charge (see below).

19.1 Faraday's Law of Induction and the Lorentz Force; Generator Voltage

In 1832 Faraday observed that a time-dependent change of magnetic flux

$$\phi_B(F) = \iint_F \boldsymbol{B} \cdot \boldsymbol{n} \mathrm{d}^2 A$$

through a current loop $\Gamma = \partial F$ gives rise to an *electromotive force* (i.e., a force by which electric charges of different sign are separated). This corresponds to a *generator voltage* which is similar to the off-load voltage between the two poles of a battery. (In the interior of a battery the current flows from the minus pole to the plus pole; only subsequently, in the external load circuit, does the current flow from plus to minus.) In Fig. 19.1 we present a sketch of the situation.

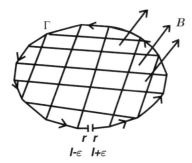

Fig. 19.1. Sketch to illustrate Faraday's law of induction. The three arrows on the r.h.s. of the figure denote a magnetic induction $\boldsymbol{B}$. An oriented loop Γ is plotted, as well as a paved surface F, which is inserted into Γ ($\Gamma = \partial F$) but which does not need to be planar as in the diagram. A change in magnetic flux $\phi_B(F) := \iint_F \boldsymbol{B} \cdot \boldsymbol{n} \mathrm{d}^2 A$ gives rise to an induced voltage $U_i(\Gamma) = \left(\oint_{\boldsymbol{r}_{1+\varepsilon}}^{\boldsymbol{r}_{1-\varepsilon}} \boldsymbol{E} \cdot \mathrm{d}\boldsymbol{r} \right)$ between two infinitesimally close points $\boldsymbol{r}_{1+\varepsilon}$ and $\boldsymbol{r}_{1-\varepsilon}$ on the loop. These two points – which are formally the initial and end points of the loop – can serve as the poles of a voltage generator (the initial point $\boldsymbol{r}_{1+\varepsilon}$ corresponds to the negative pole). The related quantitative equation is Faraday's law: $U_i(t) = -\frac{\mathrm{d}\phi_B(F)}{\mathrm{d}t}$

Faraday's law of induction states that the *induced voltage* (i.e., the *generator voltage* mentioned above)

$$ U_i = \oint_{\boldsymbol{r}_{1+\varepsilon}}^{\boldsymbol{r}_{1-\varepsilon}} \boldsymbol{E}(\boldsymbol{r}) \cdot \mathrm{d}\boldsymbol{r} $$

between (arbitrary) initial points $\boldsymbol{r}_{1+\varepsilon}$ and (almost) identical end points $\boldsymbol{r}_{1-\varepsilon}$ of an (almost) closed line[1] $\Gamma = \partial F$ obeys the following law:

$$ U_i(t) = -\frac{\mathrm{d}\phi_B(F)}{\mathrm{d}t} . \tag{19.3} $$

As already mentioned, the initial and end points of the (almost) closed loop Γ differ only infinitesimally. They correspond to the minus and plus poles of the generator, i.e., U_i is the generator voltage.

It does not matter at which position of the curve Γ the voltage is "tapped", nor does it matter whether the change of the magnetic flux results

a) from a change of Γ (i.e., form or size) *relative* to the measuring equipment,
b) from a change of the magnetic induction $\boldsymbol{B}(\boldsymbol{r}, t)$, or
c) by a combination of both effects.

---

[1] A sketch is recommended. The normal vector $\boldsymbol{n}$ of the area F should coincide with the orientation of the loop $\Gamma = \partial F$.

We thus realize that Faraday's law of induction (19.3), which appears so simple, embodies a great deal of experimental information. In addition we may already surmise at this point that Maxwell's theory is *relativistically invariant*, as we shall show later in detail.

For case b), i.e., constant Γ but variable $\boldsymbol{B}$, using Stokes's theorem, from the integral form of the law of induction,

$$\oint_{\partial F} \boldsymbol{E} \cdot \mathrm{d}\boldsymbol{r} = -\iint_F \frac{\partial \boldsymbol{B}}{\partial t} \cdot \boldsymbol{n} \mathrm{d}^2 A \,,$$

one obtains the differential form (19.1).

To derive case a) as well, we additionally use the expression for the *Lorentz force* on an electrically charged particle, which moves with velocity $\boldsymbol{v}$ in a magnetic field. This law was formulated by Hendryk A. Lorentz in Leiden decades after Faraday's discovery[2], and it states the following:

The force $\boldsymbol{F}_L$ on an electrically charged particle moving in a magnetic induction $\boldsymbol{B}(\boldsymbol{r}, t)$ with velocity

$$\boldsymbol{v}(\boldsymbol{r}, t) \quad \text{is} \quad \boldsymbol{F}_L = q\boldsymbol{v} \times \boldsymbol{B} \quad (Lorentz\ force)\,.$$

We shall now consider an infinitesimal line element $\mathrm{d}\boldsymbol{r}$ (e.g., $\mathrm{d}l\boldsymbol{e}_y$) of the current loop $\Gamma = \partial F$ and move this line element during a time interval δt with velocity $\boldsymbol{v}$ (e.g. in the x-direction), such that the line element defines a surface element

$$(\boldsymbol{v}\delta t) \times \mathrm{d}\boldsymbol{r}\,.$$

The corresponding change in magnetic flux is given by

$$\delta\phi_B = \delta t[\boldsymbol{v} \times \mathrm{d}\boldsymbol{r}] \cdot \boldsymbol{B} = -\delta t \mathrm{d}\boldsymbol{r} \cdot [\boldsymbol{v} \times \boldsymbol{B}]\,.$$

We then calculate the integral of this force along the loop, i.e.,

$$\oint_{\boldsymbol{r}_{1+\varepsilon}}^{\boldsymbol{r}_{1-\varepsilon}} \mathrm{d}\boldsymbol{r} \cdot \boldsymbol{F}_L\,.$$

But

$$\boldsymbol{F}_L = q \cdot \boldsymbol{E}\,.$$

Therefore, we finally obtain

$$U_i \equiv \oint \boldsymbol{E} \cdot \mathrm{d}\boldsymbol{r} \equiv -\frac{\delta\phi_B}{\delta t}\,.$$

Case c) follows from b) and a) by linear superposition and the product law of differentiation.

Consequently, we can state that the integral version of Faraday's law contains more than the differential version (19.1); inter alia the Lorentz force, i.e., not only b), but also a).

---

[2] Shortly before Einstein's special theory of relativity Lorentz also formulated his famous *Lorentz transformations*.

19.2 The Continuity Equation

The conservation law for total charge mentioned above, states that

$$Q_{\text{total}}(t) := \left\{ \iiint_\infty dV \varrho(\boldsymbol{r}) \right\}(t) = \text{constant} ,$$

where we integrate over the whole space $\mathcal{R}^3$. This implies that the electric charge contained in a *finite* volume V,

$$Q(V) := \int_V dV \varrho(\boldsymbol{r}) ,$$

can only increase by an influx of charge from outside, i.e.,

$$\frac{dQ(V)}{dt} = -\oiint_{\partial V} \boldsymbol{j} \cdot \boldsymbol{n} d^2 A . \tag{19.4}$$

Now for constant volume we have:

$$\frac{dQ(V)}{dt} = \iiint_V \frac{\partial \varrho}{\partial t} dV ,$$

and by Gauss's integral theorem:

$$-\oiint_{\partial V} \boldsymbol{j} \cdot \boldsymbol{n} d^2 A = -\iiint_V \text{div} \boldsymbol{j} dV .$$

As a consequence we have

$$\iiint_V \left\{ \frac{\partial \varrho}{\partial t} + \text{div} \boldsymbol{j} \right\} dV \equiv 0 ,$$

i.e. *not* $\text{div} \boldsymbol{j} \equiv 0$, but instead the *continuity equation*

$$\text{div} \boldsymbol{j} + \frac{\partial \varrho}{\partial t} \equiv 0 . \tag{19.5}$$

This equation, including its derivation, should be kept in mind, since analogous continuity equations apply to other conservation laws.

19.3 Ampère's Law with Maxwell's Displacement Current

As a result of the continuity equation, (19.5), Ampère's law

$$\text{curl} \boldsymbol{H} = \boldsymbol{j}$$

cannot remain unchanged for time-dependent electromagnetic fields, since

$$\text{div curl} \boldsymbol{H} \equiv 0 \,,$$

whereas according to (19.5) $\text{div} \boldsymbol{j} \neq 0$ applies. Maxwell found the correct solution by adding a *displacement current* $\frac{\partial \boldsymbol{D}}{\partial t}$ to the true current $\boldsymbol{j}$, to give:

$$\text{curl} \boldsymbol{H} = \boldsymbol{j} + \frac{\partial \boldsymbol{D}}{\partial t} \,.$$

Due to the permutability of partial derivatives, e.g.,

$$\frac{\partial}{\partial x} \frac{\partial}{\partial t} = \frac{\partial}{\partial t} \frac{\partial}{\partial x} \,,$$

one obtains

$$\text{div} \frac{\partial \boldsymbol{D}}{\partial t} = \frac{\partial}{\partial t} \text{div} \boldsymbol{D} \,.$$

But according to the first Maxwell equation one has $\text{div} \boldsymbol{D} = \varrho$. Thus, with the continuity equation one obtains:

$$\text{div} \left\{ \boldsymbol{j} + \frac{\partial \boldsymbol{D}}{\partial t} \right\} \equiv 0 \,,$$

as expected. The fourth Maxwell equation was therefore deduced to be:

$$\text{curl} \boldsymbol{H} = \boldsymbol{j} + \frac{\partial \boldsymbol{D}}{\partial t} \quad \text{(Maxwell IV; differential version)} \,. \qquad (19.6)$$

(By analogy with Faraday's law this equation could be termed Maxwell's law[3].)

In the corresponding integral form (see below) one considers a metal capacitor. A true current enters plate 1 of this capacitor from a lead 1, whereas from plate 2 a corresponding true current flows outwards through a lead 2.

In the space between plate 1 and plate 2 no true current flows, but the $\boldsymbol{D}$-field changes with time, since the curls of the magnetic field surrounding the true currents in the leads cannot stop in the interspace: *curls must always continue* (a diagram is recommended here!)

As with Faraday's law of induction, the integral form of the fourth Maxwell equation is again somewhat more general than the differential form (Note the time derivative in front of the integral):

$$\oint_{\partial F} \boldsymbol{H} \cdot \mathrm{d}\boldsymbol{r} = \iint_F \boldsymbol{j} \cdot \boldsymbol{n} \mathrm{d}^2 A + \frac{\mathrm{d}}{\mathrm{d}t} \iint_F \boldsymbol{D} \cdot \boldsymbol{n} \mathrm{d}^2 A \,. \qquad (19.7)$$

---

[3] The fact that on the r.h.s. of (19.1) and (19.2) the time derivatives of the electromagnetic fields $\boldsymbol{B}$ and $\boldsymbol{D}$ enter with different signs is ultimately responsible (as we shall see) for the existence of *electromagnetic waves*. The difference in signs also reminds us of a similar difference in the *canonical equations* in classical mechanics (see Part I) and gives another pointer to the existence of *matter waves* (see Part III).

19.4 Applications: Self-inductance and Mutual Inductances; Transformers; Complex Resistances; Alternating-current Resonance Circuit

a) **Inductance**

Consider a solenoid of length l and circular cross-section (radius r_0) filled with a material of relative permeability μ (typically a soft-iron core). Let l be the length of the solenoid, while r_0 is the radius of the circular cross-section and the number of windings is N_1. The total magnetic flux through all the windings of the solenoid is thus

$$\phi_B = N_1 \cdot \mu\mu_0 H \cdot \pi r_0^2 .$$

Applying Ampère's law,

$$\oint_{\partial F} \boldsymbol{H} \cdot \mathrm{d}\boldsymbol{r} = I(F) ,$$

to a long closed loop circulating around the core with all windings, implies for the magnetic field H in the interior of the solenoid:

$$H = \frac{N_1}{l} \cdot I ,$$

where I is a quasi-static, i.e., slowly (periodically) changing current in the coil. Hence we have

$$\phi_B = \mu\mu_0 \frac{N_1^2}{l} \pi r_0^2 \cdot I , \quad \text{i.e.,} \quad \phi_B \equiv L \cdot I ,$$

where L is the *(self-)inductance* of the solenoid, given by

$$L(=: L_{1,1}) = \mu\mu_0 \frac{N_1^2}{l} \pi r_0^2 . \tag{19.8}$$

The field energy of the solenoid can be written

$$E_{\mathrm{mag}} = \frac{\mu\mu_0}{2} H^2 \cdot \pi r_0^2 l .$$

With

$$H = \frac{N_1}{l} \cdot I$$

one obtains the equivalent result

$$E_{\mathrm{mag}} = \frac{L}{2} I^2 , \quad \text{i.e.,} \quad \frac{\mathrm{d}E_{\mathrm{mag}}}{\mathrm{d}t} = L \cdot I \frac{\mathrm{d}I}{\mathrm{d}t} = L \frac{\mathrm{d}I}{\mathrm{d}t} I = -U_i \cdot I ,$$

as expected.

b) **The transformer**

The index 1 above refers to the number of windings for a single solenoid. Now consider a second solenoid of equal length wrapped around the first solenoid and possessing only a marginally larger cross-sectional radius r_0, but a significantly different number of windings N_2. (This arrangement is merely designed so that the same flux of magnetic field lines passes through both coils.)

The terminals of solenoid 2 are assumed to remain *open*, i.e., no current flows through solenoid 2, in contrast to solenoid 1, through which a current I_1 flows. The magnetic field H_2 in the interior of solenoid 2 due to the current I_1 in 1 is thus

$$H_2 = \frac{N_1}{l} I_1 \ .$$

The magnetic flux through solenoid 2 is then

$$(\phi_B)_2 = L_{2,1} \cdot I_1 \ ,$$

where $L_{2,1}$ is the *mutual inductance*,

$$L_{2,1} = L_{1,1} \cdot \frac{N_2}{N_1} \ .$$

It follows that (in the case of a low-frequency[4] alternating current) the induced voltages measured at the ends of the closed solenoid 1 and the open solenoid 2 behave as the ratio of the corresponding number of windings,

$$(U_i)_2 : (U_i)_1 = N_2 : N_1 \ .$$

This is the principle of the transformer.

c) **Complex alternating-current resistances**

Consider a circuit in which an alternating current (a.c.)

$$I(t) = I^{(0)} \cdot \cos(\omega t - \alpha)$$

is generated by a voltage

$$U_G(t) = U_G^{(0)} \cdot \cos(\omega t)$$

of angular frequency ω. The phase of the *current* $I(t)$ is thus shifted with respect to that of the voltage $U_G(t)$ (for positive α: the maxima of the current are delayed w.r.t. the maxima of the generator voltage).

Making use of the Euler-Moivre relation:

$$e^{i\omega t} = \cos(\omega t) + i \sin(\omega t) \ , \tag{19.9}$$

---

[4] Only later will we see quantitatively what "low frequency" means in this context.

where $i^2 = -1$, we thus have

$$U_G(t) = \mathrm{Re}(\mathcal{U}_G e^{i\omega t})$$

and

$$I(t) = \mathrm{Re}(\mathcal{J} e^{i\omega t}) , \quad \text{where}$$
$$\mathcal{U} \equiv U_G^{(0)} \quad \text{and}$$
$$\mathcal{J} \equiv (I^{(0)} e^{-i\alpha}) .$$

By analogy with Ohm's law we then define the complex quantity $\mathcal{R}$, where

$$\mathcal{U}_G = \mathcal{R} \cdot \mathcal{I} .$$

The quantity $\mathcal{R}$ is the *complex a.c. resistance* or simply *impedance*.
The total impedance of a circuit is calculated from an appropriate combination of three types of standard elements in series or parallel, etc.

1. *Ohmic resistances* (positive and real) are represented by the well-known rectangular symbol and the letter R. The corresponding complex resistance is
$$\mathcal{R}_R = R .$$

2. *Capacitive resistances* (negatively imaginary) correspond to a pair of capacitor plates, together with the letter C. The corresponding impedance is given by
$$\mathcal{R}_C = \frac{1}{i\omega C} .$$
(A short justification: $U_C(t) = \frac{Q(t)}{C}$, i.e., $\dot{U}_C(t) = \frac{I(t)}{C}$. Thus with the *ansatz* $U_C(t) \propto e^{i\omega t}$ one obtains $\dot{U}_C(t) \equiv i\omega U_C(t)$.)

3. *Inductive resistances* (positively imaginary) are represented by a solenoid symbol, together with the letter L. The corresponding impedance is
$$\mathcal{R}_L = i\omega L .$$

(The induced voltage drop in the load results from building-up the magnetic field, according to the relation $U_L(t) = L \cdot \frac{dI(t)}{dt}$, i.e., $U_L(t) = L \cdot \dot{I}(t)$. But with the *ansatz* $I(t) \propto e^{i\omega t}$ we obtain $\dot{I}(t) \equiv i\omega I(t)$.)

One can use the same methods for mutual inductances (i.e., *transformers*; see exercises)[5].

---

[5] The input (load) voltage of the transformer is given by the relation $\mathcal{U}_{\mathrm{Tr}}^{(1)} = i\omega L_{1,2} \cdot \mathcal{J}_2$, while the output (generator) voltage is given by $\mathcal{U}_{\mathrm{Tr}}^{(2)} = -i\omega L_{2,1} \cdot \mathcal{J}_1$.

d) An a.c. resonance circuit

The following is well-known as example of *resonance phenomena*. For a series RLC circuit connected as a *load* to an alternating-voltage generator

$$U_G(t) = U_G^{(0)} \cdot \cos(\omega t) \,,$$

one has

$$\frac{J}{U} = \frac{1}{\mathcal{R}} = \left(R + \mathrm{i}\left(\omega L - \frac{1}{\omega C}\right)\right)^{-1} \,.$$

Thus we obtain

$$I(t) = I^{(0)} \cdot \cos(\omega t - \alpha) \,, \quad \text{with}$$

$$\frac{I^{(0)}}{U_G^{(0)}} = \left|\frac{1}{\mathcal{R}}\right| = \frac{1}{\sqrt{R^2 + (\omega L - \frac{1}{\omega C})^2}} \quad \text{and} \quad \tan\alpha = \frac{\omega L - \frac{1}{\omega C}}{R} \,.$$

For sufficiently small R (see below) this yields a sharp resonance at the resonance frequency

$$\omega_0 := \frac{1}{\sqrt{L \cdot C}} \,.$$

For this frequency the current and the voltage are exactly *in phase*, whereas for higher frequencies the current is *delayed* with respect to the voltage (inductive behavior) while for lower frequencies the voltage is delayed with respect to the current (capacitive behavior). At the resonance frequency ω_0 the current has a very sharp maximum of height $U_G^{(0)}/R$, and for weak damping (i.e., for sufficiently small values of R) it decays very quickly as a function of ω, for very small deviations from $\omega = \omega_0$; i.e., for

$$\omega_\pm := \omega_0 \pm \varepsilon \,, \quad \text{where} \quad \varepsilon = \frac{R}{2L} \quad \text{is} \quad \ll |\omega_0| \,,$$

the current has already decreased to 70% of the maximum (more precisely: from $1 \times I^{(0)}$ down to $\frac{1}{\sqrt{2}} \times I^{(0)}$).
The ratio

$$Q := \frac{\omega_0}{R/L}$$

is called the quality factor of the resonance; it characterizes the sharpness of the phenomenon. In fact, Q often reaches values of the order of 10^3 or more.

(Here the reader could try solving exercises 11 and 12 (file 6) from the summer term of 2002, which can be found on the internet, [2]. This can simultaneously serve as an introduction to MAPLE. In fact, it may be helpful to illustrate resonance phenomena using mathematical computer tools such as MAPLE or MATHEMATICA. See for example [12] as a recommendable presentation.)

Now consider the power loss in an a.c. circuit. By forming the derivative of the energy, we may write

$$(dE/dt)(t) = U_G(t) \cdot I(t) = U_G^{(0)} \cdot I^{(0)} \cos \omega t \cdot \cos(\omega t - \alpha)$$
$$\equiv U_G^{(0)} \cdot I^{(0)} \cdot \left[\cos \alpha \cdot (\cos \omega t)^2 + \sin \alpha \cdot (\cos \omega t \cdot \sin \omega t)\right] .$$

Averaged over a complete cycle the first term gives

$$U_G^{(0)} \cdot I^{(0)} \cdot \frac{1}{2} \cdot \cos \alpha .$$

This is the resistive part, and represents the energy dissipated. The second term, however, vanishes when averaged over a complete cycle, and is called the reactive part. Using complex quantities one must explicitly take into account the factor $\frac{1}{2}$. The resistive part may be written

$$\{\overline{(dE/dt)(t)}\} = \mathrm{Re}\left(\frac{1}{2}\mathcal{U}_G \mathcal{J}^*\right) ,$$

while the reactive part (which vanishes on average) is given by

$$\mathrm{Im}\left(\frac{1}{2}\mathcal{U}_G \mathcal{J}^*\right) .$$

Further details on alternating-current theory can be found in many standard textbooks on applied electromagnetism.

20 Maxwell's Equations II: Electromagnetic Waves

20.1 The Electromagnetic Energy Theorem; Poynting Vector

The *Poynting vector* which is defined as

$$\boldsymbol{S} := \boldsymbol{E} \times \boldsymbol{H} \tag{20.1}$$

has the meaning of "energy current density":

$$\boldsymbol{S} \equiv \boldsymbol{j}_{\text{energy}} \ .$$

Firstly we have the mathematical identity

$$\text{div}[\boldsymbol{E} \times \boldsymbol{H}] = \boldsymbol{H} \cdot \text{curl}\boldsymbol{E} - \boldsymbol{E} \cdot \text{curl}\boldsymbol{H} \ ,$$

which can be proved using the relation

$$\text{div}[\boldsymbol{E} \times \boldsymbol{H}] = \partial_i e_{i,j,k} E_j H_k = \dots \ .$$

With the Maxwell equations

$$\text{curl}\boldsymbol{E} = -\frac{\partial \boldsymbol{B}}{\partial t} \quad \text{and} \quad \text{curl}\boldsymbol{H} = \boldsymbol{j} + \frac{\partial \boldsymbol{D}}{\partial t}$$

one then obtains the continuity equation corresponding to the conservation of field energy:

$$\text{div}[\boldsymbol{E} \times \boldsymbol{H}] + \boldsymbol{E} \cdot \boldsymbol{j} + \boldsymbol{E} \cdot \frac{\partial \boldsymbol{D}}{\partial t} + \boldsymbol{H} \cdot \frac{\partial \boldsymbol{B}}{\partial t} \equiv 0 \ , \tag{20.2}$$

i.e.,

$$\text{div}\boldsymbol{j}_{\text{energy}} + \frac{\partial w_{\text{energy}}}{\partial t} = -\boldsymbol{j} \cdot \boldsymbol{E} \ , \tag{20.3}$$

where

$$\frac{\partial w_{\text{energy}}}{\partial t} := \boldsymbol{E} \cdot \frac{\partial \boldsymbol{D}}{\partial t} + \boldsymbol{H} \cdot \frac{\partial \boldsymbol{B}}{\partial t}$$

is the (formal)[1] time-derivative of the energy density and $-\boldsymbol{j} \cdot \boldsymbol{E}$ describes the *Joule losses*, i.e., one expects

$$\boldsymbol{j} \cdot \boldsymbol{E} \geq 0 \ .$$

(The losses are essentially sources of heat production, since the Ohmic behavior, $\boldsymbol{j} = \sigma \cdot \boldsymbol{E}$ (where σ is the specific conductivity) arises due to frictional processes leading to energy dissipation and heat production due to scattering of the carriers of the current, e.g., by impurities. The case of vanishing losses is called *ballistic*.)

For Ohmic behavior one has

$$\boldsymbol{j} = \frac{1}{\kappa} \boldsymbol{E} \ ,$$

where

$$\kappa = \frac{1}{\sigma}$$

is the specific resistivity[2], a constant property of the Ohmic material; i.e.,

$$-\boldsymbol{j} \cdot \boldsymbol{E} = -\frac{E^2}{\kappa} \ .$$

In the absence of electric currents the electromagnetic field energy is conserved. If the material considered shows Ohmic behavior, the field energy decreases due to Joule losses.

At this point we shall just mention two further aspects: (i) the role of the Poynting vector for a battery with attached Ohmic resistance. The Poynting vector flows radially out of the battery into the vacuum and from there into the Ohmic load. Thus the wire from the battery to the Ohmic resistance is not involved at all, and (ii) Drude's theory of electric conductivity. This theory culminates in the well-known formula

$$\boldsymbol{j}(\omega) = \sigma(\omega)\boldsymbol{E}(\omega) \ ,$$

where the alternating-current specific conductivity is given by

$$\sigma(\omega) \equiv \sigma(0)/(1 + i\omega\tau) \ , \quad \text{with} \quad \sigma(0) = n_V e^2 \tau / m_{\mathrm{e}} \ .$$

Here n_V is the volume density of the carriers, i.e., typically electrons; m_{e} is the electron mass, e the electron charge, and τ is a phenomenological relaxation time corresponding to scattering processes.

---

[1] In the case of a linear relation (e.g., between $\boldsymbol{E}$ and $\boldsymbol{D}$ or $\boldsymbol{B}$ and $\boldsymbol{H}$) the derivative is non-formal; otherwise (e.g., for the general relation $\boldsymbol{D} = \varepsilon_0 \boldsymbol{E} + \boldsymbol{P}$) this is only formally a time-derivative.

[2] The Ohmic resistance R of a wire of length l and cross-section F made from material of specific resistivity κ is thus $R = \kappa \cdot \left(\frac{l}{F}\right)$.

20.2 Retarded Scalar and Vector Potentials I: D'Alembert's Equation

For given electromagnetic fields $\boldsymbol{E}(\boldsymbol{r},t)$ and $\boldsymbol{B}(\boldsymbol{r},t)$ one can satisfy the second and third Maxwell equations, i.e., Gauss's magnetic law, $\mathrm{div}\boldsymbol{B} = 0$, and Faraday's law of induction,

$$\mathrm{curl}\boldsymbol{E} \equiv -\frac{\partial \boldsymbol{B}}{\partial t} \, ,$$

with the *ansatz*

$$\boldsymbol{B}(\boldsymbol{r},t) \equiv \mathrm{curl}\boldsymbol{A}(\boldsymbol{r},t) \quad \text{and} \quad \boldsymbol{E}(\boldsymbol{r},t) \equiv -\mathrm{grad}\phi(\boldsymbol{r},t) - \frac{\partial \boldsymbol{A}(\boldsymbol{r},t)}{\partial t} \, . \quad (20.4)$$

The scalar potential $\phi(\boldsymbol{r},t)$ and vector potential $\boldsymbol{A}(\boldsymbol{r},t)$ must now be calculated simultaneously. However, they are not unique but can be "gauged" (i.e., changed according to a gauge transformation *without* any change of the fields) as follows:

$$\boldsymbol{A}(\boldsymbol{r},t) \to \boldsymbol{A}'(\boldsymbol{r},t) := \boldsymbol{A}'(\boldsymbol{r},t) + \mathrm{grad}f(\boldsymbol{r},t),$$
$$\phi(\boldsymbol{r},t) \to \phi'(\boldsymbol{r},t) := \phi'(\boldsymbol{r},t) - \frac{\partial f(\boldsymbol{r},t)}{\partial t} \, . \quad (20.5)$$

Here the *gauge function* $f(\boldsymbol{r},t)$ in (20.5) is arbitrary (it must only be differentiable). (The proof that such gauge transformations neither change $\boldsymbol{E}(\boldsymbol{r},t)$ nor $\boldsymbol{B}(\boldsymbol{r},t)$ is again based on the fact that differentiations can be permuted, e.g., $\frac{\partial}{\partial t}\frac{\partial f}{\partial x} = \frac{\partial}{\partial x}\frac{\partial f}{\partial t}$.)

In the following we use this "gauge freedom" by choosing the so-called *Lorentz gauge* :

$$\mathrm{div}\boldsymbol{A} + \frac{1}{c^2}\frac{\partial \phi}{\partial t} \equiv 0 \, . \quad (20.6)$$

After a short calculation, see below, one obtains from the two remaining Maxwell equations (I and IV), $\mathrm{div}\boldsymbol{D} = \varrho$ and $\mathrm{curl}\boldsymbol{H} = \boldsymbol{j} + \frac{\partial \boldsymbol{D}}{\partial t}$, the so-called *d'Alembert-Poisson equations*:

$$-\left(\nabla^2 - \frac{\partial^2}{c^2\partial t^2}\right)\phi(\boldsymbol{r},t) = \varrho_E(\boldsymbol{r},t)/\varepsilon_0 \quad \text{and}$$

$$-\left(\nabla^2 - \frac{\partial^2}{c^2\partial t^2}\right)\boldsymbol{A}(\boldsymbol{r},t) = \mu_0\boldsymbol{j}_B(\boldsymbol{r},t) \, . \quad (20.7)$$

c is the velocity of light *in vacuo*, while ϱ_E and $\boldsymbol{j}_B$ are the effective charge and current density, respectively. These deviate from the true charge and true current density by polarization contributions:

$$\varrho_E(\boldsymbol{r},t) := \varrho(\boldsymbol{r},t) - \mathrm{div}\boldsymbol{P}(\boldsymbol{r}t) \quad (20.8)$$

and

$$\boldsymbol{j}_B(\boldsymbol{r},t) := \boldsymbol{j}(\boldsymbol{r},t) + \frac{\boldsymbol{J}(\boldsymbol{r},t)}{\mu_0} + \frac{\partial \boldsymbol{P}}{\partial t} \, . \quad (20.9)$$

A derivation of the two d'Alembert-Poisson equations now follows.

a)

$$\mathrm{div}\,\boldsymbol{D} = \varrho \,, \rightarrow \varepsilon_0 \mathrm{div}\,\boldsymbol{E} = \varrho - \mathrm{div}\,\boldsymbol{P} =: \varrho_E \,; \rightarrow -\nabla^2\phi - \mathrm{div}\frac{\partial \boldsymbol{A}}{\partial t} = \varrho_E/\varepsilon_0\,.$$

With

$$\mathrm{div}\frac{\partial \boldsymbol{A}}{\partial t} = \frac{\partial}{\partial t}\mathrm{div}\,\boldsymbol{A}$$

and with the Lorentz gauge (20.6) we obtain the first d'Alembert-Poisson equation.

b)

$$\mathrm{curl}\,\boldsymbol{H} = \boldsymbol{j} + \frac{\partial \boldsymbol{D}}{\partial t} \rightarrow \mathrm{curl}\,\boldsymbol{B} - \mathrm{curl}\,\boldsymbol{J} = \mu_0 \cdot \left(\boldsymbol{j} + \varepsilon_0\frac{\partial \boldsymbol{E}}{\partial t} + \frac{\partial \boldsymbol{P}}{\partial t}\right)\,;$$

$$\rightarrow \mathrm{curl\,curl}\boldsymbol{A} = \mu_0\left(\boldsymbol{j} + \mathrm{curl}\frac{\boldsymbol{J}}{\mu_0} + \frac{\partial \boldsymbol{P}}{\partial t}\right) + \mu_0\varepsilon_0\frac{\partial \boldsymbol{E}}{\partial t}\,.$$

One now inserts

$$\mathrm{curl\,curl}\boldsymbol{A} \equiv \mathrm{grad}(\mathrm{div}\,\boldsymbol{A}) - \nabla^2\boldsymbol{A} \quad \text{and} \quad \boldsymbol{E} = -\mathrm{grad}\phi - \frac{\partial \boldsymbol{A}}{\partial t}$$

and obtains with

$$\varepsilon_0\mu_0 = \frac{1}{c^2}$$

the gradient of an expression that vanishes in the Lorentz gauge. The remaining terms yield the second d'Alembert-Poisson equation.

In the next section (as for the harmonic oscillator in Part I) we discuss "free" and "fundamental" solutions of the d'Alembert equations, i.e., with vanishing r.h.s. of the equation (and $\propto \delta(\boldsymbol{r})$). Of special importance among these solutions are *planar* electromagnetic waves and *spherical waves*.

20.3 Planar Electromagnetic Waves; Spherical Waves

The operator in the d'Alembert-Poisson equations (20.7)

$$\Box := \left(\nabla^2 - \frac{\partial^2}{c^2\partial t^2}\right) \tag{20.10}$$

is called the *d'Alembert operator* or "quabla" operator.

Amongst the general solutions of the free d'Alembert equation

$$\Box\phi(\boldsymbol{r},t) \equiv 0$$

(also known simply as the wave equation) are right-moving planar waves of
the kind

$$\phi_+(\boldsymbol{r}, t) \equiv g(x - ct) .$$

Here $g(x)$ is a general function, defined on the whole x-interval, which must
be continuously differentiable twice. $g(x)$ describes the profile of the *traveling
wave*, which moves to the right here (positive x-direction) with a velocity c.
A wave traveling to the left is described by

$$\phi_-(\boldsymbol{r}, t) \equiv g(x + ct) .$$

Choosing x as the direction of propagation is of course arbitrary; in general
we could replace x by $\boldsymbol{k} \cdot \boldsymbol{r}$, where

$$\hat{\boldsymbol{k}} := \boldsymbol{k}/|\boldsymbol{k}|$$

is the direction of propagation of the planar wave.

All these relations can be evaluated directly from Maxwell's equations. In
particular, it is necessary to look for the polarization direction, especially for
the so-called *transversality*. The first two Maxwell equations,

$$\operatorname{div}\boldsymbol{E} \equiv 0 \quad \text{and} \quad \operatorname{div}\boldsymbol{B} \equiv 0 ,$$

imply (if the fields depend only on x and t) that the x-components E_x and
B_x must be constant (i.e., $\overset{!}{=} 0$, without lack of generality). Thus, only the
equations

$$\operatorname{curl}\boldsymbol{E} = -\frac{\partial \boldsymbol{B}}{\partial t} \quad \text{and} \quad \operatorname{curl}\boldsymbol{B} \equiv \mu_0\varepsilon_0\frac{\partial \boldsymbol{E}}{\partial t}$$

remain (i.e., $\equiv \frac{1}{c^2}\frac{\partial \boldsymbol{E}}{\partial t}$), which can be satisfied by

$$\boldsymbol{E} \equiv g(x - ct)\boldsymbol{e}_y \quad \text{and} \quad \boldsymbol{B} \equiv c^{-1}g(x - ct)\boldsymbol{e}_z ,$$

with one and the same arbitrary *profile function* $g(x)$. (For an electromagnetic
wave traveling *to the left* one obtains analogously $\boldsymbol{E} \equiv g(x + ct)\boldsymbol{e}_y$ and
$\boldsymbol{B} \equiv -c^{-1}g(x + ct)\boldsymbol{e}_z$.)

*For electromagnetic waves traveling to the right (or left, respectively) the
propagation direction $\hat{\boldsymbol{k}}$ and the vectors $\boldsymbol{E}$ and $c\boldsymbol{B}$ thus form a right-handed
rectangular trihedron (in both cases!), analogous to the three vectors $\pm\boldsymbol{e}_x$,
$\boldsymbol{e}_y$ and $\pm\boldsymbol{e}_z$. In particular, the amplitude functions of $\boldsymbol{E}$ and $c\boldsymbol{B}$ (in the cgs
system: those of $\boldsymbol{E}'$ and $\boldsymbol{B}'$) are always identical.*

The densities of the electromagnetic field energy are also identical:

$$w_E = \frac{\varepsilon_0}{2}E^2 \equiv w_B := \frac{B^2}{2\mu_0} .$$

The Poynting vector

$$\boldsymbol{S} := [\boldsymbol{E} \times \boldsymbol{H}]$$

is related to the total field-energy density of the wave,

$$w_{\text{total}} := w_E + w_B ,$$

as follows:

$$\boldsymbol{S} \equiv c\hat{\boldsymbol{k}} \cdot w_{\text{total}} ,$$

as expected.

Spherical waves traveling outwards,

$$\Phi_+(r,t) := \frac{g(t - \frac{r}{c})}{r} , \tag{20.11}$$

for $r > 0$ are also solutions of the free d'Alembert equation. This can easily be seen from the identity

$$\nabla^2 f(r) = r^{-1} \frac{\mathrm{d}^2 (r \cdot f(r))}{\mathrm{d}r^2} .$$

If the singular behavior of the function $1/r$ at $r = 0$ is again taken into account by exclusion of a small sphere around the singularity, one obtains from the standard definition

$$\mathrm{div} \boldsymbol{v} := \lim_{V \to 0} \left(\frac{1}{V} \oiint_{\partial V} \boldsymbol{v} \cdot \boldsymbol{n} \mathrm{d}^2 A \right)$$

for sufficiently reasonable behavior of the double-derivative $\ddot{g}(t)$ the following identity:

$$\Box \frac{g(t - \frac{r}{c})}{r} \equiv \left(\nabla^2 - \frac{1}{c^2} \frac{\partial^2}{\partial t^2} \right) \frac{g(t - \frac{r}{c})}{r} \equiv -4\pi\delta(\boldsymbol{r}) g\left(t - \frac{r}{c} \right) . \tag{20.12}$$

This corresponds to the analogous equation in electrostatics:

$$\nabla^2 \frac{1}{r} \equiv -4\pi\delta(\boldsymbol{r}) .$$

As a consequence we keep in mind that spherical waves *traveling outwards*,

$$\Phi_+(\boldsymbol{r},t) := \frac{g\left(t - \frac{r}{c} \right)}{r} ,$$

are so-called *fundamental solutions* of the d'Alembert-Poisson equations.

(The corresponding *incoming* spherical waves, $\Phi_-(r,t) := \frac{g\left(t + \frac{r}{c} \right)}{r}$, are also fundamental solutions, but in general they are *non-physical* unless one is dealing with very special initial conditions, e.g., with a pellet bombarded from all sides by intense laser irradiation, which is performed in order to initiate a thermonuclear fusion reaction.)

20.4 Retarded Scalar and Vector Potentials II: The Superposition Principle with Retardation

With equation (20.12) we are now in a position to write down the explicit solutions of the d'Alembert-Poisson equations (20.7), *viz*,

$$\phi(\boldsymbol{r}, t) = \iiint dV' \frac{\varrho_E\left(\boldsymbol{r}', t - \frac{|\boldsymbol{r}-\boldsymbol{r}'|}{c}\right)}{4\pi\varepsilon_0|\boldsymbol{r} - \boldsymbol{r}'|} \,,$$

$$\boldsymbol{A}(\boldsymbol{r}, t) = \iiint dV' \frac{\mu_0 \boldsymbol{j}_B\left(\boldsymbol{r}', t - \frac{|\boldsymbol{r}-\boldsymbol{r}'|}{c}\right)}{4\pi|\boldsymbol{r} - \boldsymbol{r}'|} \,. \tag{20.13}$$

In principle these rigorous results are very clear. For example, they tell us that the fields of single charges and currents

a) on the one hand, can be simply *superimposed*, as in the static case with Coulomb's law, while

b) on the other hand, the *retardation between cause and effect* has to be taken into account, i.e., instead of $t' = t$ (instantaneous reaction) one has to write

$$t' = t - \frac{|\boldsymbol{r} - \boldsymbol{r}'|}{c} \,,$$

i.e., the reaction is retarded, since electromagnetic signals between $\boldsymbol{r}'$ and $\boldsymbol{r}$ propagate with the velocity c.

c) *Huygens's principle*[3] of the superposition of spherical waves is also contained explicitly and quantitatively in equations (20.13).

d) According to the rigorous result (20.13) the mutual influences propagate at the *vacuum light velocity*, even in polarizable matter.

(This does not contradict the fact that stationary electromagnetic waves in dielectric and permeable matter propagate with a reduced velocity ($c^2 \to c^2/(\varepsilon\mu)$). As with the driven harmonic oscillator (see Part I) these *stationary waves* develop only after a finite *transient time*. The calculation of the transition is one of the fundamental problems solved by Sommerfeld, who was Heisenberg's supervisor.)

The explicit material properties enter the retarded potentials (20.13) only through the deviation between the "true charges" (and "true currents") and the corresponding "effective charges" (and "effective currents"). One should remember that in the rigorous equations (20.13) the *effective* quantities enter. Only *in vacuo* do they agree with the *true* quantities.

---

[3] See section on optics

20.5 Hertz's Oscillating Dipole (Electric Dipole Radiation, Mobile Phones)

It can be readily shown that the electromagnetic field generated by a time-dependent dipole $\boldsymbol{p}(t)$ at $\boldsymbol{r}' = 0$, *viz*

$$\boldsymbol{B}(\boldsymbol{r},t) = \frac{\mu_0}{4\pi}\operatorname{curl}\frac{\dot{\boldsymbol{p}}\left(t - \frac{r}{c}\right)}{r}$$

$$\boldsymbol{E}(\boldsymbol{r},t) = \frac{1}{4\pi\varepsilon_0}\operatorname{curl}\operatorname{curl}\frac{\boldsymbol{p}\left(t - \frac{r}{c}\right)}{r} \ , \tag{20.14}$$

for $t \in (-\infty,\infty)$, solves everywhere all four Maxwell equations. (If $\boldsymbol{B}(\boldsymbol{r},t)$ obeys Maxwell's equations, then so does $\boldsymbol{E}(\boldsymbol{r},t)$ and *vice versa*, in accordance with (20.14).)

A practical example of such a time-dependent electric dipole (so-called Hertz dipole) is a radio or mobile-phone antenna, driven by an alternating current of frequency ω. Since the current is given explicitly by

$$I(t) \equiv |\dot{\boldsymbol{p}}(t)| \ ,$$

the retarded vector potential has the following form, which is explicitly used in equation (20.14):

$$\boldsymbol{A}(\boldsymbol{r},t) = \frac{\mu_0}{4\pi r}\dot{\boldsymbol{p}}\left(t - \frac{r}{c}\right) \ .$$

In particular, the asymptotic behavior of the fields in both the *near-field* and *far-field range* can be derived without difficulty, as follows:

a) In the *near-field range* (for $r \ll \lambda$, where λ is the wavelength of light in a vacuum, corresponding to the frequency ω, i.e., $\lambda = \frac{\omega}{2\pi c}$) one can approximate

$$\frac{\boldsymbol{p}\left(t - \frac{r}{c}\right)}{r} \quad \text{by} \quad \frac{\boldsymbol{p}(t)}{r} \ .$$

After some elementary transformation, using the identity

$$\operatorname{curl}\operatorname{curl}\boldsymbol{v} \equiv (\operatorname{grad}\operatorname{div} - \nabla^2)\boldsymbol{v}$$

and the spherical wave equation given previously, one then obtains:

$$\boldsymbol{E}(\boldsymbol{r},t) \cong \frac{3(\boldsymbol{p}(t)\cdot\boldsymbol{r})\boldsymbol{r} - r^2\boldsymbol{p}(t)}{4\pi\varepsilon_0 r^5} \ , \tag{20.15}$$

which is the *quasi-static* result

$$\boldsymbol{E}(\boldsymbol{r},t) \propto r^{-3} \quad \text{for} \quad r \to 0$$

with which we are already familiar, while simultaneously

$$\boldsymbol{B}(\boldsymbol{r},t) \cong \frac{\mu_0}{4\pi r^3}\dot{\boldsymbol{p}}(t) \times \boldsymbol{r} \ , \quad \text{i.e.,} \quad \propto r^{-2} \quad \text{for} \quad r \to 0 \ ,$$

and is thus less strongly divergent (i.e., asymptotically negligible w.r.t. $\boldsymbol{E}(\boldsymbol{r},t)$) for $r \to 0$. Here the position-dependence of the *denominator* dominates.

b) In the *far-field range* (for $r \gg \lambda$), the position-dependence of the *numer-ator* dominates, e.g.,

$$\nabla \times \boldsymbol{p}\left(t - \frac{r}{c}\right) \cong -\frac{1}{c}\left[\hat{\boldsymbol{r}} \times \frac{\dot{\boldsymbol{p}}\left(t - \frac{r}{c}\right)}{r}\right] \, , \quad \text{with} \quad \hat{\boldsymbol{r}} = \boldsymbol{r}/r \, .$$

Here, for $r \gg \lambda$, one obtains asymptotically:

$$\boldsymbol{E}(\boldsymbol{r},t) \cong \frac{1}{4\pi\varepsilon_0 c^2 r}\left[\hat{\boldsymbol{r}} \times \left[\hat{\boldsymbol{r}} \times \ddot{\boldsymbol{p}}\left(t - \frac{r}{c}\right)\right]\right] \quad \text{and}$$

$$\boldsymbol{B}(\boldsymbol{r},t) \cong -\frac{\mu_0}{4\pi c r}\left[\hat{\boldsymbol{r}} \times \ddot{\boldsymbol{p}}\left(t - \frac{r}{c}\right)\right] \, , \tag{20.16}$$

i.e.,

$$\boldsymbol{S} = \boldsymbol{E} \times \boldsymbol{H} \cong \hat{\boldsymbol{r}}\,\frac{(\sin\theta)^2}{16\pi^2\varepsilon_0 c^3 r^2}\left|\ddot{\boldsymbol{p}}\left(t - \frac{r}{c}\right)\right|^2 \, , \tag{20.17}$$

where θ is the angle between $\ddot{\boldsymbol{p}}$ and $\hat{\boldsymbol{r}}$.

As for planar waves, the propagation vector

$$\hat{\boldsymbol{r}} := \boldsymbol{r}/r$$

and the vectors $\boldsymbol{E}$ and $c\boldsymbol{B}$ form a right-handed rectangular trihedron, where in addition the vector $\boldsymbol{E}$ in the far-field range lies asymptotically in the plane defined by $\ddot{\boldsymbol{p}}$ and $\boldsymbol{r}$.

In the far-field range the amplitudes of both $\boldsymbol{E}$ and $\boldsymbol{B}$ are $\propto \omega^2 r^{-1}\sin\theta$; the Poynting vector is therefore $\propto \omega^4 r^{-2}(\sin\theta)^2\hat{\boldsymbol{r}}$. The power integrated across the surface of a sphere is thus $\propto \omega^4$, independent of the radius! Due to the ω^4-dependence of the power, electromagnetic radiation beyond a limiting frequency range is *biologically dangerous* (e.g. *X*-rays), whereas low-frequency radiation is *biologically harmless*. The limiting frequency range (hopefully) seems to be beyond the frequency range of present-day mobile phones, which transmit in the region of 10^9 Hz.

20.6 Magnetic Dipole Radiation; Synchrotron Radiation

For vanishing charges and currents Maxwell's equations possess a symmetry which is analogous to that of the *canonical equations* of classical mechanics ($\dot{q} = -\frac{\partial\mathcal{H}}{\partial p}; \dot{p} = \frac{\partial\mathcal{H}}{\partial q}$, see Part I), called *symplectic invariance*: the set of equations does not change, if $\boldsymbol{E}$ is transformed into $c\boldsymbol{B}$, and $\boldsymbol{B}$ into

$$-c^{-1}\boldsymbol{E}\left((\boldsymbol{E},c\boldsymbol{B}) \rightarrow \begin{pmatrix} 0,+1 \\ -1,0 \end{pmatrix}\begin{pmatrix} \boldsymbol{E} \\ c\boldsymbol{B} \end{pmatrix}\right) \, .$$

To be more precise, if in equation (20.14) $\boldsymbol{B}$ is transformed into $-\boldsymbol{D}$ and $\boldsymbol{E}$ is transformed into $\boldsymbol{H}$, while simultaneously ε_0 and μ_0 are interchanged and the electric dipole moment is replaced by a magnetic one (corresponding to $B = \mu_0 H + J$, $D = \varepsilon_0 E + P$), one obtains the electromagnetic field produced by a time-dependent *magnetic* dipole. For the Poynting vector in the far-field range one thus obtains instead of (20.17):

$$S = E \times H \cong \hat{r} \frac{(\sin \theta_m)^2}{16\pi^2 \mu_0 c^3 r^2} \left| \ddot{m} \left(t - \frac{r}{c} \right) \right|^2 , \qquad (20.18)$$

where θ_m is again the angle between $\ddot{m}$ and $\hat{r}$.

We shall now show that *magnetic* dipole radiation for particles with *non-relativistic* velocities is much smaller – by a factor $\sim \frac{v^2}{c^2}$ – than *electric* dipole radiation. Consider an electron moving with constant angular velocity ω in the xy–plane in a circular orbit of radius R. The related electric dipole moment is

$$p(t) = eR \cdot (\cos(\omega t) \, \sin(\omega t), 0) .$$

On average the amplitude of this electric dipole moment is $p_0 = eR$. The corresponding magnetic dipole moment is (on average)

$$m_0 = \mu_0 \pi R^2 \frac{e\omega}{2\pi} ,$$

where we have again used – as with the calculation of the gyromagnetic ratio – the relation

$$I = \frac{e\omega}{2\pi} .$$

With

$$\omega = \frac{v}{R}$$

we thus obtain

$$m_0 = \frac{1}{2} evR .$$

Therefore, as long as $p(t)$ and $m(t)$ oscillate in their respective amplitudes p_0 and m_0 with identical frequency ω as in $\cos \omega t$, we would get on average the following ratio of the amplitudes of the respective Poynting vectors:[4]

$$\frac{|S_m|}{|S_e|} = \frac{m_0^2 \varepsilon_0}{p_0^2 \mu_0} = \frac{v^2}{4c^2} . \qquad (20.19)$$

Charged *relativistic* particles in a circular orbit are sources of intense, polarized radiation over a vast frequency range of the electromagnetic spectrum, e.g., from the infrared region up to soft X-rays. In a synchrotron, electrons travelling at almost the speed of light are forced by magnets to move in

---

[4] The factor 4 $(= 2^2)$ in the denominator of this equation results essentially from the fact that the above formula $p(T) = eR \cdot (\cos(\omega t), \, \sin(\omega t))$ can be interpreted as follows: there are in effect *two* electric dipoles (but only *one* magnetic dipole) involved.

a circular orbit. The continual acceleration of these charged particles in their circular orbit causes high energy radiation to be emitted tangentially to the path. To enhance the effectivity the electrons usually travel through special structures embedded in the orbit such as wigglers or *undulators*.

Synchrotron radiation is utilized for all kinds of physical and biophysical research at various dedicated sites throughout the world.

20.7 General Multipole Radiation

The results of this section follow directly from (20.14) for *electric* dipole radiation (Hertz dipole) and the corresponding equations for *magnetic* dipole radiation (see Sect. 20.6). We recall that a quadrupole is obtained by a limiting procedure involving the difference between two exactly opposite dipoles, one of which is shifted with respect to the other by a vector $\boldsymbol{b}(=\boldsymbol{b}_2)$; an octupole is obtained by a similar shift (with $\boldsymbol{b}(=\boldsymbol{b}_3)$) from two exactly opposite quadrupoles, etc.

As a consequence, the electromagnetic field of electric octupole radiation, for example, is obtained by application of the differential operator

$$(\boldsymbol{b}_3 \cdot \nabla)(\boldsymbol{b}_2 \cdot \nabla)$$

on the electromagnetic field of a Hertz dipole:

a) In the *near-field range*, i.e., for $r \ll \lambda$, one thus obtains the following *quasi-static* result for the $\boldsymbol{E}$-field of an electric 2^l-pole:

$$\boldsymbol{E} \cong -\mathrm{grad}\phi(\boldsymbol{r}, t) \,,$$

with

$$\phi(\boldsymbol{r}, t) \cong \frac{1}{4\pi\varepsilon_0 r^{l+1}} \sum_{m=-l}^{l} c_{l,m}(t) Y_{l,m}(\theta, \phi) \,, \qquad (20.20)$$

where $Y_{l,m}$ are *spherical harmonics*.

The coefficients of this expansion depend on the vectors $\boldsymbol{b}_1, \boldsymbol{b}_2, \ldots, \boldsymbol{b}_l$, and on time.

b) In the *far-field range*, i.e., for $r \gg \lambda$, one can approximate the expressions

$$(\boldsymbol{b}_i \cdot \nabla) \left\{ \frac{\boldsymbol{f}\left(t - \frac{r}{c}\right)}{r} \right\} \quad \text{by} \ ^5 \quad \frac{1}{r} (\boldsymbol{b}_l \cdot \hat{\boldsymbol{r}}) \left(\frac{-\partial}{c\partial t} \right) \boldsymbol{f} \left(t - \frac{r}{c} \right)$$

and one obtains for the electric 2^l-pole radiation asymptotically:

---

[5] It should be noted that the vector $\boldsymbol{f}(t - \frac{r}{c})$ is always perpendicular to $\hat{\boldsymbol{r}}$, cf. (20.16).

1)

$$E \propto \frac{1}{r} \cdot \left(\prod_{\nu=2}^{l} (b_\nu \cdot \hat{r}) \right) \cdot \left(\frac{-\partial}{c\partial t} \right)^l \left[\hat{r} \times p \left(t - \frac{r}{c} \right) \right] , \qquad (20.21)$$

2)

$$cB \cong \hat{r} \times E \quad \text{and}$$

3)

$$S \equiv E \times H , \quad \text{i.e., for} \quad p(t) = p_0 \cos \omega t$$

on average w.r.t. time:

$$S(r) \propto \frac{\omega^{2(l+1)}}{r^2} \left(\prod_{\nu=2}^{l+1} \frac{(b_\nu \cdot \hat{r})^2}{c^2} \right) |\hat{r} \times p_0|^2 \, \hat{r} . \qquad (20.22)$$

Here we have $l = 1, 2, 3$ for dipole, quadrupole and octupole radiation. *The dependence on the distance r is thus universal, i.e., $S \propto r^{-2}$ for all l; the dependence on the frequency is simple ($S \propto \omega^{2(l+1)}$); only the angular dependence of electric multipole radiation is complicated. However, in this case too, the three vectors $\hat{r}$, E and cB form a right-handed rectangular trihedron, similar to e_x, e_y, e_z. Details can be found in (20.21).[6] For magnetic multipole radiation the results are similar.*

20.8 Relativistic Invariance of Electrodynamics

We have already seen in Part I that classical mechanics had to be amended as a result of Einstein's theory of special relativity. In contrast, Maxwell's theory *is* already relativistically invariant *per se* and requires no modification (see below).

It does no harm to repeat here (see also Section 9.1) that prior to Einstein's theory of relativity (1905) it was believed that a special inertial frame existed, the so-called *aether* or *world aether*, in which Maxwell's equations had their usual form, and, in particular, where the velocity of light *in vacuo* had the value

$$c \equiv \frac{1}{\sqrt{\varepsilon_0 \mu_0}} ,$$

whereas in other inertial frames, according to the Newtonian (or Galilean) *additive behavior of velocities*, the value would be different (e.g., $c \to c+v$). In their well known experiments, Michelson and Morley attempted to measure the motion of the earth relative to the aether and thus tried to verify this behavior. Instead, they found (with great precision): $c \to c$.

---

[6] Gravitational waves obey the same theory with $l = 2$. Specifically, in the distribution of gravitational charges there are no dipoles, but only quadrupoles etc.

In fact, Hendryk A. Lorentz from Leiden had already established before Einstein that Maxwell's equations, which are not invariant under a Galilean transformation, are invariant w.r.t. a *Lorentz transformation* – as it was later called – in which space and time coordinates are "mixed" (see Section 9.1). Furthermore, the result of Michelson and Morley's experiments follows naturally from the Lorentz transformations. However, Lorentz interpreted his results only as a strange mathematical property of Maxwell's equations and (in contrast to Einstein) not as a scientific revolution with respect to our basic assumptions about *spacetime* underlying all physical events.

The relativistic invariance of Maxwell's equations can be demonstrated most clearly in terms of the Minkowski four-vectors introduced in Part I.

a) The essential point is that in addition to

$$\tilde{x} := (x, y, z, \mathrm{i}ct) \,,$$

the following two quadruplets,

1)

$$\tilde{A} := \left(\boldsymbol{A}, \mathrm{i}\frac{\varPhi}{c} \right) \quad \text{and} \,,$$

2)

$$\tilde{j} := (\boldsymbol{j}, \mathrm{i}c\varrho)$$

are Minkowski four-vectors, whereas other quantities, such as the d'Alembert operator, are Minkowski scalars (which are invariant), and the fields themselves, $\boldsymbol{E}$ plus $\boldsymbol{B}$ (six components) correspond to a skew symmetric tensor

$$F_{\mu,\nu}(= -F_{\nu,\mu})$$

generated from $\tilde{A}$, *viz*

$$F_{\mu,\nu} := \frac{\partial A_\nu}{\partial x_\mu} - \frac{\partial A_\mu}{\partial x_\nu}$$

(e.g., $F_{1,2} = -F_{2,1} = B_3$), with $x_1 := x$, $x_2 := y$, $x_3 := z$, and $x_4 := \mathrm{i}ct$.

b) One defines the "Minkowski nabla"

$$\tilde{\nabla} := \left(\frac{\partial}{\partial x}, \frac{\partial}{\partial y}, \frac{\partial}{\partial z}, \frac{\partial}{\partial \mathrm{i}ct} \right) = \left(\nabla, \frac{\partial}{\partial \mathrm{i}ct} \right) \,.$$

Similarly to Euclidian space $\mathcal{R}^3$, where the Laplace operator ∇^2 ($\equiv \varDelta$) is invariant (i.e., does not change its form) under *rotations*, in Minkowski space $\mathcal{M}_4$ the d'Alembert operator $\tilde{\nabla}^2$ ($\equiv \square$) is invariant under *pseudo-rotations*.

In addition, similar to the fact that the divergence of a vector field has a coordinate-invariant meaning with respect to rotations in $\mathcal{R}^3$, analogous results also apply for the Minkowski divergence, i.e. one has an invariant meaning of $\tilde{\nabla} \cdot \tilde{v}$ with respect to pseudo-rotations in $\mathcal{M}_4$.

c) For example, the continuity equation,

$$\mathrm{div}\boldsymbol{j} + \frac{\partial\varrho}{\partial t} = 0 \, ,$$

has a simple invariant relativistic form (which we shall use below):

$$\tilde{\nabla} \cdot \tilde{j} := \sum_{\nu=1}^{4} \frac{\partial j_\nu}{\partial x_\nu} = 0 \, . \tag{20.23}$$

d) Analogously, gauge transformations of the kind

$$\boldsymbol{A} \to \boldsymbol{A} + \mathrm{grad}g(\boldsymbol{r}, t), \Phi \to \Phi - \frac{\partial g(\boldsymbol{r}, t)}{\partial t}$$

can be combined to

$$\tilde{A}(\tilde{x}) \to \tilde{A}(\tilde{x}) + \tilde{\nabla}g(\tilde{x}) \, .$$

We are now prepared for the explicit Minkowski formulation of Maxwell's equations. As mentioned, the homogeneous equations II and III,

$$\mathrm{div}\boldsymbol{B} = 0 \, ; \quad \mathrm{curl}\boldsymbol{E} = -\frac{\partial\boldsymbol{B}}{\partial t}$$

are automatically satisfied by introducing the above *skew symmetric field tensor*

$$F_{\mu\nu} := \partial_\mu A_\nu - \partial_\nu A_\mu \, , \quad \text{with} \quad \mu, \nu = 1, \dots, 4 \, ;$$

which is analogous to the representation of $\boldsymbol{E}$ and $\boldsymbol{B}$ by a scalar potential plus a vector potential.

The remaining inhomogeneous Maxwell equations I and IV,

$$\mathrm{div}\boldsymbol{E} \equiv \varrho/\varepsilon_0 \quad \text{and} \quad \mathrm{curl}\boldsymbol{B} \equiv \mu_0\boldsymbol{j} \, ,$$

simply yield the following result, with Einstein's summation convention[7]:

$$\partial_\mu F_{\nu\mu} = \partial_\mu\partial_\nu A_\mu - \partial_\mu\partial_\mu A_\nu \equiv \mu_0 j_\nu \, .$$

With the Lorentz gauge,

$$\mathrm{div}\boldsymbol{A} + \frac{1}{c^2}\frac{\partial\phi}{\partial t} = \partial_\mu A_\mu = 0 \, , ^8$$

the first term on the l.h.s. vanishes, i.e., we again obtain the d'Alembert-Poisson equation

$$-\tilde{\nabla}^2\tilde{A} \equiv \mu_0\tilde{j} \, .$$

---

[7] Using the Einstein convention one avoids clumsy summation symbols: If an index appears twice, it is summed over.

[8] Here we again use the permutability of partial derivatives.

From the "simple" Lorentz transformations for the x- and t-components of the Minkowski four-potential $\tilde{A}$, for a transition between different inertial frames the following "more complicated" *Lorentz transformations for electromagnetic fields* result: The longitudinal components E_x and B_x remain unchanged, whereas one obtains for the transverse components:

$$B_\perp(r,t) = \frac{B'_\perp(r',t') + \frac{v}{c^2} \times E'(r',t')}{\sqrt{1 - \frac{v^2}{c^2}}}, \qquad (20.24)$$

$$E_\perp(r,t) = \frac{E'_\perp(r',t') - v \times B'(r',t')}{\sqrt{1 - \frac{v^2}{c^2}}} . \qquad (20.25)$$

These results can be used to obtain the **E**-*and* **B**-*fields of a moving point charge from the Coulomb* **E'**-*field in the co-moving frame.*[9]

---

[9] See the exercises at http://www.physik.uni-regensburg.de/forschung/krey, summer 2002, file 9.

21 Applications of Electrodynamics in the Field of Optics

21.1 Introduction: Wave Equations; Group and Phase Velocity

Firstly we shall remind ourselves of the relationship between the frequency

$$\nu = \frac{2\pi}{\omega} \quad \text{and wavelength} \quad \lambda = \frac{2\pi}{k} \quad (k = \text{wavenumber})$$

of an electromagnetic wave *in vacuo*:

$$\omega = 2\pi\nu = c \cdot k = c \cdot \frac{2\pi}{\lambda}, \quad \text{or} \quad \lambda \cdot \nu = c .$$

Secondly, electromagnetic waves cover an extremely wide spectral range. For example, radio waves have wavelengths from 1 km or more (long-wave) via 300 m (medium-wave) to about 50 m (short-wave). This range is followed by VHF (very high frequency), then the range of television and mobile-phone frequencies from $\approx 100\,\text{MHz}$ to $10\,\text{GHz}$; then we have radar, light waves, X-rays and, at very short wavelengths or high frequencies, γ-rays.

It is useful to remember that the wavelengths of visible light range from $\lambda \approx 8000\,\text{Å}$ (or 800 nm, *red*) down to $\approx 4000\,\text{Å}$ (400 nm, *violet*)[1] On the lower frequency side of the visible range come *infrared* and *far-infrared*, and on the high-frequency side *ultraviolet* and *soft-X-ray* radiation.

Thirdly, in connection with X-rays and γ-radiation, it is useful to remember that these phenomena arise from quantum transitions, see Part III (Fermi's "golden rules"), according to the formula

$$\Delta E \equiv E_i - E_f \equiv h\nu ,$$

i.e. by transitions from a higher initial energy E_i to a lower final energy E_f. The radiation may have a continuous distribution of frequencies (so-called *bremsstrahlung*, or *braking radiation*), or it may contain a discrete set of spectral "lines". The quantity h is Planck's constant:

$$h = 6.625\ldots \cdot 10^{-34}\,\text{Ws}^2 \equiv 4.136\ldots \cdot 10^{-15}\,\text{eVs} .$$

---

[1] Some readers may prefer the characteristic atomic length $1\,\text{Å}$, whereas others use units such as $1\,\text{nm}$ ($\equiv 10\,\text{Å}$). Which one is more appropriate, depends on the problem, on the method used, and on personal preferences.

X-rays have typical energies of $\Delta E \approx 10\,\mathrm{keV}$ to $\approx 1\,\mathrm{MeV}$, characteristic for the *electron shell* of atoms, whereas for γ-radiation one is dealing with excitations of *nuclei*, i.e., $\Delta E \approx 1\,\mathrm{MeV}$ up to $1\,\mathrm{GeV}$.

Fourthly, Planck's formula for black-body radiation: The total energy of the electromagnetic field contained in a volume V at Kelvin temperature T is given by

$$U(T) \equiv V \int_0^\infty d\nu\, u(\nu, T) \,,$$

with the spectral energy density

$$u(\nu, T) = \frac{8\pi\nu^2}{c^3}\,\frac{h\nu}{\mathrm{e}^{\frac{h\nu}{k_bT}} - 1} \,. \tag{21.1}$$

For the surface temperature of the sun, i.e., for $T \approx 6000\,\mathrm{K}$, the function $u(\nu, T)$ has a pronounced maximum in the *green* range, i.e., for

$$\nu = \frac{c}{\lambda} \quad \text{with} \quad \lambda \approx 6000\,\text{Å}(= 600\,\mathrm{nm}) \,.$$

See Fig. 21.1:

The vacuum velocity of electromagnetic waves (e.g., light) is

$$c_0 := (\varepsilon_0\mu_0)^{-\frac{1}{2}} \,.$$

In polarizable matter, the (stationary) velocity of electromagnetic waves is smaller:

$$c_m = \frac{c_0}{n} \,,$$

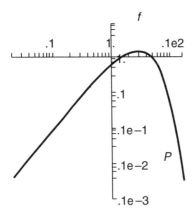

Fig. 21.1. Planck's black-body radiation formula. For the reduced frequency f ($\equiv h\nu/(k_BT)$ in the text) Planck's function $P(f) := f^3/(\exp f - 1)$ is shown as a double-logarithmic plot. It has a pronounced maximum around $f \approx 2$

where $n \; (= \sqrt{\varepsilon \cdot \mu})$ is the *refractive index*[2]. In optically dense light-trans-mitting media (e.g., glass or water) n is significantly > 1, which gives rise to the well-known phenomenon of refraction described by *Snell's law*:

$$\frac{\sin \alpha_1}{\sin \alpha_2} = \frac{n_2}{n_1} \; .$$

This expression describes the refraction of light rays at a plane boundary between a vacuum with refractive index $n = 1$ (or in practice: air) and an (optically) denser medium with $n_2 > 1$. The angle of incidence from the vacuum is α_1.

The refractive index, n, is thus related both to the (relative) dielectric constant of the material, ε_r, and also to the (relative) permeability μ_r[3]. This also means that generally n depends on the wavenumber $\boldsymbol{k}$ and frequency ω.

The above relation for dielectric and/or permeable matter follows from Maxwell's equations in the absence of charges and currents (i.e., for $\varrho \equiv \boldsymbol{j} \equiv 0$):

$$\mathrm{div}\varepsilon_r\varepsilon_0\boldsymbol{E} = \varrho \equiv 0 \; ; \mathrm{div}\boldsymbol{B} = 0 \; ;$$

$$\mathrm{curl}\boldsymbol{E} = -\frac{\partial \boldsymbol{B}}{\partial t} \; ; \mathrm{curl}\frac{\boldsymbol{B}}{\mu_r\mu_0} = \boldsymbol{j} + \varepsilon_r\varepsilon_0\frac{\partial \boldsymbol{E}}{\partial t} \; . \qquad (21.2)$$

Applying the operator curl to the third and fourth equations, together with the identity

$$\mathrm{curl}\,\mathrm{curl}\boldsymbol{v} = \mathrm{grad}\,\mathrm{div}\boldsymbol{v} - \nabla^2\boldsymbol{v}$$

we obtain the following wave equations:

$$\left(\nabla^2 - \frac{1}{c_m^2}(\partial_t)^2\right)\boldsymbol{E} = 0 \quad \text{and} \quad \left(\nabla^2 - \frac{1}{c_m^2}(\partial_t)^2\right)\boldsymbol{B} = 0 \; ,$$

where c_m (see below) is the velocity of a stationary electromagnetic wave in the considered medium.

Similar wave equations occur in other cases, e.g., for transverse sound waves (*shear waves*, $\rightarrow$ *transverse phonons* in solid state physics) it is only necessary to replace $\boldsymbol{E}$ and $c\boldsymbol{B}$ by the transverse displacements of the atoms from their rest positions and the velocity of light, c_m, by the transverse sound velocity $c_\perp^{(s)}$ (which in metals is of the order of 10^3 m/s). With these re-placements one obtains the same wave equation in totally different contexts. (There are also *longitudinal* sound waves (*compression waves*[4], $\rightarrow$ *longitudinal phonons*) with a significantly higher sound velocity $c_{\mathrm{long.}}^{(s)}$.)

---

[2] In some (mainly artificial) materials both ε and μ are negative, i.e., also n. These so-called *left-handed materials* have unusual optical properties.

[3] To avoid any misunderstanding, here we explicitly use the lower index $_r$, although hitherto we did not use this convention.

[4] In liquids and gases only *compression waves* exist, and not *shear waves*.

The relationship

$$\omega = \omega(\boldsymbol{k})$$

between frequency ω and wavenumber $\boldsymbol{k}$[5] is referred to as the *dispersion relation*.

For light waves *in vacuo* the dispersion relation is simply $\omega = c_0 \cdot |k|$.

In polarizable matter, light waves have only apparently almost the same dispersion behavior. It is true that often $\omega = c_m \cdot |k|$, but even then the light velocity in matter,

$$c_m = c_0 / \sqrt{\varepsilon_r(\omega) \cdot \mu_r(\omega)} \, ,$$

generally depends on the same frequency ω that one wishes to calculate[6], so that it makes sense to distinguish between the so-called *phase velocity*

$$\boldsymbol{v}_{\text{phase}} = \left(\frac{\omega}{k}\right) \boldsymbol{e_k}$$

and the *group velocity*

$$\boldsymbol{v}_{\text{group}} = \text{grad}_{\boldsymbol{k}}\omega(\boldsymbol{k}) \, .$$

As a consequence, the components of the group velocity $\boldsymbol{v}_{\text{group}}$ are calculated as follows:

$$(v_{\text{group}})_i = \frac{\partial \omega(\boldsymbol{k})}{\partial k_i} \, . \tag{21.3}$$

The group velocity can thus have a different direction from the wavenumber $\boldsymbol{k}$ (see below, e.g., the term *ray velocity* in the subsection on *birefringence* in crystals).

As an example for the difference between *phase velocity* and *group velocity* consider transverse sound waves propagating in an ideal crystal of cubic symmetry with lattice constant a. Let $\boldsymbol{k}$ be parallel to an edge direction. One then obtains transverse elastic plane waves with the following dispersion relation:

$$\omega(\boldsymbol{k}) = c_{\perp}^{(s)} \frac{\sin\left(\frac{k \cdot a}{2}\right)}{\left(\frac{a}{2}\right)} \, . \tag{21.4}$$

Thus, if the wavelength λ is much larger than the lattice constant a, i.e., for $k \cdot a \ll 1$, we have, as expected:

$$\omega = c_{\perp}^{(s)} k \, ,$$

and the *group velocity*,

$$|\boldsymbol{v}_{\text{group}}| \, ,$$

---

[5] $\boldsymbol{k} = \frac{2\pi}{\lambda}\boldsymbol{e_k}$, where λ is the wavelength and $\boldsymbol{e_k}$ the propagation direction of the plane wave, i.e., according to the usual *ansatz* $\Psi \propto \mathcal{R}e \exp(\text{i}(\boldsymbol{k} \cdot \boldsymbol{r} - \omega t))$.

[6] The frequency dependence of the light velocity may involve not only the amplitude, but also the direction (see below).

and *phase velocity*, $c_\perp^{(s)}$, are identical; but for $k \cdot a \to \pi$, i.e., with decreasing $\lambda \to 2a$, the *group velocity* converges to zero,

$$\frac{\partial \omega}{\partial k} \to 0 \ .$$

The meaning of "group velocity" can be illustrated by considering a wave packet generated by the superposition of monochromatic plane waves with slightly different k-vectors in the interval Δk:

$$\Psi(x,t) = \int_{\Delta k} dk a(k) e^{i(k \cdot x - \omega(k) \cdot t)} \ . \tag{21.5}$$

One obtains *beats*, which can be treated by a Taylor expansion, as follows: If K_0 is the center of the interval Δk, i.e.,

$$k = K_0 + k'$$

(with $k' \in \left[-\frac{K_0}{2}, +\frac{K_0}{2}\right]$) then we have $a(k) \approx a(K_0)$ and

$$\omega(k) \cdot t \approx \omega(K_0) \cdot t + k' \frac{d\omega}{dk} \cdot t + \dots \ .$$

Thus

$$\Psi(x,t) \approx a(K_0) e^{i(K_0 x - \omega(K_0) \cdot t)} \int_{-\frac{K_0}{2}}^{\frac{K_0}{2}} dk' e^{ik' \cdot (x - v_{\text{group}} \cdot t)} \ . \tag{21.6}$$

The factor immediately in front of the integral, which can also be written as $e^{iK_0 \cdot (x - v_{\text{phase}} t)}$ (i.e., with the phase velocity v_{phase}) describes the rapid oscillation of the amplitude of the wave. In contrast, the integral itself describes the (much slower) wave motion of an *envelope function*, which is propagated at the *group velocity* v_{group}; in particular we have

$$|\Psi(x,t)|^2 \propto |a(K_0)|^2 \cdot \left| \int_{-\frac{K_0}{2}}^{\frac{K_0}{2}} dk' e^{ik' \cdot (x - v_{\text{group}} \cdot t)} \right|^2 \ . \tag{21.7}$$

The transport of energy in a wave packet propagates with the group velocity, not the phase velocity, and according to Einstein's theory of relativity we have the constraint that $|v_{\text{group}}|$ (and not $|v_{\text{phase}}|$) must always be $\leq c$ (see below).

The difference between *phase velocity* and *group velocity* becomes clear if we consider an electromagnetic wave reflected "back and forth" between two metal plates perfectly parallel to the plane $z \equiv 0$. (These parallel plates form the simplest version of a *waveguide*.) We assume that this guided electromagnetic wave propagates in the x-direction. The *phase velocity* is given by

$$(v_{\text{phase}})_x = \frac{c}{\cos \theta} \,, \quad \text{which is} \quad > c \,.$$

In contrast, the *group velocity* is

$$(v_{\text{group}})_x = c \cdot \cos \theta \,, \quad \text{i.e.,} \quad \leq c \,,$$

as expected. Here θ is the grazing angle at which the electromagnetic wave meets the metal plates.

In the context of our treatment of *phase* and *group velocity* we should mention not only Planck's radiation formula (1900, see above) but also Einstein's "photon" hypothesis (1905, see below), which essentially resolved the problem of whether light should be considered as a particle phenomenon (Newton) or a wave phenomenon (Huygens). The answer is *"both"* ($\rightarrow$ wave-particle duality, see Part III). According to Einstein (1905, $\rightarrow$ Nobel prize 1921) electromagnetic waves result from the emission of individual relativistic *quanta* with velocity c ($\rightarrow$ "photons", see below), possessing energy $E = h\nu$, momentum

$$p = \frac{E}{c} = \frac{h\nu}{c} (= \hbar k) \,,$$

and vanishing *rest mass*. The internal energy $U(T)$ of the electromagnetic field in a large cavity of volume V at Kelvin temperature T is then (see above, and Parts III and IV) given according to *Planck's formula*:

$$U(T) = V \cdot \int_{\nu=0}^{\infty} \frac{8\pi\nu^2 \mathrm{d}\nu}{c^3} \cdot \frac{h\nu}{\exp\left(\frac{h\nu}{k_B T}\right) - 1} \,, \tag{21.8}$$

where the factor

$$V \cdot \frac{8\pi\nu^2 \mathrm{d}\nu}{c^3}$$

is the number of wave modes with frequencies $\nu \in \mathrm{d}\nu$; $h\nu$ is the energy of a single photon of wavelength $\frac{c}{\nu}$, and the thermal expectation value $\langle n \rangle_\nu$ of the number of these photons is

$$\langle n \rangle_\nu = \frac{\sum_{n=0}^{\infty} n \cdot \mathrm{e}^{-n\frac{h\nu}{k_B T}}}{\sum_{n=0}^{\infty} \mathrm{e}^{-n\frac{h\nu}{k_B T}}} = \frac{1}{\exp\left(\frac{h\nu}{k_B T}\right) - 1} \,.$$

(k_B is Boltzmann's constant.)

If in this formula the integration variable

$$\frac{8\pi\nu^2 d\nu}{c^3}$$

is replaced by the more general expression

$$e_f \cdot \frac{d^3 k}{(2\pi)^3} \ ,$$

where for photons the degeneracy factor e_f has the value 2 and the function

$$\omega(\boldsymbol{k}) = c \cdot k$$

has to be used, then Planck's formula can be generalized.

For example, for the excitations of so-called *quasi-particles* resembling photons, e.g. for "phonons" or "magnons" (the quanta of sound waves and of spin waves, respectively) or for "plasmons" (quanta corresponding to oscillations of the charge density in solids), one has similar properties as for photons, but different dispersion relations (e.g. $\omega(k) = D \cdot k^2$ for *magnons in ferromagnets* and

$$\omega(k) = \omega_p + b \cdot k^2$$

for *plasmons*, where ω_p is the *plasma frequency* and b a positive factor.)

Thus in analogous manner to that for photons one obtains different results for the *internal energy* $U(T)$ of the quasi-particle gas considered. One sees here that the dispersion relation $\omega(\boldsymbol{k})$ of a wave plays an important role not only in optics but also in many other branches of physics.

21.2 From Wave Optics to Geometrical Optics; Fermat's Principle

By *avoiding abrupt changes* (e.g., by neglecting such quantities as

$$|\lambda \cdot \mathrm{grad} A(\boldsymbol{r})| \ ,$$

see below) one can start with *wave optics* and arrive at the field of *geometrical or ray optics* as follows.

Commencing with the wave equation

$$\left(\nabla^2 - \left(\frac{1}{c_m}\right)^2 \partial_{tt}\right)\Psi(\boldsymbol{r},t) = 0 \ , \tag{21.9}$$

where Ψ is one of the Cartesian components of the electromagnetic fields or one of the equivalent wave quantities considered[7], we try to solve this

---

[7] Neglecting the vectorial character of the electromagnetic field is already an approximation, the so-called *scalar approximation*. This can already be essential, which should not be forgotten.

equation with the usual stationary ansatz. Thus we are led to the following wave equation:

$$\left(\nabla^2 + k^2(\boldsymbol{r})\right) \psi(\boldsymbol{r}) \overset{!}{=} 0 \,, \tag{21.10}$$

with

$$k(\boldsymbol{r}) = \frac{\omega}{c_m} = k_0 \cdot n(\boldsymbol{r}) \,,$$

where $k_0 = \frac{\omega}{c_0}$ and $n(\boldsymbol{r})$ is the refractive index, which is a real quantity (see above). For cases involving dielectric losses or for Ohmic behavior, one should replace $n(\boldsymbol{r})$ by the complex quantity

$$\tilde{n} := n + \mathrm{i}\kappa \,, \quad \text{where} \quad \kappa^{-1}$$

is the absorption length.

The solution is now

$$\psi(\boldsymbol{r}) := A(\boldsymbol{r}) \cdot \exp \mathrm{i}k_0 \cdot S(\boldsymbol{r}) \,,$$

with the so-called *eikonal* $S(\boldsymbol{r})$. In spite of its strange name, which has mainly historical reasons, this function S is most important for *geometrical optics*; it has the dimensionality of an effective length, *viz* the minimum distance between two equivalent wave fronts.[8]

One now assumes that with the formation of the second derivatives the terms $\propto k_0^2$ dominate, e.g.,

$$\frac{\partial^2 \psi}{\partial x^2} \approx -k_0^2 \left(\frac{\partial S}{\partial x}\right)^2 \psi + \dots \,,$$

where the dots represent neglected terms, which are not proportional to k_0^2, but only to the first or zeroth power of k_0.[9] In this way the wave equation (21.10) is systematically replaced by the so-called *eikonal equation*[10]

$$\left(\mathrm{grad}S(\boldsymbol{r})\right)^2 \approx n(\boldsymbol{r})^2 \,. \tag{21.11}$$

Here the surfaces $S(\boldsymbol{r}) = constant$ describe the *wave fronts*, and their gradients describes the *ray directions*.

The eikonal approximation of a scalar wave equation, derived from Maxwell's theory, is the basis of geometrical (or ray) optics. In particular one can derive from it *Fermat's principle* of the "shortest optical path":

$$\int_{\boldsymbol{r}_1}^{\boldsymbol{r}_2} \mathrm{d}l n(\boldsymbol{r}) \overset{!}{=} \min \tag{21.12}$$

---

[8] The *eikonal* function $S(\boldsymbol{r})$ should of course not be confused with the *Poynting vector* $\boldsymbol{S}$ or the *entropy* $S(T)$ of statistical physics (Part IV). In each of these cases the same letter S is used.

[9] Note that k_0 is not small, but large ($k_0 = \frac{2\pi}{\lambda_0}$, where λ_0 is the wavelength).

[10] A similar approximation leads from Schrödinger's wave equation of quantum mechanics to the Hamilton-Jacobi equations of classical mechanics (see below).

for the real path of a ray of light. Due to lack of space, further details on Fermat's principle will not be described here. We only mention that not only Snell's law of refraction (see above), but essentially the whole of *lense optics*, including optical microscopy, can be derived from it.[11]

The *eikonal approximation* is significant because it contains the transition from (i) the *wave picture* of light, explored by Huygens and Young (with the basic issue of the ability to *interfere*; → *holography*, see below) to the (ii) *particle representation*.

In the particle representation, light rays without the ability to interfere are interpreted as particle trajectories, while the wave fronts $S(\boldsymbol{r})$ are interpreted as purely fictitious mathematical quantities. This is analogous to the transition from quantum mechanics to classical mechanics in *semiclassical theories*: one tries to solve the Schrödinger equation

$$-\frac{\hbar^2}{2m}\nabla^2\psi + V(\boldsymbol{r})\psi = E\psi ,$$

which plays the role of a *matter-wave equation*, by a kind of *eikonal ansatz*,

$$\psi(\boldsymbol{r}) := A(\boldsymbol{r}) \cdot \exp\frac{i}{\hbar}S(\boldsymbol{r}) .$$

(Here $S(\boldsymbol{r})$ has the dimensionality of *action*, i.e., the same dimensionality as Planck's constant h ($\hbar$ is Planck's constant divided by (2π).) An expansion in powers of $\frac{1}{\hbar}$ (taking into account only the dominant terms analogously to the *eikonal ansatz*) leads to the so-called Hamilton-Jakobi equations of classical mechanics, *viz*

$$\frac{(\mathrm{grad}S)^2}{2m} + V(\boldsymbol{r}) \equiv E .$$

The Hamilton-Jakobi equation contains the totality of classical mechanics, e.g., the Hamilton function, the canonical equations and also *Hamilton's principle of least action*, which ultimately corresponds to *Fermat's principle*.[12]

---

[11] See, e.g., Fliessbach's book on electrodynamics, problem 36.2

[12] The equivalence can be rather easily shown: Hamilton's principle says that $\int_{t_1}^{t_2} dt\mathcal{L} \overset{!}{\equiv}$ extremal for variation of all virtual orbits in the space of generalized coordinates q_i, for fixed initial and final coordinates. (The momenta $p_i \equiv \partial\mathcal{L}/\partial\dot{q}_i$ result implicitly). Using a Legendre transformation, one replaces $\mathcal{L}(q_i,\dot{q}_i) \rightarrow \sum_i p_i\dot{q}_i - \mathcal{H}$, with the Hamilton function $\mathcal{H}(p_i,q_k)$, where the generalized coordinates and momenta are varied independently. Meanwhile one substitutes $\mathcal{H}$ by the constant E, i.e., one does not vary all orbits in phase space, but only those with constant $\mathcal{H} \equiv E$. In this way one obtains the so-called Maupertuis principle (→ Landau-Lifshitz I, [13], Chap. 44) of classical mechanics. In the field of optics, it corresponds exactly to Fermat's principle.

21.3 Crystal Optics and Birefringence

For (i) fluids (gases and liquids), (ii) polycrystalline solids, (iii) amorphous substances and (iv) cubic crystals, the dielectric displacement $\boldsymbol{D}$ is *proportional* to $\boldsymbol{E}$, i.e., as assumed hitherto:

$$\boldsymbol{D} = \varepsilon_r \varepsilon_0 \boldsymbol{E} \ .$$

In contrast, for non-cubic crystals plus all solid systems under strong uniaxial tension or compression, the behavior is more complicated. For such systems we have (for $i = 1, 2, 3$)

$$D_i = \sum_{k=1}^{3} \varepsilon_{i,k} \varepsilon_0 E_k \ , \tag{21.13}$$

where the *dielectric constants* $\varepsilon_{i,k}$ now form a *tensor*.

In the following we shall assume (i) that we are dealing with nonmagnetic material, such that the relative permeability is 1, and (ii) that there are no magnetic fields, such that

$$\varepsilon_{i,k} \equiv \varepsilon_{k,i}$$

The tensor $\varepsilon_{i,k}$ can be diagonalized by a suitable rotation, i.e., in the new orthogonal basis it has the diagonal representation

$$\varepsilon_{\alpha,\beta} = \begin{pmatrix} \varepsilon_1 \ , \ 0 \ , \ 0 \\ 0 \ , \ \varepsilon_2 \ , \ 0 \\ 0 \ , \ 0 \ , \ \varepsilon_3 \end{pmatrix} \ . \tag{21.14}$$

The quadratic form

$$w_{\mathrm{e}} = \frac{1}{2} \boldsymbol{E} \cdot \boldsymbol{D} \equiv \frac{\varepsilon_0}{2} \sum_{i,k=1}^{3} \varepsilon_{i,k} E_i E_k$$

is thus diagonal in the eigenvector basis, i.e., there are analogous relations as for the mechanical rotational energy of a rigid body (see Part I).

The relation

$$w_{\mathrm{e}} = \frac{1}{2} \boldsymbol{E} \cdot \boldsymbol{D}$$

for the energy density of the electric field corresponds in fact to the mechanical relation

$$T_{\mathrm{Rot}} = \frac{1}{2} \boldsymbol{\omega} \cdot \boldsymbol{L} \ ,$$

and the mechanical relation

$$L_i = \sum_{k=1}^{3} \Theta_{i,k} \omega_k \ ,$$

with the components $\Theta_{i,k}$ of the *inertia tensor*, corresponds to the electric relation

$$D_i/\varepsilon_0 = \sum_k \varepsilon_{i,k} E_k \; ;$$

i.e., $\boldsymbol{\omega}$, $\boldsymbol{L}$, and the *inertia ellipsoid* of classical mechanics correspond to the quantities $\boldsymbol{E}$, $\boldsymbol{D}$ and the $\varepsilon_{i,k}$-*ellipsoid* of crystal optics.

However, as in *mechanics*, where in addition to the *inertia ellipsoid* ($\boldsymbol{\omega}$-ellipsoid) there is also an equivalent second form, the so-called *Binet ellipsoid* ($\boldsymbol{L}$-ellipsoid), which can be used to express the rotational energy in the two equivalent ellipsoid forms

$$2 \cdot T_{\mathrm{Rot}} \overset{(i)}{=} \Theta_1 \omega_1^2 + \Theta_2 \omega_2^2 + \Theta_3^2 \omega_3^2 \overset{(ii)}{=} \frac{L_1^2}{\Theta_1} + \frac{L_2^2}{\Theta_2} + \frac{L_3^2}{\Theta_3} \;,$$

in the field of crystal optics it is also helpful to use two methods. One may prefer either (i) the $\boldsymbol{E}$-ellipsoid or (ii) the equivalent $\boldsymbol{D}$-ellipsoid, as follows: (i) if $\boldsymbol{E}$ and/or the Poynting vector $\boldsymbol{S}(= \boldsymbol{E} \times \boldsymbol{H})$ are preferred, then one should use the *Fresnel ellipsoid* ($\boldsymbol{E}$-ellipsoid). On the other hand, (ii), if $\boldsymbol{D}$ and/or the propagation vector $\boldsymbol{k}$ are involved, then one should use the *index ellipsoid* ($\boldsymbol{D}$-ellipsoid). These two ellipsoids are:

$$w_e \overset{(i)}{=} \frac{\varepsilon_0}{2} \cdot \left(\varepsilon_1 E_1^2 + \varepsilon_2 E_2^2 + \varepsilon_3^2 E_3^2 \right) \overset{(ii)}{=} \frac{1}{2\varepsilon_0} \cdot \left(\frac{D_1^2}{\varepsilon_1} + \frac{D_2^2}{\varepsilon_2} + \frac{D_3^2}{\varepsilon_3} \right) \;. \qquad (21.15)$$

For a given $\boldsymbol{E}$, the direction of the vector $\boldsymbol{D}$ is obtained in a similar way to that in the mechanics of rigid bodies. In that case the direction of the vector $\boldsymbol{L}$ for a given $\boldsymbol{\omega}$ must be determined by means of a *Poinsot construction*: $\boldsymbol{L}$ has the direction of the *normal* to the tangential plane belonging to $\boldsymbol{\omega}$, tangential to the inertia ellipsoid. Analogously, in the present case, the direction of $\boldsymbol{D}$ has to be determined by a Poinsot construction w.r.t. $\boldsymbol{E}$, i.e., by constructing the normal to the tangential plane of the $\boldsymbol{E}$-ellipsoid, and *vice versa*. In general, the vector $\boldsymbol{E}$ is thus rotated with respect to the direction of $\boldsymbol{D}$. (The vector $\boldsymbol{S}$ is rotated with respect to the vector $\boldsymbol{k}$ by exactly the same amount about the same rotation axis, $\boldsymbol{H}$, see below.)

In fact, the three vectors $\boldsymbol{S}$, $\boldsymbol{E}$ and $\boldsymbol{H}$ form a right-handed trihedron ($\boldsymbol{S} \equiv \boldsymbol{E} \times \boldsymbol{H}$), similar to the three vectors $\boldsymbol{k}$, $\boldsymbol{D}$ and $\boldsymbol{H}$. These statements follow from Maxwell's equations: the monochromatic *ansatz*

$$\boldsymbol{D} \propto \boldsymbol{D_k} e^{\mathrm{i}(\boldsymbol{k} \cdot \boldsymbol{r} - \omega t)} \;,$$

plus analogous assumptions for the other vectors, implies with

$$\mathrm{div}\,\boldsymbol{D} = 0 \quad \text{that} \quad \boldsymbol{k} \cdot \boldsymbol{D_k} \equiv 0 \;.$$

Similarly

$$\mathrm{div}\,\boldsymbol{B} = 0 \quad \text{implies that} \quad \boldsymbol{k} \cdot \boldsymbol{H_k} \equiv 0 \;.$$

Furthermore the relation

$$\operatorname{curl} \boldsymbol{H} = \frac{\partial \boldsymbol{D}}{\partial t} \quad \text{implies that} \quad \boldsymbol{k} \times \boldsymbol{H_k} = -\omega \boldsymbol{D_k} \; ;$$

and with

$$\operatorname{curl} \boldsymbol{E} = -\frac{\partial \boldsymbol{B}}{\partial t} \quad \text{it follows that} \quad \boldsymbol{k} \times \boldsymbol{E_k} = \mu_0 \omega \boldsymbol{H_k} \; .$$

The four vectors $\boldsymbol{D}$, $\boldsymbol{E}$, $\boldsymbol{k}$ and $\boldsymbol{S}$ thus all lie in the same plane perpendicular to $\boldsymbol{H}$. The vectors $\boldsymbol{D}$ and $\boldsymbol{k}$ (as well as $\boldsymbol{E}$ and $\boldsymbol{S}$) are perpendicular to each other. $\boldsymbol{E}$ originates from $\boldsymbol{D}$ (and $\boldsymbol{S}$ from $\boldsymbol{k}$) by a rotation with the same angle about the common rotation axis $\boldsymbol{H}$[13].

One therefore distinguishes between (i) the *phase velocity* $\boldsymbol{v}_{\text{phase}}$, which is immediately related to the wave-propagation vector $\boldsymbol{k}$, and (ii) the so-called *ray velocity* $\boldsymbol{v}_{\text{ray}}$, which is instead related to the energy-current density vector $\boldsymbol{S}$. (These different velocities correspond to the *phase* and *group* velocities discussed above.)

In most cases *one* of the three principal axes of the *dielectric tensor*, the *c-axis*, is crystallographically distinguished, and the other two orthogonal axes are equivalent, i.e., one is dealing with the symmetry of an *ellipsoid of revolution.*

This is true for tetragonal, hexagonal and trigonal[14] crystal symmetry. Only for orthorhombic[15], monoclinic and triclinic crystal symmetry can all three eigenvalues of the dielectricity tensor be different. (This exhausts all crystal classes). In the first case one is dealing with so-called *optically uniaxial* crystals, in the latter case with *optically biaxial* crystals.

We have now arrived at the phenomenon of *birefringence*, and as a simple example we shall consider a linearly polarized monochromatic electromagnetic wave with wavenumber $\boldsymbol{k}$, incident at right angles from a vacuum onto the surface of a non-cubic crystal. In the interior of the crystal, $\boldsymbol{k}$ must also be perpendicular to the surface. For fixed $\boldsymbol{k}$ we should use the $\boldsymbol{D}$-ellipsoid (*index ellipsoid*) which is assumed (by the symmetry of the crystal) in general to be *inclined* to the direction of incidence, i.e., the propagation vector $\boldsymbol{k}$ is not necessarily parallel to a *principal axis* of the *dielectric tensor*.

The values of $\boldsymbol{D_k}$ corresponding to the vector $\boldsymbol{k}$ are all located on an *elliptical section* of the *index ellipsoid* with the *plane perpendicular to $\boldsymbol{k}$* through the origin.

---

[13] With this (somewhat rough) formulation we want to state that the *unit vector* $\hat{E}$ originates from $\hat{D}$ by the above-mentioned rotation.

[14] This corresponds to the symmetry of a distorted cube stressed along one space diagonal; the *c*-axis is this diagonal, and in the plane perpendicular to it one has trigonal symmetry.

[15] This corresponds to the symmetry of a cube, which is differently strained in the three edge directions.

On the one hand, the vector $\boldsymbol{E}$ should be perpendicular to the tangential plane of the *index ellipsoid viz* by the *Poinsot construction*; on the other hand it should belong to the plane defined by $\boldsymbol{D}$ and $\boldsymbol{k}$. As one can show, these two conditions can only be satisfied if the direction of $\boldsymbol{D}$ is a principal direction of the above-mentioned *section*. This allows only two (orthogonal) polarization directions of $\boldsymbol{D}$; thus the two corresponding sets of dielectric constants are also fixed. In general *they are different from each other* and the corresponding phase velocities,

$$c_P = \frac{c_0}{\sqrt{\varepsilon(\boldsymbol{D})}} \ ,$$

differ as well. (In addition, in general, the *ray velocities* ($\hat{=}$ *group velocities*) are different from the *phase velocities*, see above; i.e., two different *ray velocities* also arise.)

Usually the incident wave has contributions from both polarizations. As a consequence, even if in vacuo the wave has a unique linear polarization direction (not parallel to a principal axis of the dielectric tensor), in the interior of the crystal generally a superposition of two orthogonal linearly polarized components arises, which propagate with different velocities.

The phenomenon becomes particularly simple if one is dealing with *optically uniaxial* systems. In this case the index ellipsoid is an *ellipsoid of revolution*, i.e., with two identical dielectric constants $\varepsilon_1 \equiv \varepsilon_2$ and a different value ε_3. Under these circumstances one of the two above-mentioned polarization directions of the vector $\boldsymbol{D}$ can be stated immediately, *viz* the direction of the plane corresponding to

$$\boldsymbol{k} \times \boldsymbol{e}_3 \ .$$

For this polarization one has simultaneously $\boldsymbol{E} \sim \boldsymbol{D}$ (i.e., also $\boldsymbol{S} \sim \boldsymbol{k}$), i.e., one is dealing with totally usual relations as in a vacuum (the so-called *ordinary beam*). In contrast, for orthogonal polarization the vectors $\boldsymbol{E}$ and $\boldsymbol{D}$ (and $\boldsymbol{S}$ and $\boldsymbol{k}$) have different directions, so that one speaks of an *extraordinary beam*.

If the phase-propagation vector $\boldsymbol{k}$ is, e.g., in the (x_1, x_3)-plane under a general angle, then the *in-plane polarized* wave is *ordinary*, whereas the wave polarized perpendicular to the plane is *extraordinary*. In the limiting cases where (i) $\boldsymbol{k}$ is $\sim \boldsymbol{e}_3$ both waves are *ordinary*, whereas if (ii) $\boldsymbol{k} \sim \boldsymbol{e}_1$ both polarizations would be *extraordinary*.

For *optically biaxial crystals* the previous situation corresponds to the general case.

It remains to be mentioned that for *extraordinary* polarizations not only the directions but also the *magnitudes* of *phase* and *ray velocities* are different, *viz*

$$\boldsymbol{v}_{\mathrm{phase}} \propto \boldsymbol{k} \ , \quad \text{with} \quad |\boldsymbol{v}_{phase}| = \frac{c_0}{\sqrt{\varepsilon(\boldsymbol{D})}}$$

in the first case, and

$$v_S \propto S \quad \text{(for} \quad S = E \times H), \quad \text{with} \quad |v_S| = \frac{c_0}{\sqrt{\varepsilon(E)}}$$

in the second case. In the first case we have to work with the D-ellipsoid (index ellipsoid), in the second case with the E-ellipsoid (Fresnel ellipsoid).

21.4 On the Theory of Diffraction

Diffraction is an important wave-optical phenomenon. The word alludes to the fact that it is not always possible to keep rays together. This demands a mathematically precise description. We begin by outlining Kirchhoff's law, which essentially makes use of Green's second integral theorem, a variant of Gauss's integral theorem. It starts from the identity:

$$\{u(r)\nabla^2 v(r) - v(r)\nabla^2 u(r)\} \equiv \nabla \cdot (u\nabla v - v\nabla u),$$

and afterwards proceeds to

$$\iiint_V dV \{u\nabla^2 v - v\nabla^2 u\} \equiv \oiint_{\partial V} d^2 A\, n \cdot (u\nabla v - v\nabla u). \qquad (21.16)$$

This expression holds for continuous real or complex functions $u(r)$ and $v(r)$, which can be differentiated at least twice, and it even applies if on the l.h.s. of (21.16) the operator ∇^2 is replaced by similar operators, e.g.,

$$\nabla^2 \to \nabla^2 + k^2.$$

Kirchhoff's law (or "2nd law", as it is usually called) is obtained by substituting

$$u(r) = \frac{\exp(ik|r - r'|)}{|r - r'|} \quad \text{and} \quad v(r) := \psi(r)$$

into (21.16). Here r' *is an arbitrary point in the interior of the (essentially hollow) volume V* (see below). As a consequence we have

$$\left(\nabla^2 + k^2\right) v(r) \equiv 0 \quad \text{and} \quad \left(\nabla^2 + k^2\right) u(r) \equiv -4\pi\delta(r - r').$$

Thus *Kirchhoff's 2nd law* states rigorously that

$$\psi(r') \equiv \oiint_{\partial V} \frac{d^2 A}{4\pi} n(r) \cdot \left(\frac{e^{ik|r-r'|}}{|r - r'|}\nabla\psi(r) - \psi(r)\nabla\frac{e^{ik|r-r'|}}{|r - r'|}\right). \qquad (21.17)$$

We assume here that the volume V is illuminated externally and that ∂V contains an opening (or aperture) plus a "wall". Only a small amount of the light is *diffracted* from the aperture to those regions within the interior of V in the geometrical shadow. Therefore it is plausible to make the following approximations of (21.17)[16]:

---

[16] a prerequisite is that the involved distances are $\gg \lambda$.

a) only the *aperture* contributes to the integral in equation (21.17), and
b) within the *aperture* one can put

$$\psi(\boldsymbol{r}) \approx \frac{\mathrm{e}^{\mathrm{i}k|\boldsymbol{r}-\boldsymbol{r}_Q|}}{|\boldsymbol{r}-\boldsymbol{r}_Q|} \,.$$

Here $\boldsymbol{r}_Q$ is the position of a *point* light source outside the volume, which illuminates the *aperture*. This is the case of so-called *Fresnel diffraction*. (In the field of applied seismics, for example, one may be dealing with a point source of seismic waves produced by a small detonation. Fresnel diffraction is explained in Fig. 21.2 below.)
We then have

$$\psi(\boldsymbol{r}_P) \approx \frac{-\mathrm{i}k}{2\pi} \iint_{\text{aperture}} \mathrm{d}^2 A \frac{\mathrm{e}^{\mathrm{i}k r_Q}}{r_Q} \cdot \frac{\mathrm{e}^{\mathrm{i}k r_P}}{r_P} \,. \qquad (21.18)$$

Here r_Q and r_P are the distances between the source point $\boldsymbol{r}_Q$ on the left and the integration point $\boldsymbol{r}$, and between integration point $\boldsymbol{r}$ and observation point

$$r_P \equiv r' \,.$$

(Note that $\nabla_{\boldsymbol{r}} = -\nabla_{\boldsymbol{r}_Q}$).
c) The situation becomes particularly simple when dealing with a planar *aperture* illuminated by a *plane wave*

$$\propto \mathrm{e}^{\mathrm{i}\boldsymbol{k}_0 \cdot \boldsymbol{r}}$$

(so-called *Fraunhofer diffraction*). We now have

$$|\boldsymbol{k}_0| \stackrel{!}{=} k \,,$$

and for a point of observation

$$r' = r_P$$

behind the boundary ∂V (i.e., not necessarily behind the *aperture*, but possibly somewhere in the shadow behind the "wall"):

$$\psi(\boldsymbol{r}_P) \approx f \mathrm{e}^{\mathrm{i}\boldsymbol{k}_0 \cdot \boldsymbol{r}} \iint_{\text{aperture}} \mathrm{d}^2 A \frac{\exp(\mathrm{i}k|\boldsymbol{r}-\boldsymbol{r}_P|)}{|\boldsymbol{r}-\boldsymbol{r}_P|} \qquad (21.19)$$

Here f is a (noninteresting) factor.

Equation (21.19) is an explicit and particularly simple form of *Huygens' principle*: Every point of the *aperture* gives rise to spherical waves, whose effects are superimposed.

Two standard problems, which are special cases of (21.19) and (21.18), should now be mentioned:

a) *Fraunhofer diffraction at a single slit* (and with *interference*, at a *double slit*). This case, which is discussed in almost all textbooks on optics, will be treated later.
b) *Fresnel diffraction at an edge*. This problem is also important in the field of reflection seismology when, for example, there is an abrupt shift in the rock layers at a *fault*.

21.4.1 Fresnel Diffraction at an Edge; Near-field Microscopy

In the following we shall consider *Fresnel diffraction at an edge*.

Fresnel diffraction means that one is dealing with a point source. The surface in shadow is assumed to be the lower part of a semi-infinite vertical plane, given as follows:

$$x \equiv 0 , \quad y \in (-\infty, \infty) , \quad z \in (-\infty, 0] .$$

This vertical half-plane, with a sharp edge $z \equiv 0$, is illuminated by a point source

$$\boldsymbol{r}_Q := (-x_Q, y_Q, 0)$$

from a position perpendicular to the plane at the height of the edge, i.e., x_Q is assumed to be the (positive) perpendicular distance from the illuminating point to the edge. Additionally we assume

$$y_Q^2 \ll x_q^2 , \quad \text{while} \quad y_P \equiv 0 .$$

The point of observation behind the edge is

$$\boldsymbol{r}_P := (x_P, 0, z_P) ,$$

where we assume $x_P > 0$, whereas z_P can be negative. In this case the point of observation would be in the shadow; otherwise it is directly illuminated. All distances are assumed to be $\gg \lambda$. The situation is sketched in Fig. 21.2.

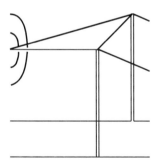

Fig. 21.2. Schematic diagram to illustrate Fresnel diffraction. In the diagram (which we have intentionally drawn without using coordinates; see the text), rays starting on the l.h.s. from a point source $\boldsymbol{Q}$ (e.g., $\boldsymbol{Q} = (-x_Q, y_Q, z_Q)$) are diffracted at the edge of a two-dimensional half-plane (e.g., $(0, y, z)$, with $z \leq z_Q$), from where waves proceed to the observation point $\boldsymbol{P}$, which belongs to the three-dimensional space behind the plane (e.g., (x_P, y_P, z_P), with $x_P > 0$), for example, into the shadow region (e.g., $z_P < z_Q$). In the directly illuminated region one observes so-called *fringes*, as explained in the text

We then have for small values of y^2 and z^2 :

$$\psi(\boldsymbol{r}_P) \sim \int\limits_{-\infty}^{\infty} \mathrm{d}y \int\limits_{0}^{\infty} \mathrm{d}z \exp\left[\mathrm{i}k \cdot \left(x_Q + \frac{(y - y_Q)^2 + z^2}{2x_q} + x_P \right.\right.$$
$$\left.\left. + \frac{y^2 + (z - z_P)^2}{2x_p} \right)\right] , \qquad (21.20)$$

i.e., apart from a constant complex factor

$$\psi(\boldsymbol{r}_P) \sim \int\limits_{0}^{\infty} \mathrm{d}z \exp\left[\mathrm{i}k \cdot \left(\frac{z^2}{2x_Q} + \frac{(z - z_P)^2}{2x_P} \right)\right] . \qquad (21.21)$$

By substitution, this result can be written (again apart from a complex factor[17] of order of magnitude 1)

$$\psi(\boldsymbol{r}_P) \sim \int\limits_{-w}^{\infty} \mathrm{d}\eta \, e^{\mathrm{i}\eta^2} , \qquad (21.22)$$

with the *w-parameter*

$$w := z_P \cdot \sqrt{\frac{k \cdot x_q}{2x_P(x_P + x_Q)}} . \qquad (21.23)$$

One thus obtains for the intensity at the point of observation:

$$I = \frac{I_0}{2} \left| \sqrt{\frac{2}{\pi}} \int\limits_{-w}^{\infty} \mathrm{d}\eta \, e^{\mathrm{i}\eta^2} \right|^2 . \qquad (21.24)$$

The real and imaginary parts of the integral $I(w)$ appearing in (21.24) define the *Fresnel integrals* $C(w)$ and $S(w)$:

$$C(w) + \frac{1}{2} := \sqrt{\frac{2}{\pi}} \int\limits_{-w}^{\infty} \cos(\eta^2)\mathrm{d}\eta , \quad S(w) + \frac{1}{2} := \sqrt{\frac{2}{\pi}} \int\limits_{-w}^{\infty} \sin(\eta^2)\mathrm{d}\eta , \quad \text{or}$$

$$C(w) = \sqrt{\frac{2}{\pi}} \int\limits_{0}^{w} \cos(\eta^2)\mathrm{d}\eta , \qquad S(w) = \sqrt{\frac{2}{\pi}} \int\limits_{0}^{w} \sin(\eta^2)\mathrm{d}\eta ,$$

and the closely related *Cornu spiral*, which is obtained by plotting $S(w)$ over $C(w)$, while w is the line parameter of the spiral; cf. Sommerfeld, [14], or Pedrotti *et al.*, Optics, [15], Fig. 18.17[18].

---

[17] Landau-Lifshitz II (*Field theory* (sic)), chapter 60
[18] Somewhat different, but equivalent definitions are used by Hecht in [16].

An asymptotic expansion of (21.24), given, e.g., in the above-mentioned volume II of the textbook series by Landau and Lifshitz, [8], yields

$$\frac{I(w)}{I_0} \cong \begin{cases} \frac{1}{4\pi w^2} & \text{for} \quad w \ll 0 \\ 1 + \sqrt{\frac{1}{\pi}} \frac{\sin(w^2 - \frac{\pi}{4})}{w^2} & \text{for} \quad w \gg 0 \end{cases} . \qquad (21.25)$$

A more detailed calculation also yields intermediate behavior, i.e., a smooth function that does not jump discontinuously from 1 to 0, when the geometrical shadow boundary is crossed, but which increases monotonically from $(I/I_0) = 0$ for $w = -\infty$ (roughly $\sim 1/w^2$) up to a maximum amplitude

$$(I/I_0) \approx 1.37 \; for \; w \approx \sqrt{\frac{3\pi}{4}}$$

(The characteristic length scale for this monotonic increase is of the order of half a wavelength), and then oscillates about the asymptotic value 1, with decreasing amplitude and decreasing period: In this way *fringes* appear near the shadow boundary on the positive side (cf. Fig. 12 in Chap. 60 of volume II of the textbook series by Landau and Lifshitz, [8]), i.e. with an envelope-decay length Δz_P which obeys the equation

$$\frac{(\Delta z_P)^2 x_Q}{\lambda 2 x_P \cdot (x_P + x_Q)} \equiv 1 \,,$$

and which is therefore not as small as one might naively believe (in particular it can be significantly larger than the characteristic length $\frac{\lambda}{2}$ for the above-mentioned monotonic increase), but which is

$$\Delta z_P := \sqrt{2\lambda \cdot (x_Q + x_P) \cdot \frac{x_P}{x_Q}} \,.$$

Thus, even if the edge of the shadowing plane were atomically sharp, the optical image of the edge would not only be *unsharp* (as naively expected) on the scale of a typical "decay length" of the light (i.e., approximately on the scale of $\frac{\lambda}{2}$, where λ is the wavelength) but also on the *scale of unsharpness* Δz_P, which is significantly larger[19]. The signal would thus be *alienated* and *disguised* by the above long-wavelength fringes. In our example from seismology, however, a disadvantage can be turned into an advantage, because from the presence of fringes a fault can be discovered.

---

[19] This enlargement of the scale of unsharpness through the oscillating sign of the fringes is a similar effect to that found in statistics, where for homogeneous cases one has a $1/N$-behavior of the error whereas for random signs this changes to $1/\sqrt{N}$-behavior, which is significantly larger.

The fact that the accuracy Δx of optical mappings is roughly limited to $\frac{\lambda}{2}$ follows essentially from the above relations. In expressions of the form

$$\exp(\mathrm{i}k \cdot \Delta x)$$

the phases should differ by π, if one wants to resolve two points whose positions differ by Δx. With

$$k = \frac{2\pi}{\lambda} \quad \text{this leads to} \quad \Delta x \approx \frac{\lambda}{2} \,.$$

This limitation of the accuracy in optical microscopy is essentially based on the fact that in microscopy usually only the far-field range of the electromagnetic waves is exploited.

Increased accuracy can be gained using *near-field microscopy* (SNOM $\equiv$ *Scanning Near Field Optical Microscopy*). This method pays for the advantage of better resolution by severe disadvantages in other respects[20], i.e., one has to scan the surface point by point with a sharp micro-stylus: electromagnetic fields evolve from the sharp point of the stylus. In the far-field range they correspond to electromagnetic waves of wavelength λ, but in the near-field range they vary on much shorter scales.

21.4.2 Fraunhofer Diffraction at a Rectangular and Circular Aperture; Optical Resolution

In the following we shall treat Fraunhofer diffraction, at first very generally, where we want to show that in the transverse directions a *Fourier transformation* is performed. Apart from a complex factor, (for distances $\gg \lambda$) the equality (21.19) is identical with

$$\psi(\boldsymbol{r}_P) \propto \iint_{\text{aperture}} \mathrm{d}^2 r \, \mathrm{e}^{\mathrm{i}(\boldsymbol{k}_0 - \boldsymbol{k}_P) \cdot \boldsymbol{r}} \,. \tag{21.26}$$

(If the aperture, analogous to an eye, is filled with a so-called "pupil function" $P(\boldsymbol{r})$, instead of (21.26) one obtains a slightly more general expression:

$$\psi \propto \iint_{\text{aperture}} \mathrm{d}^2 r \, P(\boldsymbol{r}) \mathrm{e}^{\mathrm{i}(\boldsymbol{k}_0 - \boldsymbol{k}_P) \cdot \boldsymbol{r}} \,.)$$

Here $\boldsymbol{k}_P$ is a vector of *magnitude* k and *direction* $\boldsymbol{r}_P$, i.e., it is true for $r_P \gg r$ that

$$\exp(\mathrm{i}k|\boldsymbol{r} - \boldsymbol{r}_P|) \cong \exp[\mathrm{i}k r_P - \mathrm{i}(\boldsymbol{r}_p \cdot \boldsymbol{r})/r_P] \,,$$

---

[20] There is some kind of *conservation theorem* involved in these and other problems, i.e. again the theorem of *conserved effort*.

such that apart from a complex factor, the general result (21.19) simplifies to the Fourier representation (21.26).

For the special case of a perpendicularly illuminated *rectangular* aperture in the (y, z)-plane one sets

$$\mathbf{k}_0 = k \cdot (1, 0, 0)$$

and also

$$\mathbf{k}_P = k \cdot \left(\sqrt{1 - \sin^2 \theta_2 - \sin^2 \theta_3}, \quad \sin \theta_2, \quad \sin \theta_3 \right),$$

and obtains elementary integrals of the form

$$\int_{-a_j/2}^{a_j/2} \mathrm{d}y_j e^{-i(\sin \theta_j) \cdot y_j}.$$

In this way one finds

$$\psi(\mathbf{r}_P) \propto a_2 a_3 \cdot \prod_{j=2}^{3} \frac{(\sin \theta_j) \cdot \frac{a_j \pi}{\lambda}}{\frac{a_j \pi}{\lambda}}. \tag{21.27}$$

The intensity is obtained from $\psi \cdot \psi^*$.

For a *circular* aperture with radius a one obtains a slightly more complicated result:

$$\psi(\mathbf{r}_P) \propto \pi a^2 \cdot \frac{2 J_1 \left[2\pi \cdot (\sin \theta) \cdot \frac{a}{\lambda} \right]}{2\pi \cdot (\sin \theta) \cdot \frac{a}{\lambda}}, \tag{21.28}$$

where $J_1[x]$ is a Bessel function. In this case the intensity has a sharp maximum at $\sin \theta = 0$ followed by a first minimum at

$$\sin \theta = 0.61 \frac{\lambda}{a},$$

so that the angular resolution for a telescope with an aperture a is limited by the *Abbé result*

$$\sin \theta \geq 0.61 \frac{\lambda}{a}.$$

In this case too, diffraction effects limit the resolution to approximately $\frac{\lambda}{2}$.

21.5 Holography

Hitherto we have not used the property that electromagnetic waves can *interfere* with each other, i.e., the property of *coherence*:

$$\left| \sum_j \psi_j \right|^2 \equiv \left| \sum_{j,k} \psi_j^* \psi_k \right| , \quad \text{and not simply} \quad \equiv \sum_j |\psi_j|^2 .$$

(The last expression – addition of the *intensities* – would (in general) only be true, if the phases and/or the complex amplitudes of the terms ψ_j were *uncorrelated random numbers*, such that from the double sum $\sum_{j,k} \psi_j^* \psi_k$ only the diagonal terms remained.)

If the spatial and temporal correlation functions of two wave fields,

$$C_{1,2} := \left\langle \left(\psi^{(1)} \right)^* (\boldsymbol{r}_1, t_1) \psi^{(2)} (\boldsymbol{r}_2, t_2) \right\rangle$$
$$\equiv \sum_{\boldsymbol{k}_1, \boldsymbol{k}_2} \left(c_{\boldsymbol{k}_1}^{(1)} \right)^* e^{-i(\boldsymbol{k}_1 \cdot \boldsymbol{r}_1 - \omega t_1)} c_{\boldsymbol{k}_2}^{(2)} e^{i(\boldsymbol{k}_2 \cdot \boldsymbol{r}_2 - \omega t_2)} , \qquad (21.29)$$

decay exponentially with increasing spatial or temporal distance, *viz*

$$C_{1,2} \propto e^{-\left(\frac{|\boldsymbol{r}_1 - \boldsymbol{r}_2|}{l_c} + \frac{(t_1 - t_2)}{\tau_c} \right)} ,$$

then the decay length l_c and decay time τ_c are called the *coherence length* and *coherence time*, respectively.

Due to the invention of the laser one has gained light sources with macroscopic coherence lengths and coherence times. (This subject cannot be treated in detail here.)

For photographic records the blackening is proportional to the *intensity*, i.e., one loses the information contained in the *phase* of the wave. Long before the laser was introduced, the British physicist Dennis Gabor (in 1948) found a way of keeping the phase information intact by *coherent superposition of the original wave with a reference wave*, i.e., to *reconstruct* the *whole* ($\to$ holography) original wave field $\psi_G(\boldsymbol{r}, t)$[21] from the intensity signal recorded by photography. However, it was only later (in 1962) that the physicists Leith and Upatnieks at the university of Michigan used laser light together with the "off-axis technique" (i.e., *oblique illumination*) of the reference beam, which is still in use nowadays for conventional applications[22].

---

[21] This means that $\psi_G(\boldsymbol{r}, t)$ is obtained by illuminating the object with *coherent light*.

[22] Nowadays there are many conventional applications, mainly in the context of security of identity cards or banknotes; this is treated, e.g., in issue 1, page 42, of the German "Physik Journal", 2005, see [17]

The conventional arrangement for recording optical holograms is presented, for example, in Fig. 13.2 of the textbook by Pedrotti and coworkers. The hologram, using no lenses at all (!), measures the blackening function of a photographic plate, which is illuminated simultaneously by an *object wave* and a *reference wave*.

As a consequence, the intensity I corresponds to the (coherent) superposition of (i) the object wavefunction ψ_G and (ii) the reference wave

$$\sim e^{i \boldsymbol{k}_0 \cdot \boldsymbol{r}} , \quad \text{i.e.,} \quad I \propto |\psi_G + a_0 \exp{(i(\boldsymbol{k}_0 \cdot \boldsymbol{r}))}|^2 .$$

In particular, the intensity I does not depend on time, since all waves are

$$\propto e^{-i\omega t} .$$

If the terms $\propto a_0$, $\propto a_0^*$ and $\propto |a_0|^2$ dominate, one thus obtains

$$I \propto |a_0|^2 + a_0 e^{i \boldsymbol{k}_0 \cdot \boldsymbol{r}} \psi_G(\boldsymbol{r}) + a_0^* e^{-i \boldsymbol{k}_0 \cdot \boldsymbol{r}} \psi_G^*(\boldsymbol{r}) (\to I_{\text{hologram}}) , \qquad (21.30)$$

where additionally one has to multiply with the blackening function of the photographic emulsion.

The second term on the r.h.s. of (21.30) is just the object wave field while the third term is some kind of "conjugate object field" or "twin field", which yields a strongly disguised picture related to the object (e.g., by transition to complex-conjugate numbers, $e^{i\Phi(\boldsymbol{r})} \to e^{-i\Phi(\boldsymbol{r})}$, points with lower coordinates and points with higher coordinates are interchanged).

It is thus important to *view* the hologram in such a way that out of the photograph of the *whole* intensity field I_{hologram} the second term, i.e., the *object field*, is reconstructed. This is achieved, e.g., by illuminating the hologram with an additional so-called *reconstruction wave*

$$\propto e^{-i \boldsymbol{k}_1 \cdot r}$$

(e.g., approximately opposite to the direction of the reference wave, $\boldsymbol{k}_1 \approx \boldsymbol{k}_0$).

In the coherent case one thus obtains

$$I_{\text{view}} \propto |a_0|^2 e^{-i \boldsymbol{k}_1 \cdot \boldsymbol{r}} + a_0 \psi_G(\boldsymbol{r}) e^{i (\boldsymbol{k}_0 - \boldsymbol{k}_1) \cdot \boldsymbol{r}} + a_0^* \psi_G^*(\boldsymbol{r}) e^{-i (\boldsymbol{k}_0 + \boldsymbol{k}_1) \cdot \boldsymbol{r}} . \quad (21.31)$$

For suitable viewing one mainly sees the second term on the r.h.s., i.e., exactly the object wave field.

It is clear that not all possibilities offered by holography have been exploited systematically as yet (see, for example, the short article, in German, in the *Physik Journal* p. 42 (2005, issue 1) already mentioned in [17]; certainly in other journals similar articles in a different language exist). Some of the present-day and future applications of holography (without claiming completeness) are: color holography, volume holograms, distributed information, filtering, holographic data-storage and holographic pattern recognition. These new methods may be summarized as potential applications of *analogue optical quantum computing*[23].

---

[23] In *quantum computation* (see Part III) one exploits the *coherent* superposition of Schrödinger's matter waves, where to date mainly quantum-mechanical *two-level systems* are being considered. In the field of optics, in principle one is further ahead, with the invention of coherent light sources, *the laser*, and with the methods of *holography*.

22 Conclusion to Part II

In this part of the book we have outlined the foundations of theoretical electrodynamics and some related aspects of optics (where optics has been essentially viewed as a branch of "applied electrodynamics"). If in Maxwell's lifetime Nobel prizes had existed, he would certainly have been awarded one. His theory, after all, was really revolutionary, and will endure for centuries to come. Only the quantum mechanical aspects (e.g., light quanta), due to the likes of Einstein and Planck etc. ($\rightarrow$ Part III) are missing. It was not coincidental that Einstein's views on relativity turned out to be in accord with Maxwell's theory of electrodynamics (from which he was essentially inspired). In fact, without realizing it, Maxwell had overturned the Newtonian view on space and time in favor of Einstein's theory. Actually, as mentioned in the Introduction, much of our present culture (or perhaps "lack of it") is based on (the) electrodynamics (of a Hertzian dipole)!

Part III

Quantum Mechanics

23 On the History of Quantum Mechanics

Quantum mechanics first emerged after several decades of experimental and theoretical work at the end of the nineteenth century on the physical laws governing *black-body radiation*. Led by industrial applications, such as the improvement of furnaces for producing iron and steel, physicists measured the flux of energy of thermal radiation emitted from a cavity,[1] finding that at moderate frequencies it almost perfectly follows the classical theory of Rayleigh and Jeans:

$$dU(\nu, T) = V \cdot \frac{8\pi\nu^2}{c^3} d\nu \cdot k_{\mathrm{B}} T , \tag{23.1}$$

where $dU(\nu, T)$ is the spectral energy density of electromagnetic waves in the frequency interval between ν and $\nu + d\nu$, and V is the volume of the cavity. The factor $k_{\mathrm{B}} T$ is the usual expression for the average energy $\langle \varepsilon \rangle_T$ of a classical harmonic oscillator at a Kelvin temperature T and k_B is Boltzmann's constant. Rayleigh and Jeans' law thus predicts that the energy of the radiation should increase indefinitely with frequency – which does not occur in practice. This failure of the classical law at high frequency has been dubbed "the ultraviolet catastrophe".

However, in 1896 the experimentalist Wilhelm Wien had already deduced that for sufficiently hiqh frequencies, i.e.,

$$h\nu \gg k_{\mathrm{B}} T ,$$

where h is a constant, the following behavior should be valid:

$$dU(\nu, T) = V \cdot \frac{8\pi\nu^2}{c^3} d\nu \cdot h\nu \cdot \exp\left(-\frac{h\nu}{k_{\mathrm{B}} T}\right) . \tag{23.2}$$

Here we have adopted the terminology already used in 1900 by *Max Planck* in his paper in which he introduced the quantity h, which was later named after him Planck's *constant*

$$h = 6.25 \ldots \cdot 10^{-34} \, \mathrm{Ws}^2 .$$

(In what follows we shall often use the reduced quantity $\hbar = \frac{h}{2\pi}$.)

---

[1] A method named *bolometry*.

Planck then effectively *interpolated* between (23.1) and (23.2) in his famous *black-body radiation formula*:

$$dU(\nu, T) = V \cdot \frac{8\pi\nu^2}{c^3} \, d\nu \cdot \frac{h\nu}{\exp\left(\frac{h\nu}{k_B T}\right) - 1} . \tag{23.3}$$

In order to derive (23.3) Planck postulated that the energy of a harmonic oscillator of frequency ν is *quantized*, and given by

$$E_n = n \cdot h\nu , \tag{23.4}$$

where[2] $n = 0, 1, 2, \ldots$.

Five years after Planck's discovery came the *annus mirabilis*[3] of Albert Einstein, during which he not only published his *special theory of relativity* (see Parts I and II) but also introduced his *light quantum hypothesis*:

– Electromagnetic waves such as light possess both wave properties (e.g., the ability to interfere with other waves) and also particle properties; they appear as single *quanta* in the form of massless relativistic particles, so-called *photons*, of velocity c, energy $E = h\nu$ and momentum

$$|\boldsymbol{p}| = \frac{E}{c} = \frac{h\nu}{\lambda} = \hbar|\boldsymbol{k}|$$

(where $\nu \cdot \lambda = c$ and $|\boldsymbol{k}| = \frac{2\pi}{\lambda}$; λ is the wavelength of a light wave (*in vacuo*) of frequency ν.)

According to classical physics, the simultaneous appearance of wave and particle properties would imply a contradiction, but as we shall see later, this is *not* the case in quantum mechanics, where the concept of *wave – particle duality* applies (see below).

By postulating the existence of photons, Einstein was then able to explain the experiments of Philipp Lenard on the photoelectric effect; i.e., it became clear why the *freqency* of the light mattered for the onset of the effect, and not its *intensity*. Later the Compton effect (the scattering of light by electrons) could also be explained conveniently[4] in terms of the impact between particles

---

[2] The correct formula, $E_n = (n + \frac{1}{2}) \cdot h\nu$, also leads to the result (23.3). The addition of the zero-point energy was derived later after the discovery of the formalism of *matrix mechanics* (Heisenberg, see below).

[3] In this one year, 1905, Einstein published five papers, all in the same journal, with revolutionary Nobel-prize worthy insight into three topics, i.e., (i) he presented *special relativity*, [5]; (ii) stated the *light quantum hypothesis*, [18], which in fact gained the Nobel prize in 1921; and (iii) (a lesser well-known work) he dealt with *Brownian motion*, [35], where he not only explained the phenomenon atomistically, but proposed a basic relation between diffusion and friction in thermal equilibrium.

[4] The Compton effect can also be explained (less conveniently, but satisfactorily) in a wave picture.

(governed by the conservation of energy and momentum). Indirectly due to Einstein's hypothesis, however, the *particle aspects* of quantum mechanics were initially placed at the center of interest, and *not* the *wave aspects* of matter, which were developed later by de Broglie and Schrödinger (see below).

In fact, in 1913, following the pioneering work of Ernest Rutherford, *Niels Bohr* proposed his atomic model, according to which the electron in a hydrogen atom can only orbit the nucleus on discrete circular paths of radius $r_n = n \cdot a_0$ (with $n = 1, 2, 3, \ldots$, the *principal quantum number*[5] and $a_0 = 0.529 \ldots$ Å, the so-called *Bohr radius*[6]), and where the momentum of the electron is quantized:

$$\oint \boldsymbol{p} \cdot \mathrm{d}\boldsymbol{q} = \int_0^{2\pi} p_\varphi \cdot r \, \mathrm{d}\varphi \overset{!}{=} n \cdot h \,. \tag{23.5}$$

Thus, according to Bohr's model, in the ground state of the H-atom the electron should possess a *finite* angular momentum

$$\hbar \left(= \frac{h}{2\pi} \right) \,.$$

Later it turned out that this is one of the basic errors of Bohr's model since actually the angular momentum of the electron in the ground state of the H-atom is zero. (*A frequent examination question*: Which are other basic errors of Bohr's model compared to Schrödinger's wave mechanics?)

At first Bohr's atomic model seemed totally convincing, because it appeared to explain all essential experiments on the H-atom (e.g., the *Rydberg formula* and the corresponding spectral series) not only qualitatively or approximately, but even quantitatively. One therefore tried to explain the spectral properties of other atoms, e.g., the He atom, in the same way; but without success.

This took more than a decade. Finally in 1925 came the decisive breakthrough in a paper by the young PhD student Werner Heisenberg, who founded a theory (the first fully correct quantum mechanics) which became known as *matrix mechanics*, [19,20]. Heisenberg was a student of Sommerfeld in Munich; at that time he was working with Born in Göttingen.

Simultaneously (and independently) in 1924 the French PhD student Louis *de Broglie*[7], in his PhD thesis *turned around* Einstein's light-quantum hypothesis of *wave-matter correspondence* by complementing it with the proposition of a form of *matter-wave correspondence*:

---

[5] We prefer to use the traditional atomic unit 1Å = 0.1 nm.

[6] Later, by Arnold Sommerfeld, as possible particle orbits also *ellipses* were considered.

[7] L. de Broglie, Ann. de physique (4) **3** (1925) 22; Thèses, Paris 1924

a) (*de Broglie's hypothesis of "material waves"*):
   Not only is it true to say that an electrodynamic wave possesses particle
   properties, but conversely it is also true that a particle possesses wave
   properties (*matter waves*). Particles with momentum p and energy E
   correspond to a complex wave function

   $$\psi(\boldsymbol{r},t) \propto e^{i(\boldsymbol{k}\cdot\boldsymbol{r}-\omega t)} ,$$

   with

   $$\boldsymbol{k} = \frac{\boldsymbol{p}}{\hbar} + e \cdot \operatorname{grad} f(\boldsymbol{r},t) \quad \text{and} \quad \omega = \frac{E}{\hbar} - e \cdot \frac{\partial f}{\partial t} .$$

   (Here the real function $f(\boldsymbol{r},t)$ is arbitrary and usually set $\equiv 0$, if no
   electromagnetic field is applied. This is a so-called *gauge function*, i.e., it
   does not influence local measurenents; e is the charge of the particle.)
b) De Broglie's hypothesis of *matter waves*, which was directly confirmed in
   1927 by the crystal diffraction experiments of Davisson and Germer, [21],
   gave rise to the development of *wave mechanics* by *Erwin Schrödinger*,
   [22]. Schrödinger also proved in 1926 the equivalence of his "wave mechanics" with Heisenberg's "matrix mechanics".
c) Finally, independently and almost simultaneously, quantum mechanics
   evolved in England in a rather abstract form due to *Paul M. Dirac*[8]. All
   these seemingly different formulations, which were the result of considerable direct and indirect contact between many people at various places,
   are indeed equivalent, as we shall see below.

Nowadays the standard way to present the subject – which we shall adhere
to – is (i) to begin with Schrödinger's *wave mechanics*, then (ii) to proceed
to Dirac's more abstract treatment, and finally (iii) – quasi *en passant*, by
treating the different quantum mechanical "aspects" or "representations" (see
below) – to present Heisenberg's *matrix mechanics*.

---

[8] In 1930 this became the basis for a famous book, see [23], by John von Neumann,
born in 1903 in Budapest, Hungary, later becoming a citizen of the USA, deceased
(1957) in Washington, D.C., one of the few universal geniuses of the 20th century
(e.g., the "father" of information technology).

24 Quantum Mechanics: Foundations

24.1 Physical States

Physical states in quantum mechanics are described by equivalence classes of vectors in a *complex Hilbert space* (see below). The equivalence classes are so-called "rays", i.e., one-dimensional subspaces corresponding to the Hilbert vectors. In other words, state functions can differ from each other by a constant complex factor[1], similar to eigenvectors of a matrix in linear algebra.

Unless otherwise stated, we shall generally choose representative vectors with unit magnitude,

$$\langle \psi, \psi \rangle = 1 \; ;$$

but even then these are not yet completely defined: Two unit vectors, differing from each other by a constant complex factor of magnitude 1 ($\psi \to e^{i\alpha}\psi$, with real α) represent the same state.

Furthermore, the states can depend on time. Vectors distinguished from each other by different time-dependent functions typically do *not* represent the same state [2].

In the position representation corresponding to Schrödinger's wave mechanics, a physical state is described by a complex function

$$\psi = \psi(\boldsymbol{r}, t) \quad with \quad \int dV |\psi(\boldsymbol{r}, t)|^2 \overset{!}{=} 1 \; ,$$

where the quantity

$$|\psi(\boldsymbol{r}, t)|^2 \cdot dV$$

represents the instantaneous probability that the particle is found in the infinitesimally small volume element dV.

This is still a preliminary definition, since we have not yet included the concept of *spin* (see below).

---

[1] Two Hilbert vectors which differ by a constant complex factor thus represent the same physical state.

[2] At least not in the Schrödinger picture, from which we start; see below.

24.1.1 Complex Hilbert Space

To be more specific, the Hilbert vectors of wave mechanics are square-integrable complex functions $\psi(\mathbf{r})$, defined for $\mathbf{r} \in V$, where V is the available volume of the system and where (without lack of generality) we assume normalization to 1. The function $\psi(\mathbf{r})$ is also allowed to depend on a time parameter t. For the *scalar product* of two vectors in this Hilbert space we have by definition:

$$\langle \psi_1 | \psi_2 \rangle \equiv \int dV \psi_1(\mathbf{r}, t)^* \psi_2(\mathbf{r}, t) ,$$

where ψ_1^* is the complex conjugate of ψ_1. (Unfortunately mathematicians have slightly different conventions[3]. However, we shall adhere to the convention usually adopted for quantum mechanics by physicists.)

As in linear algebra the scalar product is *independent* of the basis, i.e., on a change of the basis (e.g., by a rotation of the basis vectors) it must be transformed covariantly (i.e., the new basis vectors are the rotated old ones). Moreover, a scalar product has *bilinear properties* with regard to the addition of a finite number of vectors and the multiplication of these vectors by complex numbers. By the usual "square-root" definition of the *distance* between two vectors, one obtains (again as in linear algebra) a so-called *unitary vector space* (or *pre-Hilbert space*), which by *completion* wrt (= *with respect to*) the *distance*, and with the postulate of the existence of at least one countably-infinite basis (the postulate of so-called *separability*) becomes a Hilbert space ($\mathcal{HR}$).

If one is dealing with a countable *orthonormal* basis (orthonormality of a countable basis can always be assumed, essentially because of the existence of the Erhardt-Schmidt orthogonalization procedure), then every element $|\psi\rangle$ of the Hilbert space can be represented in the form

$$|\psi\rangle = \sum_i c_i |u_i\rangle \quad \text{with} \quad c_i = \langle u_i | \psi \rangle$$

where

$$\langle \psi_1 | \psi_2 \rangle = \int dV \psi_1^*(\mathbf{r}, t) \psi_2(\mathbf{r}, t) = \sum_i \left(c_i^{(1)} \right)^* \cdot c_i^{(2)} . \tag{24.1}$$

This is similar to what is known from linear algebra; the main difference here is that countably-infinite sums appear, but for all elements of the Hilbert space convergence of the sums and (Lebesgue) integrals in (24.1) is assured[4].

---

[3] Mathematicians are used to writing $\langle \psi_1 | \psi_2 \rangle = \int dV \psi_1(\mathbf{r}, t)(\overline{\psi_2(\mathbf{r}, t)})$, i.e., (i) in the definition of the scalar product they would not take the complex conjugate of the *first* factor but of the *second* one, and (ii) instead of the "star" symbol they use a "bar", which in physics usually represents an average.

[4] by definition of $\mathcal{HR}$ and from the properties of the Lebesgue integral

24.2 Measurable Physical Quantities (Observables)

Measurable quantities are represented by Hermitian operators[5] in Hilbert space, e.g., in the position representation the coordinates "x" of Hamiltonian mechanics give rise to multiplication operators[6],

$$\psi(\mathbf{r},t) \to \psi'(\mathbf{r},t) := (\hat{x}\psi)(\mathbf{r},t) = x \cdot \psi(\mathbf{r},t) \,,$$

while the momenta are replaced by differential operators

$$\psi(\mathbf{r},t) \to \psi'(\mathbf{r},t) := (\hat{p}_x\psi)(\mathbf{r},t) := (\hbar/\mathrm{i})(\partial\psi(\mathbf{r},t)/\partial x) \,.^{[7]}$$

Therefore one writes compactly

$$\hat{p}_x = (\hbar/\mathrm{i})(\partial/\partial x) \,.$$

More precisely, one assumes that the operator corresponding to an *observable* $\hat{A}$ is not only Hermitian, i.e.,

$$\langle\psi_1|\hat{A}\psi_2\rangle = \langle\hat{A}\psi_1|\psi_2\rangle$$

for all ψ_1 and ψ_2 belonging to the range of definition of the operator $\hat{A}$, but that $\hat{A}$, if necessary after a subtle widening of its definition space, has been enlarged to a so-called *self-adjoint* operator: Self-adjoint operators are (i) Hermitian and (ii) additionally possess a *complete system* of square-integrable (so-called *proper*) and square-nonintegrable (so-called *improper*) eigenvectors $|\psi_j\rangle$ and $|\psi_\lambda\rangle$, respectively (see (24.3)).

The corresponding *eigenvalues* a_j and $a(\lambda)$ (point spectrum and continuous spectrum, respectively) are real, satisfying the equations:

$$\hat{A}|\psi_j\rangle = a_j|\psi\rangle \quad , \quad \hat{A}|\psi_\lambda\rangle = a(\lambda)|\psi_\lambda\rangle \,. \tag{24.2}$$

Here the position-dependence of the states has not been explicitly written down (e.g., $|\psi_\lambda\rangle \hat{=} \psi_\lambda(\mathbf{r},t)$) to include Dirac's more abstract results. Moreover, using the square-integrable (*proper*) and square-nonintegrable (*improper*) eigenvectors one obtains an *expansion theorem* ($\hat{=}$ so-called *spectral resolution*). Any Hilbert vector $|\psi\rangle$ can be written as follows, with complex coefficients c_i and square-integrable complex functions[8] $c(\lambda)$:

$$|\psi\rangle \equiv \sum_i c_i|\psi_i\rangle + \int \mathrm{d}\lambda c(\lambda)|\psi_\lambda\rangle \,. \tag{24.3}$$

---

[5] To be mathematically more precise: *Self-adjoint* operators; i.e., the operators must be Hermitian plus complete (see below).

[6] Operators are represented by a hat-symbol.

[7] Mathematically these definitions are restricted to dense subspaces of $\mathcal{HR}$.

[8] $c(\lambda)$ exists and is square-integrable, if in (24.4) ψ_λ is an improper vector and $\psi \in \mathcal{HR}$ (*weak topology*). Furthermore, in (24.3) we assume that our basis does not contain a so-called "singular continuous" part, see below, but only the usual "absolute continuous" one. This is true in most cases.

However the l.h.s. of (24.3) for $\psi(\boldsymbol{r}, t)$ does not necessarily agree pointwise with the r.h.s. for every $\boldsymbol{r}$, but the identity is only valid "almost everywhere" in the following sense (so-called *strong topology*):

$$(|\psi_n\rangle \to |\psi\rangle \Longleftrightarrow \int \mathrm{d}V |\psi_n(\boldsymbol{r}) - \psi(\boldsymbol{r})|^2 \to 0) \ .$$

The coefficients c_i and $c(\lambda)$ are obtained by *scalar multiplication* from the left with $\langle\psi_i|$ and $\langle\psi_\lambda|$, i.e.,

$$c_i = \langle\psi_i|\psi\rangle \quad , \quad c(\lambda) = \langle\psi_\lambda|\psi\rangle \ . \tag{24.4}$$

In these equations the following orthonormalisation is assumed:

$$\langle\psi_i|\psi_j\rangle = \delta_{i,j} \quad , \quad \langle\psi_{\lambda'}|\psi_\lambda\rangle = \delta(\lambda' - \lambda) \quad , \quad \langle\psi_i|\psi_\lambda\rangle = 0 \ , \tag{24.5}$$

with the Kronecker delta $\delta_{i,j} = 1$ for $i = j$; $\delta_{i,j} = 0$ otherwise (i.e., $\sum_j \delta_{i,j} f_j = f_i$ for all complex vectors f_i), and the *Dirac δ function* $\delta(x)$, a so-called generalized function (distribution), which is represented (together with the limit[9] $\varepsilon \to 0$) by a set $\{\delta_\varepsilon(x)\}_\varepsilon$ of increasingly narrow and at the same time increasingly high bell-shaped functions (e.g., Gaussians) with

$$\int_{-\infty}^{\infty} \mathrm{d}x \delta_\varepsilon(x) \equiv 1 \ ),$$

defined in such a way that for all "test functions" $f(\lambda) \in \mathcal{T}$ (i.e., for all arbitrarily often differentiable complex functions $f(\lambda)$, which decay for $|\lambda| \to \infty$ faster than any power of $1/|\lambda|$) one has the property (see Part II):

$$\int_{-\infty}^{\infty} \mathrm{d}\lambda \delta(\lambda' - \lambda) \cdot f(\lambda) \equiv f(\lambda'), \quad \forall f(\lambda) \in \mathcal{T} \ . \tag{24.6}$$

This implies the following expression (also an extension of linear algebra!) for the scalar product of two vectors in Hilbert space after expansion in the basis belonging to an arbitrary observable $\hat{A}$ (consisting of the orthonormal proper and improper eigenvectors of $\hat{A}$):

$$\left\langle \psi^{(1)} \middle| \psi^{(2)} \right\rangle = \sum_i \left(c_i^{(1)}\right)^* \cdot c_i^{(2)} + \int \mathrm{d}\lambda \left(c^{(1)}(\lambda)\right)^* \cdot c^{(2)}(\lambda) \ . \tag{24.7}$$

For simplicity it is assumed below, unless otherwise stated, that we are dealing with a pure point spectrum, such that in (24.7) only summations appear.

---

[9] The limit $\varepsilon \to 0$ must be performed *in front of* the integral.

a) However, there are important observables with a purely continuous spectrum (e.g., the *position operator* $\hat{x}$ with improper eigenfunctions

$$\psi_\lambda(x) := \delta(x - \lambda)$$

and the *momentum operator* $\hat{p}_x$ with improper eigenfunctions

$$\psi_\lambda(x) := (2\pi\hbar)^{-1/2} \exp(i\lambda \cdot x/\hbar) \ ;$$

the eigenvalues appearing in (24.2) are then $a(\lambda) = x(\lambda) = p(\lambda) = \lambda)$.

b) In rare cases a third spectral contribution, the *singular-continuous contribution*, must be added, where it is necessary to replace the integral $\int d\lambda \ldots$ by a Stieltje's integral

$$\int dg(\lambda) \ldots \ ,$$

with a continuous and monotonically nondecreasing, but nowhere differentiable function $g(\lambda)$ (the usual above-mentioned continuous contribution is obtained in the *differentiable case* $g(\lambda) \equiv \lambda$).

For a pure *point spectrum* one can thus use *Dirac's abstract bra-ket formalism*:

a) "observables" are represented by *self-adjoint*[10] operators. In *diagonal representation* they are of the form

$$\hat{A} = \sum_i a_i |\psi_i\rangle\langle\psi_i| \ , \tag{24.8}$$

with real eigenvalues a_i and orthonormalized eigenstates $|\psi_i\rangle$,

$$\langle\psi_i|\psi_k\rangle = \delta_{ik} \ ,$$

and

b) the following statement is true (which is equivalent to the expansion theorem (24.5)):

$$\hat{1} = \sum_i |\psi_i\rangle\langle\psi_i|. \tag{24.9}$$

Equation (24.9) is a so-called "resolution of the identity operator $\hat{1}$" by a sum of *projections*

$$\hat{P}_i := |\psi_i\rangle\langle\psi_i| \ .$$

The action of these projection operators is simple:

$$\hat{P}_i|\psi\rangle = |\psi_i\rangle \langle\psi_i|\psi\rangle = |\psi_i\rangle c_i \ .$$

Equation (24.9) is often applied as

$$|\psi\rangle \equiv \hat{1}|\psi\rangle \ .$$

---

[10] The difference between hermiticity and self-adjointness is subtle, e.g. instead of vanishing values at the boundaries of an interval one only demands periodic behavior. In the first case the differential expression for $\hat{p}_x$ has no eigenfunctions for the interval, in the latter case it has a complete set.

24.3 The Canonical Commutation Relation

In contrast to classical mechanics (where *observables* correspond to arbitrary real functions $f(r, p)$ of position and momentum, and where for products of x_i and p_j the sequential order does not matter), in quantum mechanics two *self-adjoint* operators representing observables typically *do not commute*. Instead, the following so-called canonical commutation relation holds[11]:

$$[\hat{p}_j, \hat{x}_k] := \hat{p}_j \hat{x}_k - \hat{x}_k \hat{p}_j \equiv \frac{\hbar}{\mathrm{i}} \delta_{jk} \ . \tag{24.10}$$

The canonical commutation relation does not depend on the representation (see below). It can be derived in the wave mechanics representation by applying (24.10) to an arbitrary function $\psi(r)$ (belonging of course to the maximal intersection $\mathcal{I}_{\max}$ of the regions of definition of the relevant operators). Using the product rule for differentiation one obtains

$$\frac{\hbar}{\mathrm{i}} \left\{ \partial(x_k \psi(r))/\partial x_j - x_k \partial \psi(r)/\partial x_j \right\} \equiv \frac{\hbar}{\mathrm{i}} \delta_{jk} \psi(r) \ , \quad \forall \psi(r) \in \mathcal{I}_{max} \subseteq \mathcal{HR} \ . \tag{24.11}$$

This result is identical to (24.10).

24.4 The Schrödinger Equation; Gauge Transformations

Schrödinger's equation describes the time-development of the wave function $\psi(r, t)$ between two measurements. This fundamental equation is

$$-\frac{\hbar}{\mathrm{i}} \frac{\partial \psi}{\partial t} = \hat{H} \psi \ . \tag{24.12}$$

$\hat{H}$ is the so-called *Hamilton operator* of the system, see below, which corresponds to the classical Hamiltonian, insofar as there is a relationship between classical and quantum mechanics (which is not always the case [12]). This important self-adjoint operator determines the *dynamics* of the system.

Omitting *spin* (see below) one can obtain the Hamilton *operator* directly from the Hamilton *function* (*Hamiltonian*) of classical mechanics by replacing the classical quantities r and p by the corresponding *operators*, e.g.,

$$\hat{x}|\psi\rangle \rightarrow x \cdot \psi(r) \ ; \quad \hat{p}_x|\psi\rangle \rightarrow \frac{\hbar}{\mathrm{i}} \frac{\partial}{\partial x} \psi(r) \ .$$

---

[11] Later we will see that this commutation relation is the basis for many important relations in quantum mechanics.

[12] For example the spin of an electron (see below) has no correspondence in classical mechanics.

For example, the classical Hamiltonian $H(\boldsymbol{r}, \boldsymbol{p}, t)$ determining the motion of a particle of mass m and electric charge e in a conservative force field due to a potential energy $V(\boldsymbol{r})$ plus electromagnetic fields $\boldsymbol{E}(\boldsymbol{r}, t)$ and $\boldsymbol{B}(\boldsymbol{r}, t)$, with corresponding scalar electromagnetic potential $\Phi(\boldsymbol{r}, t)$ and vector potential $\boldsymbol{A}(\boldsymbol{r}, t)$, i.e., with

$$\boldsymbol{B} = \operatorname{curl}\boldsymbol{A} \quad \text{and} \quad \boldsymbol{E} = -\operatorname{grad}\Phi - \frac{\partial \boldsymbol{A}}{\partial t} \ ,$$

is given by

$$H(\boldsymbol{r}, \boldsymbol{p}, t) = \frac{(\boldsymbol{p} - e\boldsymbol{A}(\boldsymbol{r}, t))^2}{2m} + V(\boldsymbol{r}) + e\Phi(\boldsymbol{r}, t) \ . \tag{24.13}$$

The corresponding Schrödinger equation is then

$$-\frac{\hbar}{\mathrm{i}} \frac{\partial \psi(\boldsymbol{r}, t)}{\partial t} = \hat{H}\psi = \frac{1}{2m} \left(\frac{\hbar}{\mathrm{i}} \nabla - e \cdot \boldsymbol{A}(\boldsymbol{r}, t) \right)^2 \psi(\boldsymbol{r}, t)$$
$$+ \{V(\boldsymbol{r}) + e \cdot \Phi(\boldsymbol{r}, t)\}\, \psi(\boldsymbol{r}, t) \ . \tag{24.14}$$

The corresponding Newtonian equation of motion is the equation for the Lorentz force (see Parts I and II)[13]

$$m\frac{\mathrm{d}\boldsymbol{v}}{\mathrm{d}t} = -\nabla V(\boldsymbol{r}) + e \cdot (\boldsymbol{E} + \boldsymbol{v} \times \boldsymbol{B}) \ . \tag{24.15}$$

Later we come back to these equations in connection with *spin* and with the *Aharonov-Bohm effect*.

In (24.13) and (24.14), one usually sets $\boldsymbol{A} \equiv 0$, if $\boldsymbol{B}$ vanishes everywhere. But this is neither necessary in electrodynamics nor in quantum mechanics. In fact, by analogy to electrodynamics, see Part II, a slightly more complex *gauge transformation* can be defined in quantum mechanics, by which certain "nonphysical" (i.e., unmeasurable) functions, e.g., the probability amplitude $\psi(\boldsymbol{r}, t)$, are non-trivially transformed without changes in measurable quantities.

To achieve this it is only necessary to perform the following simultaneous changes of $\boldsymbol{A}$, Φ and ψ into the corresponding *primed* quantities[14]

$$\boldsymbol{A}'(\boldsymbol{r}, t) = \boldsymbol{A}(\boldsymbol{r}, t) + \nabla f(\boldsymbol{r}, t) \tag{24.16}$$

$$\Phi'(\boldsymbol{r}, t) = \Phi(\boldsymbol{r}, t) - \frac{\partial f(\boldsymbol{r}, t)}{\partial t} \tag{24.17}$$

$$\psi'(\boldsymbol{r}, t) = \exp\left[+\mathrm{i}e \cdot \frac{f(\boldsymbol{r}, t)}{\hbar} \right] \cdot \psi(\boldsymbol{r}, t) \tag{24.18}$$

---

[13] With the Hamiltonian $\mathcal{H} = \frac{(\boldsymbol{p} - e\boldsymbol{A})^2}{2m}$ one evaluates the so-called *canonical equations* $\dot{x} = \partial\mathcal{H}/\partial p_x$, $\dot{p}_x = -\partial\mathcal{H}/\partial x$, where it is useful to distinguish the *canonical momentum* $\boldsymbol{p}$ from the *kinetic momentum* $m\boldsymbol{v} := \boldsymbol{p} - e\boldsymbol{A}$.

[14] Here we remind ourselves that in classical mechanics the *canonical momentum* $\boldsymbol{p}$ must be gauged. In Schrödinger's wave mechanics one has instead $\frac{\hbar}{\mathrm{i}}\nabla\psi$, and only ψ must be gauged.

with an arbitrary real function $f(\mathbf{r}, t)$. Although this transformation changes both the Hamiltonian $\hat{H}$, see (24.14), and the probability amplitude $\psi(\mathbf{r}, t)$, all *measurable* physical quantities, e.g., the electromagnetic fields $\mathbf{E}$ and $\mathbf{B}$ and the probability density $|\psi(\mathbf{r}, t)|^2$ as well as the probability-current density (see below, (25.12)) do *not* change, as can be shown.

24.5 Measurement Process

We shall now consider a general state $|\psi\rangle$. In the basis belonging to the observable $\hat{A}$ (*i.e., the basis is formed by the complete set of orthonormalized eigenvectors $|\psi_i\rangle$ and $|\psi_\lambda\rangle$ of the self-adjoint ($\hat{=}$ hermitian plus complete) operator $\hat{A}$), this state has complex expansion coefficients

$$c_i = \langle \psi_i | \psi \rangle \quad \text{and} \quad c(\lambda) = \langle \psi_\lambda | \psi \rangle$$

(obeying the δ-conditions (24.5)). If in such a state *measurements* of the observable $\hat{A}$ are performed, then

a) only the values a_i and $a(\lambda)$ are obtained as the result of a single measurement, and
b) for the probability $W(\hat{A}, \psi, \Delta a)$ of finding a result in the interval

$$\Delta a := [a_{\min}, a_{\max}) ,$$

the following expression is obtained:

$$W(\hat{A}, \psi, \Delta a) = \sum_{a_i \in \Delta a} |c_i|^2 + \int_{a(\lambda) \in \Delta a} \mathrm{d}\lambda |c(\lambda)|^2 . \qquad (24.19)$$

In this way we obtain what is known as the *quantum mechanical expectation value*, which is equivalent to a fundamental experimental value, *viz* the average over an infinitely long series of measurements of the observable $\hat{A}$ in the state ψ:

$$(\overline{A})_\psi \overset{(i)}{=} \sum_i a_i |c_i|^2 + \int \mathrm{d}\lambda a(\lambda)|c(\lambda)|^2 \overset{(ii)}{=} \langle \psi | \hat{A} | \psi \rangle . \qquad (24.20)$$

Analogously one finds that the operator

$$\left(\delta \hat{A} \right)^2 := \left(\hat{A} - (\overline{A})_\psi \right)^2$$

corresponds to the *variance* (the square of the mean variation) of the values of a series of measurements around the average.

For the product of the variances of two series of measurements of the observables $\hat{A}$ and $\hat{B}$ we have, with the commutator

$$[\hat{a}, \hat{b}] := \hat{A}\hat{B} - \hat{B}\hat{A} ,$$

Heisenberg's uncertainty principle:

$$\langle\psi| \left(\delta\hat{A}\right)^2 |\psi\rangle \cdot \langle\psi| \left(\delta\hat{B}\right)^2 |\psi\rangle \geq \frac{1}{4} \left|\langle\psi| \left[\hat{A}, \hat{B}\right] |\psi\rangle\right|^2 . \qquad (24.21)$$

Note that this relation makes a very precise statement; however one should also note that it does *not* deal with single measurements but with expectation values, which depend, moreover, on $|\psi\rangle$.

Special cases of this important relation (which is not hard to derive) are obtained for

$$\hat{A} = \hat{p}_x , \quad \hat{B} = \hat{x} \quad \text{with} \quad \left[\hat{A}, \hat{B}\right] = \frac{\hbar}{i} ,$$

and for the orbital angular moments[15]

$$\hat{A} = \hat{L}_x \quad \text{and} \quad \hat{B} = \hat{L}_y \quad \text{with} \quad \left[\hat{A}, \hat{B}\right] = i\hbar\hat{L}_z .$$

On the other hand permutable operators have identical sets of eigenvectors (but different eigenvalues).

Thus in quantum mechanics a measurement generally has a finite influence on the state (e.g., $|\psi\rangle \to |\psi_1\rangle$; state reduction), and two series of measurements for the same state $|\psi\rangle$, but non-commutable observables $\hat{A}$ and $\hat{B}$, typically (this depends on $|\psi\rangle$!) cannot simultaneously have vanishing expectation values of the variances[16].

24.6 Wave-particle Duality

In quantum mechanics this important topic means that

a) (on the one hand) the complex probability amplitudes (and not the probabilities themselves) are linearly superposed, in the same way as field *amplitudes* (not *intensities*) are superposed in coherent optics, such that interference is possible (i.e., $|\psi_1 + \psi_2|^2 = |\psi_1|^2 + |\psi_2|^2 + 2\mathcal{R}e(\psi_1^* \cdot \psi_2)$); whereas

---

[15] and also for the spin momenta (see below).

[16] Experimentalists often prefer the "short version" "...cannot be measured simultaneously (with *precise results*)"; unfortunately this "shortening" gives rise to many misunderstandings. In this context the relevant section in the "Feynman lectures" is recommended, where it is demonstrated by construction that for a single measurement (but of course not on average) even $\hat{p}_x$ and $\hat{x}$ can simultaneously have precise values.

b) (on the other hand) measurement and interaction processes take place
with single particles such as photons, for which the usual fundamental
conservation laws (conservation of *energy* and/or *momentum* and/or *angular momentum*) apply per single event, and not only on average. (One
should mention that this important statement had been proved experimentally even in the early years of quantum mechanics!)

For example, photons are the "particles" of the electromagnetic wave field,
which is described by Maxwell's equations. They are realistic objects, i.e.
(massless) relativistic particles with energy

$$E = \hbar\omega \quad \text{and momentum} \quad \boldsymbol{p} = \hbar\boldsymbol{k} \; .$$

Similarly, (nonrelativistic) electrons are the quanta of a "Schrödinger field"[17],
i.e., a "matter field", where for the matter field the Schrödinger equation plays
the role of the Maxwell equations .

*The solution of the apparent paradox of wave-particle duality in quan-
tum mechanics can thus be found in the probabilistic interpretation of the
wave function ψ. This is the so-called "Copenhagen interpretation" of quantum mechanics, which dates back to Niels Bohr (in Copenhagen) and Max
Born (in Göttingen). This interpretation has proved to be correct without
contradiction, from Schrödinger's discovery until now – although initially the
interpretation was not undisputed, as we shall see in the next section.*

24.7 Schrödinger's Cat: Dead *and* Alive?

Remarkably Schrödinger himself fought unsuccessfully against the Copenhagen interpretation [18] of quantum mechanics, with a question, which we
paraphrase as follows: "*What is the 'state' of an unobserved cat confined in
a box, which contains a device that with a certain probability kills it immediately?*"

Quantum mechanics tends to the simple answer that the cat is either in
an "alive" state ($|\psi\rangle = |\psi_1\rangle$), a "dead" state($|\psi\rangle = |\psi_2\rangle$) or in a (coherently)
"superposed" state ($|\psi\rangle \equiv c_1|\psi_1\rangle + c_2|\psi_2\rangle$).

With his question Schrödinger was in fact mainly casting doubt on the
idea that a system could be in a state of coherent quantum mechanical superposition with nontrivial probabilities of states that are *classically mutually*

---

[17] Relativistic electrons would be the quanta of a "Dirac field", i.e., a matter field,
where the Dirac equation, which is *not* described in this book, plays the role of
the Maxwell equations of the theory.

[18] Schrödinger preferred a "charge-density interpretation" of $e|\psi(\boldsymbol{r})|^2$. But this
would have necessitated an addition $\delta\mathcal{V}$ to the potential energy, i.e., classically:
$\delta\mathcal{V} \equiv e^2 \iint dV\, dV' |\psi(\boldsymbol{r})|^2 |\psi(\boldsymbol{r}')|^2 /(8\pi\varepsilon_0|\boldsymbol{r}-\boldsymbol{r}'|)$, which – by the way – after a systematic quantization of the corresponding classical field theory (the so-called
"2nd quantization") leads back to the usual quantum mechanical single-particle
Schrödinger equation *without* such an addition, see [24].

exclusive ("alive" and "dead" simultaneously!). Such states are nowadays called "Schrödinger cat states", and although Schrödinger's objections were erroneous, the question led to a number of important insights. For example, in practice the necessary coherence is almost always *destroyed* if one deals with a *macroscopic* system. This gives rise to corresponding quantitative terms such as the *coherence length* and *coherence time*. In fact there are many other less *spectacular* "cat states", e.g. the state describing an object which is simultaneously in the vicinity of two different places,

$$\psi = c_1\psi_{x \approx x_1} + c_2\psi_{x \approx x_2} \ .$$

Nowadays one might update the question for contemporary purposes. For example we could assume that Schrödinger's proverbial cat carries a bomb attached to its collar[19], which would not only explode spontaneously with a certain probability, but also *with certainty* due to any external interaction process ("measurement with interaction"). The serious question then arises as to whether or not it would be possible to verify by means of a "quantum measurement without interaction", i.e., without making the bomb explode, that a suspicious box is empty or not.

This question is treated below in Sect. 36.5; the answer to this question is actually *positive*, i.e., there *is* a possibility of performing an "interaction-free quantum measurement", but the probability for an interaction ($\Rightarrow$ explosion), although reduced considerably, does not vanish completely. For details one should refer to the above-mentioned section or to papers such as [31].

---

[19] If there are any cat lovers reading this text, we apologize for this thought experiment.

25 One-dimensional Problems in Quantum Mechanics

In the following we shall deal with *stationary states*. For these states one can make the *ansatz*:
$$\psi(\boldsymbol{r}, t) = u(\boldsymbol{r}) \cdot e^{-i\frac{Et}{\hbar}} \ .$$

As a consequence, for stationary states, the expectation values of a constant observable $\hat{A}$ is also constant (w.r.t. time):

$$\langle \psi(t)|\hat{A}|\psi(t)\rangle \equiv \langle u|\hat{A}|u\rangle \ .$$

The related differential equation for the amplitude function $u(\boldsymbol{r})$ is called the *time-independent Schrödinger equation*. In one dimension it simplifies for vanishing electromagnetic potentials, $\varPhi = \boldsymbol{A} \equiv 0$, to:

$$u'' = -k^2(x)u(x) \ , \quad \text{with} \quad k^2(x) := \frac{2m}{\hbar^2}(E - V(x)) \ . \qquad (25.1)$$

This form is useful for values of

$$x \quad \text{where} \quad E > V(x) \ , \quad \text{i.e., for} \quad k^2(x) > 0 \ .$$

If this not the case, then it is more appropriate to write (25.1) as follows

$$u'' = +\kappa^2(x)u(x) \ , \quad \text{with} \quad \kappa^2(x) := \frac{2m}{\hbar^2}(V(x) - E) \ . \qquad (25.2)$$

For a potential energy that is constant w.r.t. time, we have the following general solution:

$$u(x) = A_+ \exp(ik \cdot x) + A_- \cdot \exp(-ik \cdot x) \quad \text{and}$$
$$u(x) = B_+ \exp(+\kappa \cdot x) + B_- \cdot \exp(-\kappa \cdot x) \ .$$

Using

$$\cos(x) := (\exp(ix) + \exp(-ix))/2 \quad \text{and}$$
$$\sin(x) := (\exp(ix) - \exp(-ix))/(2i)$$

we obtain in the first case

$$u(x) = C_+ \cdot \cos(\kappa \cdot x) + C_- \cdot \sin(\kappa) , \quad \text{where}$$
$$C_+ = A_+ + A_- \quad \text{and}$$
$$C_- = i(A_+ - A_-) .$$

The coefficients A_+, A_- etc. are real or complex numbers, which can be determined by consideration of the boundary conditions.

25.1 Bound Systems in a Box (Quantum Well); Parity

Assume that

$$V(x) = 0 \quad \text{for} \quad |x| \geq a , \quad \text{whereas}$$
$$V(x) = -V_0(< 0) \quad \text{for} \quad |x| < a .$$

The potential is thus an *even* function,

$$V(x) \equiv V(-x) , \quad \forall x \in \mathcal{R} ,$$

cf. Fig. 25.1. Therefore the corresponding *parity* is a "good quantum number" (see below).

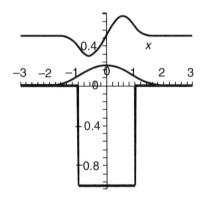

Fig. 25.1. A "quantum well" potential and a sketch of the two lowest eigenfunctions. A symmetrical quantum-well potential of width $\Delta x = 2$ and depth $V_0 = 1$ is shown as a function of x. In the two upper curves the qualitative behavior of the lowest and 2nd-lowest stationary wave functions ($\rightarrow$ even and odd parity) is sketched, the uppermost curve with an offset of 0.5 units. Note that the quantum mechanical wave function has exponential tails in the external region which a classical bound particle never enters. In fact, where the classical bound particle has a *point of return* to the center of the well, the quantum mechanical wave function only has a *turning point*, i.e., only the curvature changes sign

Next we assume $E < 0$ (i.e., for *bound states*). Then we have the following solutions of the Schrödinger equation

$$u(x) = B_-^{(1)} \exp(-\kappa|x|) + B_+^{(1)} \exp(+\kappa|x|) \,, \quad -x \geq a \,, \tag{25.3}$$
$$u(x) = C_1 \cos(k \cdot x) + C_2 \sin(k \cdot x) \,, \qquad\qquad |x| < a \,, \tag{25.4}$$
$$u(x) = B_-^{(2)} \exp(-\kappa x) + B_+^{(2)} \exp(+\kappa x) \,, \quad +x \geq a \,. \tag{25.5}$$

a) Firstly, we recognise that the coefficents $B_+^{(1)}$ and $B_+^{(2)}$ must vanish, since otherwise it would not be possible to satisfy the condition

$$\int_{-\infty}^{\infty} dx |u(x)|^2 \overset{!}{=} 1 \,.$$

b) The remaining coefficients are determined (apart from a common factor, where only the magnitude is fixed by the normalization condition) from the continuity conditions for u and u' at the potential steps, $x = \pm a$.

The calculation is thus much easier for symmetrical potentials, $V(x) = V(-x)$, since then all solutions can be divided into two different classes:

even parity, i.e., $\quad u(-x) = u(x) \,, \quad (\Rightarrow B_-^{(1)} = B_-^{(2)} \,; \quad C_2 = 0) \quad$ and

odd parity, i.e., $\quad u(-x) = -u(x) \,, \quad (\Rightarrow B_-^{(1)} = -B_-^{(2)} \,; \quad C_1 = 0) \,.$

One then only needs *one* continuity condition, i.e. the one for u'/u at $x = +a$. From this condition (see below) one also obtains the discrete energy values $E = E_n$, for which continuity is possible (for $k > 0$ and $\kappa > 0$):

$$\frac{\kappa(E)}{k(E)} = \tan(k(E) \cdot a) \,, \quad \text{for even parity,} \tag{25.6}$$

$$-\frac{k(E)}{\kappa(E)} = \tan(k(E) \cdot a) \,, \quad \text{for odd parity} \,. \tag{25.7}$$

These equations can be solved *graphically* (this is a typical exercise), by plotting all branches of $\tan(k \cdot a)$ as a function of $k \cdot a$ (these branches intercept the x-axis at $k \cdot a = n \cdot \pi$, where n is integer, and afterwards they diverge to $+\infty$ from $-\infty$, at $k_\pm \cdot a = (2n \pm 1) \cdot \pi/2 \mp 0^+$). Then one can determine the intersections of this multi-branched curve with the line obtained by plotting the l.h.s. of (25.6) or (25.7) as a function of $k(E) \cdot a$.

In this way one obtains the following general statements which are true for a whole class of similar problems:

Existence I:

There is always at least one bound state. (This statement is true for similar problems in one and two dimensions, but *not in three dimensions*[1].

---

[1] In d=3 dimensions one can show, see below, that for so-called *s*-states, i.e., if the state does not depend on the angular coordinates ϑ and φ, the wave-function

For example, for the analogous three-dimensional "potential box model" for the mutual binding of a neutron and proton in the deuteron nucleus the depth V_0 is just deep enough to generate a bound state, whereas for a "di-neutron" it is just *not* deep enough.)

Nodal theorem: The ground state, ψ_0, has no "nodes" (i.e., no zeroes) at all (between the interval limits, i.e., here for $-\infty < x < \infty$). In contrast, an eigenstate ψ_n, for $n = 1, 2, \ldots$, if existent, has exactly n nodes.

If parity is a "good" quantum number, i.e., for symmetric potentials, $V(x) \equiv V(-x)$, the following principle is additionally true:

Alternating parity: The ground state, $\psi_{(n=0)}$, has *even* parity, the first excited state *odd* parity, the 2nd excited state again *even* parity, etc..

Existence II:

Quantitatively one finds that the nth bound state, $n = 1, 2, \ldots$, exists iff the quantum well is *sufficiently deep and broad*, i.e., for the present case iff

$$\sqrt{\frac{2m|V_0|a^2}{\hbar^2}} > n \cdot \frac{\pi}{2} \, .$$

25.2 Reflection and Transmission at Steps in the Potential Energy; Unitarity

For simplicity we assume firstly that

$$V(x) \equiv 0 \quad \text{for} \quad x < 0 \quad \text{and} \quad \equiv \Delta V(x)(> 0) \quad \text{for all} \quad x \geq 0 \, ,$$

with a barrier in a finite range including $x = 0$. Consider the reflection and transmission of a monochromatic wave traveling from the left. We assume below that E is sufficiently high (e.g., $E > V(\infty)$ in Fig. 25.2, see below). Otherwise we have *total reflection*; this case can be treated separately.)

We thus have, with $\omega := E/\hbar$:

$$\psi(x, t) = A \cdot \left(e^{i(k_- x - \omega t)} + r \cdot e^{i(-k_- x - \omega t)} \right) \quad \text{for} \quad x < 0 \, , \quad \text{and}$$

$$= A \cdot t \cdot e^{i(k_+ x - \omega t)} \quad \text{for} \quad x > a \, ; \tag{25.8}$$

is quasi one-dimensional in the following sense: The auxiliary quantity $w(r) := r \cdot \psi(r)$ satisfies the same "quasi one-dimensional" Schrödinger equation as noted above, see the three equations beginning with (25.3). As a consequence, one only needs to put $x \to r$ and $u(x) \to w(r)$, and can thus transfer the above "one-dimensional" results to three dimensions. But now one has to take into account that negative r values are not allowed and that $w(0) \overset{!}{=} 0$ (remember: $w(r) = r \cdot \psi(r)$). The one-dimensional solutions with even parity, i.e. $u(x) \propto \cos kx$, are thus unphysical for a "three-dimensional quantum box", i.e. for $V(r) = -V_0$ for $r \leq a$, $V \equiv 0$ otherwise. In contrast, the solutions of odd parity, i.e., $w(r) = r \cdot \psi(r) \propto \sin kr$, transfer to d=3. – This is a useful tip for similar problems in written examinations.

k_- and k_+ are the wave numbers on the l.h.s. and r.h.s. of the barrier (see below).

The amplitude A is usually replaced by 1, which does not lead to any restriction. The complex quantities r and t are the *coefficients of reflection* and *transmission* (not yet the *reflectivity* R and *transmittivity* T, see below).

The coefficients r and t follow in fact from the two continuity conditions for $\psi(x)$ and $\frac{\mathrm{d}\psi(x)}{\mathrm{d}x}$. The *reflectivity* $R(E)$ and the *transmittivity* $T(E)$ themselves are functions of $r(E)$ and $t(E)$, i.e.,

$$R = |r|^2, \quad T = \frac{k_+}{k_-}|t|^2 . \tag{25.9}$$

The fraction $\frac{k_+}{k_-}$ in the formula for T is the ratio of the velocities on the r.h.s. and l.h.s. of the barrier in the potential energy; i.e.,

$$E = \frac{\hbar^2 k_-^2}{2m} \equiv \frac{\hbar^2 k_+^2}{2m} + \Delta V. \tag{25.10}$$

T can be directly calculated from R using the so-called unitarity relation[2]

$$R + T \equiv 1. \tag{25.11}$$

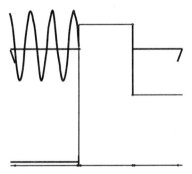

Fig. 25.2. Scattering of a plane wave by a barrier (schematically). A wave $\propto e^{\mathrm{i}k \cdot x}$ traveling from $-\infty$ with a certain energy meets a rectangular barrier, where it is partially reflected and transmitted (indicated by the *straight lines with arrows*, i.e., the corresponding complex amplitudes r and t are also associated with cosine-like behavior). The conditions determining r and t are that the wave function and its derivative are continuous. The velocities on each side are different. The energy is also allowed to be higher than that of the barrier

---

[2] The name *unitarity relation* follows by addition of an incoming wave from the *right*, i.e., the *incoming* wave is now a two-component vector with indices l and r (representing *left* and *right*, respectively). This is also the case with the *outgoing* wave. The incoming and outgoing waves are related, as can be shown, by a *unitary matrix* (the so-called *S-matrix*), which generalizes (25.11).

25.3 Probability Current

All these statements follow explicitly (with $\hat{v} = m^{-1}(\hat{p} - e\hat{A})$) from a gauge invariant relation for the *probability current density*:

$$j_w(r, t) := \mathcal{R}e\left\{\psi^*(r, t)\hat{v}\psi(r, t)\right\} = \frac{\hbar}{2im}\left(\psi^*\nabla\psi - \psi\nabla\psi^*\right)$$
$$-\frac{e}{m}A|\psi|^2 . \tag{25.12}$$

Together with the scalar probability density

$$\varrho_w(r, t) = |\psi(r, t)|^2 ,$$

the current density j_w satisfies (as one can show) the *continuity equation*

$$\frac{\partial \varrho_w(r, t)}{\partial t} + \mathrm{div}\, j_w(r, t) \equiv 0 . \tag{25.13}$$

As shown in Part II in the context of electrodynamics, this continuity equation is equivalent to the conservation theorem of the total probability:

$$\int_\infty \mathrm{d}^3 r \varrho_w(r, t) \equiv 1, \forall t .$$

Ultimately it is this fundamental conservation theorem, which stands behind the unitarity relation (25.11).

For a *series* of different steps in potential energy the complex coefficients r_n and t_n may be calculated sequentially. This gives rise to a so-called "transfer matrix".

25.4 Tunneling

In this section the probability of tunneling through a symmetrical rectangular barrier of width a and height $V_0 (> 0)$ will be considered. Assume that $V(x) = 0$ for $x < 0$ and $x > a$, but $V(x) = V_0$ for $0 \le x \le a$; furthermore assume that the energy E is smaller than the barrier height, i.e., $0 < E < V_0$ (but see below!). Classically, in such a situation a particle will be elastically reflected at the barrier. In contrast, quantum mechanically, with the methods outlined above, it is straightforward to show that one obtains a finite tunneling probability, given by

$$T(E) = \frac{1}{1 + \frac{1}{4}(\frac{\kappa}{k} + \frac{k}{\kappa})^2 \sinh^2(\kappa a)} . \tag{25.14}$$

Here

$$k(E) = \left(\frac{2m}{\hbar^2} E\right)^{1/2}, \quad \text{and} \quad \kappa(E) = \left(\frac{2m}{\hbar^2} (V_0 - E)\right)^{1/2}.$$

For barriers with $\kappa a \gg 1$ the tunneling transmittivity is therefore exponentially small, $T \ll 1$, but finite, as can be seen from

$$\sinh x = (1/2)(e^x + e^{-x}) \cong e^x/2$$

for $x \gg 1$:

$$T(E) \cong \frac{16}{\left(\frac{k}{\kappa} + \frac{\kappa}{k}\right)^2} \cdot e^{-2\kappa a}. \tag{25.15}$$

The factor in front of the exponential in (25.15) is of the order of $\mathcal{O}(4)$, if k and κ are comparable. Therefore the following result for a *non-rectangular* tunneling barrier is plausible (here we assume that $V(x) > E$ only in the interval $a < x < b$). Then one obtains (as long as the result is $\ll 1$):

$$T(E) \cong \mathcal{O}(4) \cdot \exp\left\{-2 \int_a^b dx \sqrt{\frac{2m}{\hbar^2}(V(x) - E)}\right\}. \tag{25.16}$$

(The exponent in this expression is essentially proportional to the product of the *width and the square-root of the height* of the barrier. This yields a rough but systematic approximation for tunnelling through a barrier.)

Furthermore, the factor 2 in the exponent and the factor $\mathcal{O}(4)$ in (25.16) have an obvious meaning. They result from the correspondence $T \propto |\psi|^2$; (i), the factor 2 in the exponent in front of the integral immediately follows from the exponent 2 in $|\psi|^2$, and, (ii), the prefactor $\mathcal{O}(4)$ is obtained from the relation $4 = 2^2$ by the fact that the wave-function must decay exponentially on *both sides* of the barrier. Furthermore, the reciprocal length appearing in the exponent of (25.16) is, as expected, proportional to $\hbar$, i.e., this decay length *vanishes* in the classical limit $\hbar \to 0$.

Quantum mechanical reflection at a *depression* in the potential energy, i.e., at a barrier with *negative* sign (e.g., a "quantum well" as in figure 25.1) is also of interest: We now assume an incoming plane wave with $E > 0$ (instead of the problem of bound states, $E < 0$); in any case (as already mentioned) for $V(x)$ we have the same situation as in Fig. 19.1, i.e. $V \equiv 0$ for $x < 0$ and $x > a$, but $V(x) \equiv -|V_0|$ for $0 \leq x \leq a$. For such a "quantum well" (as already stated) there is at least one bound state for $E < 0$. On the other hand, classically an electron with $E > 0$ does not "see" the quantum well. Quantum mechanically, however, even in this case there is a finite reflection probability. In fact the transmittivity T for this case is analogous to (25.14), again without proof:

$$T(E) = \frac{1}{1 + \frac{1}{4}\left(\frac{k_0}{k} + \frac{k}{k_0}\right)^2 \sin^2(k_0 \cdot a)}, \tag{25.17}$$

where

$$k_0(E) = \sqrt{\frac{2m}{\hbar^2}(E + |V_0|)} \ .$$

Complete transmission ($T = 1$) is only obtained for the special energy values $k_0 \cdot a = n\pi$ (where n is an integer); otherwise $T < 1$, i.e., there is a finite reflectivity.

26 The Harmonic Oscillator I

This is one of the most important problems of quantum mechanics. The Hamilton operator is

$$\hat{\mathcal{H}} = \frac{\hat{p}^2}{2m} + \frac{m\omega_0^2\hat{x}^2}{2} \ . \tag{26.1}$$

The harmonic oscillator is important *inter alia* because a potential energy $V(x)$ in the vicinity of a local minimum can almost always be approximated by a parabola (which is the characteristic potential energy of the harmonic oscillator),

$$V(x) = V_0 + \frac{1}{2}V''(x_0)(x - x_0)^2 + \ldots .$$

We thus have

$$m\omega_0^2 \equiv V''(x_0) \ ,$$

and it is assumed that anharmonicities, i.e., the correction terms of higher order, which are denoted by the dots, $+\ldots$, can be neglected. In contrast, the assumptions $x_0 = 0$ and $V_0 = 0$ impose no restrictions on generality.

In the position representation (*wave mechanics*), we thus have to solve the Schrödinger equation

$$-\frac{\hbar}{\mathrm{i}}\frac{\partial\psi(x,t)}{\partial t} = \hat{\mathcal{H}}\psi(x,t) \ .$$

Using the *ansatz* for stationary states

$$\psi(x,t) \overset{!}{=} u(x) \cdot e^{-\mathrm{i}Et/\hbar} \ ,$$

we obtain the following time-independent Schrödinger equation:

$$u''(x) \overset{!}{=} \left(\frac{2m}{\hbar^2}\cdot\frac{m\omega_0^2x^2}{2} - \frac{2m}{\hbar^2}E\right)\cdot u(x) = \left[\left(\frac{m\omega_0}{\hbar}\right)^2 - \frac{2m}{\hbar^2}E\right]\cdot u(x) \tag{26.2}$$

It is now advantageous to use *reduced variables* without physical dimension, i.e.,

$$\varepsilon := E/(\hbar\omega_0/2), \xi := x/\sqrt{\hbar/(m\omega_0)} \quad \text{and}$$

$$\tilde{u}(\xi) := u(x)\sqrt{\frac{\hbar}{m\omega_0}} \ .$$

Equation (26.2) is thus simplified to

$$\frac{d^2 \tilde{u}(\xi)}{d\xi^2} = (\xi^2 - \varepsilon) \cdot \tilde{u}(\xi) \,, \tag{26.3}$$

and the normalization condition

$$\int_{\infty}^{\infty} |u(x)|^2 \, dx \stackrel{!}{=} 1 \quad \text{becomes}$$

$$\int_{-\infty}^{\infty} |\tilde{u}(\xi)|^2 \, d\xi \stackrel{!}{=} 1 \,.$$

Using the *ansatz*

$$\tilde{u}(\xi) =: v(\xi) \cdot e^{-\xi^2/2}$$

the (asymptotically dominating) exponential behavior $\sim e^{-\xi^2/2}$ can now be separated out; the differential equation for $v(\xi)$ (the so-called Hermitian differential equation) is solved by means of a power series,

$$v(\xi) = \sum_{\nu=0,1,\ldots}^{\infty} a_\nu \xi^\nu \,.$$

Finally after straightforward calculations, for the coefficients a_ν the following recursion relation is obtained:

$$\frac{a_{\nu+2}}{a_\nu} = \frac{2\nu + 1 - \varepsilon}{(\nu + 1)(\nu + 2)} \,. \tag{26.4}$$

Although Schrödinger's differential equation (26.3) is now satisfied, one must additionally demand that the recursion relation *terminates* at a finite

$$\nu(= \nu_0) \,, \quad \text{i.e.,} \quad \varepsilon = 2\nu_0 + 1 \,.$$

If it were not terminated, i.e., if for all non-negative integers ν the identity $\varepsilon = 2\nu + 1$ were violated, then $v(\xi)$ would diverge for $|\xi| \gg 1$ (for even parity, i.e., for $[a_0 \neq 0, a_1 = 0]$, the divergency would be asymptotically as $\sim e^{+\xi^2}$, whereas for odd parity, i.e., for $[a_1 \neq 0, a_0 = 0]$, $v(\xi) \sim \xi \cdot e^{+\xi^2}$). This would violate the condition that

$$u(\xi) \left(= v(\xi) \cdot e^{-\frac{\xi^2}{2}} \right)$$

must be square-integrable. In contrast, if there is *termination* at a finite ν, then the question of the convergence of the ratio $a_{\nu+2}/a_\nu$ becomes obsolete.

In conclusion, iff the power series for $v(\xi)$ terminates at a finite n, i.e. iff the reduced energy

$$\varepsilon = E / \left(\frac{\hbar}{\omega_0/2} \right)$$

is identical to one of the eigenvalues $2n + 1$ *with* $n = 0, 1, 2, \ldots,$

$$\tilde{u}(\xi) \left(= v(\xi) \cdot e^{-\xi^2/2}\right)$$

is square-integrable.

In this way the following energy levels for the harmonic oscillator are obtained:

$$E_n = \hbar\omega_0 \cdot (n + 1/2) , \tag{26.5}$$

for $n = 0, 1, 2, \ldots$, with the eigenfunctions

$$\tilde{u}_n(\xi) \sim H_n(\xi) e^{-\xi^2/2} . \tag{26.6}$$

The $H_n(\xi)$ are the so-called *Hermite polynomials*; e.g.,

$$H_0(\xi) = 1 , \quad H_1(\xi) = \xi , \quad H_2(\xi) = 1 - 2\xi^2 \quad \text{and} \quad H_3(\xi) = \xi \cdot \left(1 - \frac{2}{3}\xi^2\right)$$

(only the first and second should be kept in mind). The missing normalization factors in (26.6) are also unimportant.

Eigenfunctions of a Hermitian operator, corresponding to different eigenvalues of that operator, are necessarily *orthogonal*, as can easily be shown. We thus have

$$\langle \tilde{u}_i | \tilde{u}_j \rangle = \langle u_i | u_j \rangle = \delta_{i,j} ,$$

as expected.

The *completeness*[1], i.e., the *basis* property of the function system $\{u_n\}$, is also given, if all polynomial degrees $n = 0, 1, 2, \ldots$ are taken into account[2]. In fact, the probability that an oscillating particle of a given energy E_n is found outside the "inner" region which is *classically* allowed, is very small, i.e., $\propto e^{-x^2/x_0^2}$, but *finite* even for large n.

As we shall see in a later section ($\rightarrow$ 28.1), the harmonic oscillator can also be treated in a purely abstract (i.e., algebraic) way.

---

[1] We remember that the *completeness* property is not satisfied for the bound functions of an arbitary quantum well.

[2] Here the following examination questions suggest themselves: (i) What do the eigenfunctions $u_n(x)$ of a harmonic oscillator look like (qualitatively!) for the following three cases: $n = 0$; $n = 1$; and $n = 256$? (ii) What would be obtained, classically and quantum mechanically, for the probability of a harmonically oscillating particle of energy E to be found in a small interval Δx?

27 The Hydrogen Atom according to Schrödinger's Wave Mechanics

27.1 Product Ansatz; the Radial Function

The electron in a hydrogen atom is treated analogously to the previous cases. The Hamilton operator is written

$$\hat{\mathcal{H}} = \frac{\hat{\boldsymbol{p}}^2}{2m} + V(r) \; . \tag{27.1}$$

The explicit expression for the potential energy,

$$V(r) = -\frac{Z\,\mathrm{e}^2}{4\pi\varepsilon_0 r} \; ,$$

is not immediately important: one only requires that $V(r)$ should be rotationally invariant.

Next, the usual *ansatz* for stationary states is made:

$$\psi(\boldsymbol{r}, t) = u(\boldsymbol{r}) \cdot \mathrm{e}^{-\mathrm{i}Et/\hbar} \; .$$

Then, in spherical coordinates a *product ansatz* (*separation of variables*) follows:

$$u(\boldsymbol{r}) = R(r) \cdot Y_{lm}(\theta, \varphi) \; ,$$

with the *spherical harmonics* $Y_{lm}(\theta, \varphi)$ (which have already been introduced in Part II[1]).

This product ansatz can be made essentially because the operators $\hat{\boldsymbol{L}}^2$ and $\hat{L}_z$ commute with each other and (as conserved quantities) also with $\hat{\mathcal{H}}$, such that all three operators can be diagonalised simultaneously.

As for the Y_{lm}, here we only need the property that they are eigenfunctions of the (orbital) angular momentum operators $\hat{\boldsymbol{L}}^2$ and $\hat{L}_z$, i.e.,

$$\hat{\boldsymbol{L}}^2 Y_{lm}(\theta, \varphi) = \hbar^2 l \cdot (l+1) Y_{lm}(\theta, \varphi) \; , \tag{27.2}$$
$$\hat{L}_z Y_{lm}(\theta, \varphi) = \hbar m Y_{lm}(\theta, \varphi) \; . \tag{27.3}$$

---

[1] It is important to remind ourselves of other instances where the same or similar ideas or equations are used.

The characteristic values for the orbital angular momentum, $l = 0, 1, 2, 3, \ldots$[2] correspond, in "chemical language", to s, p, d, f, ... orbitals (we also have the relation $l \leq n - 1$, where $n(= 1, 2, 3, \ldots)$ is the principal quantum number, see below). The "magnetic" quantum number m assumes the allowed values $m = -l, -l + 1, \ldots, +l$ ($\rightarrow$ quantization of the orbital angular momentum; we do not yet consider the spin momentum which will be treated below).

In addition we need to express the Laplace operator in spherical coordinates:

$$\nabla^2 f = \frac{1}{r} \frac{\partial^2}{\partial r^2}(rf) - \frac{\hat{\boldsymbol{L}}^2 f}{\hbar^2 r^2} , \qquad (27.4)$$

where[3] the square of the orbital angular momentum can be written as

$$-\frac{\hat{\boldsymbol{L}}^2}{\hbar^2} f(\theta, \varphi) := \frac{1}{\sin \theta} \frac{\partial}{\partial \theta} \left(\sin \theta \frac{\partial f}{\partial \theta} \right) + \frac{1}{\sin^2 \theta} \frac{\partial^2 f}{\partial \varphi^2} . \qquad (27.5)$$

For the radial function $R(r)$ one obtains (using an auxiliary function $w(r) := r \cdot R(r)$) the following one-dimensional differential equation (N.B. this depends on l, but not on m):

$$-\frac{\hbar^2}{2m} w(r)'' + \left[\frac{\hbar^2 l(l+1)}{2mr^2} + V(r) \right] \cdot w(r) = E \cdot w(r) . \qquad (27.6)$$

Additionally, the following boundary conditions must be satisfied: $w(0) \overset{!}{=} 0$ [4] and

$$\int_0^\infty r^2 |R(r)|^2 \, \mathrm{d}r = \int_0^\infty \mathrm{d}r |w(r)|^2 \overset{!}{=} 1$$

since $\mathrm{d}^3 r = r^2 \, \mathrm{d}r \, \mathrm{d}\Omega$ and $\oint \mathrm{d}\Omega |Y_{lm}(\theta, \varphi)|^2 := \int\limits_{\theta:=0}^{\pi} \sin \theta \mathrm{d}\theta \int\limits_{\varphi=0}^{2\pi} \mathrm{d}\varphi |Y_{lm}(\theta, \varphi)|^2 = 1$). From now on we concentrate on a Coulomb potential

$$V(r) = -|\mathrm{e}|^2 / (4\pi\varepsilon_0 r) ,$$

i.e., to an A/r-potential, which has already been shown to be special in classical mechanics (remember that for A/r-potentials the Runge-Lenz vector is conserved ($\rightarrow$ *absence of perihelion motion of the Kepler ellipses and hyperbolae*)).

---

[2] It can be shown algebraically that the commutation relations of angular momenta can also be satisfied for half-integral numbers, 1/2, 3/2, 5/2 etc., which corresponds to the possibility of *non-orbital angular momenta*, e.g., the *spin*, see below.

[3] The relation (27.4) has (since $\hat{p}^2 = -\hbar^2 \nabla^2$) the following classical pendant: $\boldsymbol{p}^2 = p_r^2 + r^{-2}[\boldsymbol{r} \times \boldsymbol{p}]^2 = p_r^2 + \boldsymbol{L}^2/r^2$; here $\boldsymbol{L} = \boldsymbol{r} \times \boldsymbol{p}$ is the orbital angular momentum vector.

[4] Remember the footnote on the quasi one-dimensional behavior of bound s states in a 3d quantum box in the previous Chap. 25.

By analogy with the harmonic oscillator, reduced lengths and energies are also defined, i.e. the variables $\varrho := r/a_0$ and $\varepsilon := -E/E_R$, where

$$a_0 = \frac{\hbar^2}{4\pi\varepsilon_0\,e^2}(\approx 0.529\,\text{Å}(\equiv 0.0529\,\text{nm}))$$

is called the Bohr radius and

$$E_R := \frac{2m}{\hbar^2}\frac{e^4}{(4\pi\varepsilon_0)^2}(\approx 13.59\,\text{eV})$$

the Rydberg energy.

Finally the differential equation for $w(\varrho)$ thus obtained can again be simplified by an *ansatz* separating the dominating asymptotic behavior for $\varrho \ll 1$ and $\gg 1$ by two factorizations, i.e. by

$$w(\varrho) =: \varrho^{l+1} \cdot F(\varrho) \cdot e^{-\sqrt{\varepsilon}\varrho} \ .$$

Using this ansatz one obtains a rather simple differential equation (*Laguerre's equation*) for $F(\varrho)$, which is again solved by a power series

$$F(\varrho) = \sum_{\nu=0,1,2,\ldots}^{\infty} b_\nu \varrho^\nu \ .$$

In this way one is lead to the recursion relation

$$\frac{b_{\nu+1}}{b_\nu} = 2\frac{\sqrt{\varepsilon}\cdot(l+1+\nu)-1}{(\nu+1)(\nu+2l+2)} \ . \tag{27.7}$$

27.1.1 Bound States ($E < 0$)

These states correspond to negative energies E (positive values of ε). As for the harmonic oscillator, the condition of square-integrability of

$$w(\varrho)\ , \quad \int_0^\infty d\varrho|w(\varrho)|^2 \overset{!}{=} 1\ ,$$

is again only satisfied if the recursion relation (27.7) *terminates*. Again a positive integer n must exist, the "principal quantum number" (characterising the electron *shell*), such that the recursion relation (27.7) results in $b_{\nu+1} \equiv 0$, as soon as ν has reached the integer $n - l - 1$.

The energy E is then $= -\frac{E_R}{n^2}$, i.e., it does *not* depend on l, although the radial function $R(\varrho)(= R_{nl}(\varrho))$ does.

The fact that the binding energy of the H-atom only depends on the principal quantum number n (and not on the orbital quantum number l)

is a special property of the Coulomb potential[5], i.e., this "accidental degeneracy", as it it often called somewhat misleadingly, is *not* valid for general $V(r)$.

The product ansatz

$$u(\boldsymbol{r}) = R_{nl}(r) \cdot Y_{lm}(\theta, \varphi) \ ,$$

however, applies generally for *rotationally invariant potentials* $V(r)$, and the energy of bound states depends generally both on n and l.

27.1.2 The Hydrogen Atom for Positive Energies ($E > 0$)

This case is important *inter alia* because of its relation to the solar atmosphere and corona. The analysis corresponds to the Kepler problem, but with hyperbolic orbits, i.e., ε is now imaginary. Thus the recursion relation (27.7) no longer terminates, since

$$\sqrt{\varepsilon} = \mathrm{i}\sqrt{|\varepsilon|} \ ,$$

and one obtains radial functions $R_{E,l}(r)$, which are (analogously to monochromatic plane waves) only *almost* square integrable, i.e., although not being square integrable themselves, they can be superposed to square integrable wave packets.

The corresponding wave functions are

$$u_{E;l,m}(\boldsymbol{r}) = Y_{lm}(\theta, \varphi) \cdot R_{E,l}(r) \ ,$$

where the radial functions can be orthonormalized as

$$\langle R_{E,l} | R_{E',l'} \rangle = \int_0^\infty \mathrm{d}r r^2 R_{E,l}(r)^* R_{E',l'}(r) = \delta_{ll'} \delta(E - E') \ . \qquad (27.8)$$

The asymptotic behavior of the $R_{E,l}(r)$ for $r \to \infty$ is given by

$$R_{E,l}(r) \propto \frac{\sin\left(k_E \cdot r + \kappa_E \ln(2k_E r) + \eta_{l,E} - \frac{l\pi}{2}\right)}{r} \qquad (27.9)$$

with

$$k_E := \sqrt{\frac{2m}{\hbar^2} E} \ , \quad \kappa_E := \frac{1}{\sqrt{|\varepsilon|}}$$

and with a real so-called Coulomb phase $\eta_{l,E}$.

---

[5] In classical mechanics the corresponding speciality is the conservation of the so-called Runge-Lenz vector for the Kepler problem (i.e., planetary motion), which (as already mentioned) corresponds to the absence of any *perihelion rotation* in the ideal case, i.e., for a perfect A/r-potential.

27.2 Spherical Harmonics

The spherical harmonics $Y_{l,m}(\vartheta, \varphi)$ appearing in the above product ansatz, and in (27.2) and (27.3), are eigenfunctions of the operators $\hat{\boldsymbol{L}}^2$ and $\hat{L}_z$ with eigenvalues

$$\hbar^2 l \cdot (l+1) \quad \text{and} \quad m\hbar \,,$$

respectively (here, for convenience, we are using the index "m" (not mass) instead of "m_l"). The facts concerning $\hat{\boldsymbol{L}}^2$ and $\hat{L}_z$ are most important at this point and related to properties already known from electrodynamics, see Part II, since products

$$r^l \cdot Y_{l,m}(\vartheta, \varphi)$$

are harmonic, i.e.,

$$\nabla^2 \left[r^l \cdot Y_{l,m}(\vartheta, \varphi) \right] \equiv 0 \,,$$

which yields a simple proof of the above relation (27.2).

Concerning the second relation we note that the functions $Y_{lm}(\vartheta, \varphi)$ are defined as

$$Y_{lm}(\theta, \varphi) = c_{l,|m|} \cdot P_{l,|m|}(\cos\theta) \cdot e^{im\varphi} \,, \tag{27.10}$$

where the normalization factors $c_{l,|m|}$ are not of interest (in particular, because there are different conventions, distinguished by a complex factor of magnitude unity, a fact that one should consider when discovering apparent mistakes or inconsistencies in some formulae). The $P_{l,|m|}$ are called "associated Legendre polynomials"; for $m = 0$ one has the "genuine" Legendre polynomial.

The fact that $Y_{lm}(\vartheta, \varphi)$ is an eigenfunction of the z-component $\hat{L}_z$, with eigenvalue $\hbar m$, follows directly from the definition (27.10), since for

$$\hat{L}_z = x\hat{p}_y - y\hat{p}_x \,,$$

by application to a function of r, ϑ and φ, we have the simple identity

$$\hat{L}_z = \frac{\hbar}{i} \frac{\partial}{\partial\varphi} \,,$$

which can easily be proved (usually as an exercise).

In addition, the relation

$$\hat{\boldsymbol{L}}^2 Y_{lm} = \hbar^2 l(l+1) Y_{lm} \,,$$

which is equivalent to

$$\nabla^2 \left(r^l \cdot Y_{lm} \right) = 0 \,,$$

follows easily from the expression (27.4) for the ∇^2 operator in spherical coordinates.

The normalization coefficients (see below) are chosen in such a way that one obtains a complete orthonormalized system of angular functions on the surface of a sphere, i.e.

$$\langle Y_{lm}|Y_{l'm'}\rangle : = \int_0^\pi d\theta \sin\theta \int_0^{2\pi} d\varphi Y_{lm}(\theta,\varphi)^* Y_{l'm'}(\theta,\varphi)$$
$$= \delta_{ll'}\delta_{mm'}, \tag{27.11}$$

such that a generalized Fourier expansion of an arbitrary function $f(\vartheta,\varphi)$ is possible.

Important spherical harmonics are listed in the following table (we repeat that the normalization factors are not important. The bracketed terms on the r.h.s. show explicitly the relation to harmonic functions.)

$$Y_{0,0}(\theta,\varphi) = \sqrt{\frac{1}{4\pi}}, \tag{27.12}$$

$$Y_{1,0}(\theta,\varphi) = \sqrt{\frac{3}{4\pi}}\cos\theta \quad \left(\propto \frac{z}{r}\right) \tag{27.13}$$

$$Y_{1,\pm1}(\theta,\varphi) = \sqrt{\frac{3}{8\pi}}\sin\theta \cdot e^{\pm i\varphi} \quad \left(\propto \frac{x \pm iy}{r}\right) \tag{27.14}$$

$$Y_{2,0}(\theta,\varphi) = \sqrt{\frac{5}{16\pi}}(3\cos^2\theta - 1) \quad \left(\propto \frac{3z^2 - r^2}{r^2}\right) \tag{27.15}$$

$$Y_{2,\pm1}(\theta,\varphi) = \sqrt{\frac{15}{16\pi}}\cos\theta\sin\theta\,e^{\pm i\varphi} \quad \left(\propto \frac{z\cdot(x \pm iy)}{r^2}\right) \tag{27.16}$$

$$Y_{2,\pm2}(\theta,\varphi) = \sqrt{\frac{15}{32\pi}}\sin^2\theta\,e^{\pm 2i\varphi} \quad \left(\propto \frac{(x \pm iy)^2}{r^2}\right). \tag{27.17}$$

A nodal theorem is also valid for the spherical harmonics: The real and imaginary parts of Y_{lm} possess $l - |m|$ *polar* zero-lines ($\vartheta = $ constant) and additionally $|m|$ *azimuthal* zero-lines ($\varphi = $ constant). The radial functions $R_{n,l}(r)$ have $n - l - 1$ radial nodal surfaces. The sum of the radial, polar and azimuthal nodal surfaces of the product functions

$$R_{nl}(r) \cdot \mathcal{Re}[Y_{lm}(\vartheta,\varphi)]$$

(and similarly for the imaginary part) is thus $n - 1$, as expected.

As a consequence, in the hydrogen atom, the probability of finding the electron outside a radius $r_0 \gg n \cdot a_0$ is exponentially small,

$$\propto e^{-2r_0/(n\cdot a_0)} \, 6.$$

---

[6] i.e., not as small as for the harmonic oscillator, where it is $\propto e^{-x^2/x_0^2}$, see above.

28 Abstract Quantum Mechanics (Algebraic Methods)

28.1 The Harmonic Oscillator II: Creation and Destruction Operators

Heisenberg was the first to recognise[1] that for a given Hamiltonian the important point is to represent the observables by algebraic entities (e.g., matrices within his formalism; operators in Schrödinger's wave mechanics) which satisfy the *canonical commutation relations* (24.10)[2]. Then everything else follows purely algebraically (sometimes after a certain amount of perturbation theory) without the aid of differential operators.

This approach will now be used to treat the harmonic oscillator (although even the *hydrogen atom* was treated completely algebraically by Wolfgang Pauli[3], before Schrödinger invented his wave mechanics).

For the harmonic oscillator the Hamilton operator $\hat{\mathcal{H}}$ written as a function of the reduced length

$$\xi := x / \sqrt{\hbar/(m\omega_0)} \, ,$$

see Sect. 26, and reduced momentum

$$\hat{p}_\xi := \frac{1}{\mathrm{i}} \frac{\partial}{\partial \xi}$$

has the form:

$$\hat{\mathcal{H}} = \frac{\hbar\omega_0}{2} \left(\hat{p}_\xi^2 + \hat{\xi}^2 \right) \, .$$

The canonical commutation relations are equivalent to

$$\left[\hat{p}_\xi, \hat{\xi} \right] = \frac{1}{\mathrm{i}} \, .$$

---

[1] in his famous publication where he created *matrix mechanics* and in a subsequent publication with the co-authors Max Born and Pascual Jordan.

[2] This generalizes the *Poisson brackets* of classical mechanics, see Part I.

[3] See the book on quantum mechanics, [24], by W. Döring, Göttingen 1962, who also presents Pauli's algebraic approach for the H atom.

By complex rotation one now introduces the following operators, which are called the *destruction operator* and the *creation operator*, respectively:

$$\hat{b} := \frac{1}{\sqrt{2}}\left(\hat{\xi} + i\hat{p}_\xi\right) \quad \text{and} \quad \hat{b}^+ := \frac{1}{\sqrt{2}}\left(\hat{\xi} - i\hat{p}_\xi\right) . \tag{28.1}$$

They are *mutually adjoint* to each other,

$$\left\langle \hat{b}^+\psi_1 \middle| \psi_2 \right\rangle = \left\langle \psi_1 \middle| \hat{b}\psi_2 \right\rangle$$

(if $|\hat{b}\psi_2\rangle \neq 0$); i.e. they are clearly *non-Hermitian*, but instead they have a number of remarkable properties, *inter alia*:

If $|\psi_n\rangle$ ($n = 0, 1, 2, \ldots$) is a (normalized) eigenstate of $\hat{\mathcal{H}}$ then one has for $n = 0, 1, 2, \ldots$, and for $n = 1, 2, 3, \ldots$, respectively:

$$b^+|\psi_n\rangle = \sqrt{n+1}|\psi_{n+1}\rangle \quad \text{and} \quad \hat{b}|\psi_n\rangle = \sqrt{n}|\psi_{n-1}\rangle , \tag{28.2}$$

as well as $\hat{b}|\psi_0\rangle = 0$.

Thus $\hat{b}^+$ *increases* the number of excited quanta by 1, whereas $\hat{b}$ *decreases* this number by 1, which explains their names. The Hamilton operator itself can be written as:

$$\mathcal{H} = \frac{\hbar\omega_0}{2} \cdot (\hat{b}^+\hat{b} + \hat{b}\hat{b}^+) = \hbar\omega_0 \cdot \left(\hat{b}^+\hat{b} + \frac{1}{2}\right) . \tag{28.3}$$

In (28.3) we have used the commutation relation[4]

$$\left[\hat{b}, \hat{b}^+\right] = 1$$

(which follows from $[\hat{p}, \hat{x}] = \frac{\hbar}{i}$). The number operator $\hat{n} := \hat{b}^+\hat{b}$ is Hermitian, with eigenvalues $n = 0, 1, 2, \ldots$.

To prove the above statements we firstly calculate

$$|\Phi\rangle := \hat{b}^+\hat{b}\left(\hat{b}^+|\psi_n\rangle\right)$$

and find with the canonical commutation relation

$$\hat{b}\hat{b}^+ = 1 + \hat{b}^+\hat{b} \quad \text{that} \quad |\Phi\rangle = \hat{b}^+\left(1 + \hat{b}^+\hat{b}\right)|\psi_n\rangle .$$

Since $|\psi\rangle \equiv |\psi_n\rangle$ *is* already a (normalized) eigenstate of the number operator $\hat{n} = \hat{b}^+\hat{b}$ with eigenvalue n, we find that the state

$$|\psi'\rangle := b^+|\psi_n\rangle$$

must *also* be eigenstate of $\hat{n}$, *viz* with eigenvalue $n + 1$.

---

[4] The operator $\hat{b}$ is thus also not unitary, since for unitary operators $\hat{U}$ one would have $\hat{U}\hat{U}^+ = 1$, i.e., *without* commutator bracket.

The length of $|\psi'\rangle$ is calculated with

$$\langle \psi'|\psi'\rangle = \left\langle \hat{b}^+\psi_n \middle| \hat{b}^+\psi_n \right\rangle = \langle \psi_n|\hat{b}\hat{b}^+\psi_n\rangle \; ,$$

and with the canonical commutation relations we find

$$(1+n)\,\langle \psi_n|\psi_n\rangle \, (\equiv 1+n) \; .$$

Thus

$$||\psi'\rangle|$$

must be $\sqrt{n+1}$. In addition, if one assumed that the number operator $\hat{n}$ had eigenvalues *different* from the non-negative integer numbers, one would be led into contradictions.

28.2 Quantization of the Angular Momenta; Ladder Operators

One can proceed similarly with the angular momenta. Firstly, it follows from the *canonical commutation relations* for $\hat{\boldsymbol{r}}$ and $\hat{\boldsymbol{p}}$ that for the orbital angular momentum

$$\boldsymbol{\mathcal{L}} := \hat{\boldsymbol{r}} \times \hat{\boldsymbol{p}}$$

canonical angular momentum commutation relations are valid, i.e.,

$$\left[\hat{J}_x, \hat{J}_y\right] = \mathrm{i}\hbar J_z$$

(etc., by cyclic permutation of the three Cartesian components of an angular momentum vector operator $\hat{\boldsymbol{J}}$, which correspond to Hermitian operators $\hat{J}_k$, with $k = x, y, z$).

In the following, the algebra of the angular commutation relations is assumed to apply more generally, i.e., not only for orbital moments, $\boldsymbol{J} = \boldsymbol{\mathcal{L}}$, where it can be derived from the canonical commutation relations for $\hat{\boldsymbol{r}}$ and $\hat{\boldsymbol{p}}$, but without derivation right from the beginning. For example, on purely mathematical reasoning we consider the possibility that there may be, formally or not, other angular momenta or quasi-momenta[5] than those of the orbital motion.

The generalization of what is known for the harmonic oscillator is based on the following facts: the *ladder operators*,

$$\hat{J}^{\pm} := \hat{J}_x \pm \mathrm{i}\hat{J}_y \; ,$$

increase (or decrease) the eigenvalue of $\hat{J}_z$ by one unit of $\hbar$ from $m \cdot \hbar$ to $(m \pm 1) \cdot \hbar$, as far as possible, so that the following statements are true and can be proved purely algebraically ($\rightarrow$ exercises):

---

[5] e.g., the so-called isospin.

244 28 Abstract Quantum Mechanics (Algebraic Methods)

a) The operators $\hat{\boldsymbol{J}}^2$ and $\hat{J}_z$ commute with each other, and are thus simultaneously diagonalizable.

b) The orthonormalized abstract eigenvectors $|\psi_{J,M_J}\rangle$ of $\hat{\boldsymbol{J}}^2$ and $\hat{J}_z$ (i.e., of this simultaneous diagonalization) satisfy the relations

$$\hat{\boldsymbol{J}}^2|\psi_{J,M_J}\rangle = \hbar^2 J(J+1)|\psi_{J,M_J}\rangle , \qquad (28.4)$$

with *integral* or *half-integral* (!) value of J $(= 0, \frac{1}{2}, 1, \frac{3}{2}, \ldots)$, while

$$\hat{J}_z|\psi_{J,M_J}\rangle = \hbar M_J|\psi_{J,M_J}\rangle , \qquad (28.5)$$

with

$$M_J = -J, -J+1, \ldots, J .$$

c) Under these constraints (and the usual phase conventions concerning square roots of a complex number) one has

$$\hat{J}^{\pm}|\psi_{J,M_J}\rangle = \hbar\sqrt{J(J+1) - M_J(M_J \pm 1)}|\psi_{J,M_J\pm 1}\rangle , \qquad (28.6)$$

i.e.,

$$\hat{J}^{\pm}|\psi_{J,\pm J}\rangle = 0 .$$

Thus, in addition to the orbital angular momentum $(J = 0, 1, 2, \ldots)$, algebraically a second kind of angular momentum

$$\left(J = \frac{1}{2}, \frac{3}{2}, \ldots\right)$$

should exist, which has no classical analogue. This comprises (in addition to other cases) the so-called intrinsic or spin angular momentum of an electron[6], for which $J = \frac{1}{2}$.

In fact, for $J = 1/2$ the above-mentioned *canonical angular momentum commutation relations* can be implemented as follows:

$$\hat{\boldsymbol{J}} = \frac{\hbar}{2}\boldsymbol{\sigma} ,$$

with the *Pauli matrices*

$$\sigma_x = \begin{pmatrix} 0 , 1 \\ 1 , 0 \end{pmatrix} , \quad \sigma_y = \begin{pmatrix} 0 , -i \\ i , 0 \end{pmatrix} , \quad \sigma_z = \begin{pmatrix} 1 , 0 \\ 0 , -1 \end{pmatrix} . \qquad (28.7)$$

The corresponding states (complex two-component "spinors") can be generated by the following basic states:

$$\downarrow := \left|\psi_{\frac{1}{2}, -\frac{1}{2}}\right\rangle = \begin{pmatrix} 0 \\ 1 \end{pmatrix} , \quad \uparrow := \left|\psi_{\frac{1}{2}, \frac{1}{2}}\right\rangle = \begin{pmatrix} 1 \\ 0 \end{pmatrix} . \qquad (28.8)$$

---

[6] For other particles the spin may be different.

28.3 Unitary Equivalence; Change of Representation

All the above statements are not only independent of the basis functions used in the calculations but are also independent of the dynamical "aspect" used in the formulation of quantum mechanics.

One usually begins with the first of the following three "aspects" (or "pictures", or "representations") (i) the *Schrödinger aspect*, (ii) the *Heisenberg aspect*, and (iii) the *interaction (or Dirac) aspect*.

Changes in the basis functions and/or "representations"[7] are implemented by means of *unitary operators* $\hat{U}$. These are special linear operators defined on the total Hilbert space[8], which can be interpreted as (complex) rotations of the Hilbert space, since they leave scalar products invariant.

We thus assume that for all $|\psi\rangle$, $|\psi_1\rangle$ and $|\psi_2\rangle$ in Hilbert space the following equalities are valid:

$$|\psi'\rangle = \hat{U}|\psi\rangle \quad \text{or} \quad \langle\psi_1'|\psi_2'\rangle = \left\langle\hat{U}\psi_1\Big|\hat{U}\psi_2\right\rangle = \left\langle\psi_1\Big|\hat{U}^+\,\hat{U}\psi_2\right\rangle = \langle\psi_1|\psi_2\rangle \;.$$
(28.9)

For unitary transformations one therefore demands that

$$\hat{U}^+\hat{U} = \hat{U}\hat{U}^+ = \hat{1}\;, \quad \text{or} \quad \hat{U}^+ = \hat{U}^{-1}\;.$$

Such operators, if they do not depend on time, can always be written as $\hat{U} = e^{i\hat{B}}$ with *self-adjoint* (Hermitian plus complete) $\hat{B}$. The transition from position functions to momentum functions, or the transitions between various matrix representations, can be written in this way. However, unitary operators can also depend nontrivially on time. For example, the generalized *gauge transformation* (24.16), which was treated in an earlier section, corresponds to a (generally time-dependent) unitary transformation.

In addition to the invariance of all scalar products a unitary transformation also preserves the quantum mechanical expectation values, probability statements, commutation relations etc., i.e., all measurable physical statements. It is only necessary that all observables $\hat{A}$ appearing in these statements are transformed *covariantly* with the states, i.e. as follows:

$$|\psi\rangle \to |\psi'\rangle := \hat{U}|\psi\rangle\;, \quad \hat{A} \to \hat{A}' := \hat{U}^+\hat{A}\hat{U}\;.$$
(28.10)

It is easy to show that such transformations conserve all expectation values:

$$\left\langle\psi\Big|\hat{A}\psi\right\rangle \equiv \left\langle\psi'\Big|\hat{A}'\psi'\right\rangle\;.$$
(28.11)

---

[7] „Darstellungs- bzw. Bildwechsel" in German texts

[8] We remind ourselves that usually the Hamilton operator $\hat{H}$ is not defined in the total Hilbert space, since the functions ψ must be differentiable twice. In contrast, $\exp[-i\hat{H}\cdot t/\hbar]$, by a subtle limiting procedure, is defined in the total Hilbert space.

Up till now we have used Schrödinger's "picture". The transitions to the other representations, (ii) the Heisenberg or (iii) the Dirac (= interaction) representations, are described in the following. The essential point is that one uses the unitary transformations corresponding to the so-called *time-displacement operators* $\hat{U}(t, t_0)$. In addition we shall use indices S, H and I, which stand for "Schrödinger", "Heisenberg" and "interaction", respectively.

a) In the *Schrödinger picture* the states are time-dependent, but in general *not* the operators (e.g., position operator, momentum operator,...). We thus have

$$|\psi_S(t)\rangle \equiv \hat{U}(t, t_0)|\psi_S(t_0)\rangle .$$

Here the time-displacement operator $\hat{U}(t, t_0)$ is uniquely defined according to the equation

$$-\frac{\hbar}{i}\frac{\partial \hat{U}(t, t_0)}{\partial t} = \hat{\mathcal{H}}_S \hat{U}(t, t_0)$$

(i.e., the Schrödinger equation) and converges $\rightarrow \hat{1}$ for $t \rightarrow t_0$. Here t_0 is fixed, but arbitrary.

b) In the *Heisenberg picture* the state vectors are constant in time,

$$|\psi'\rangle \equiv |\psi_H\rangle := |\psi_S(t_0)\rangle ,$$

whereas now the operators depend explicitly on time, even if they do *not* in the Schrödinger picture: i.e., we have in any case:

$$\hat{A}_H(t) := \hat{U}^+(t, t_0)\hat{A}_S(t)\hat{U}(t, t_0) . \tag{28.12}$$

If this formalism is applied to a matrix representation

$$(A_H)_{j,k}(t) := \langle (\psi_H)_j | \hat{A}_H(t)(\psi_H)_k \rangle ,$$

one obtains, quasi "by the way", Heisenberg's *matrix mechanics*.

c) In Dirac's *interaction picture*, the Hamilton operator

$$\hat{\mathcal{H}}(= \hat{\mathcal{H}}_S)$$

is decomposed into an "unperturbed part" and a "perturbation":

$$\hat{\mathcal{H}}_S = \hat{H}_0 + \hat{V}_S ,$$

where $\hat{H}_0$ does not explicitly depend on time, whereas V_S *can* depend on t.

One then introduces as "unperturbed time-displacement operator" the unitary operator

$$\hat{U}_0(t, t_0) := e^{-i\hat{H}_0(t-t_0)/\hbar} .$$

and transforms all operators only with U_0, i.e., with the definition

$$A_I(t) := U_0^+(t, t_0)\hat{A}_S(t)\hat{U}_0(t, t_0) \, . \tag{28.13}$$

($A_I(t)$ *is thus a more or less trivial modification of $A_S(t)$, although the time dependence is generally different.*)

In contrast, the time-displacement of the state vectors in the interaction picture is more complicated, although it is already determined by (28.13) plus the postulate that the physical quantities, e.g., the expectation values, should be independent of which aspect one uses. In fact, for

$$|\psi_I(t)\rangle := \hat{U}_0^+(t, t_0)|\psi_S(t)\rangle$$

a modified Schrödinger equation is obtained, given by

$$-\frac{\hbar}{i}\frac{\partial|\psi_I(t)\rangle}{\partial t} = \hat{V}_I(t)|\psi_I(t)\rangle \, . \tag{28.14}$$

The unperturbed part of the Hamilton operator, $\hat{H}_0$, is thus "transformed away" from (28.14) in the interaction picture; but one should note that one has some kind of "conservation of effort" theorem: i.e., in (28.14) one must use $\hat{V}_I(t)$ instead of $\hat{V}_S$.

The formal solution of (28.14) is[9] the the following perturbation series (Dyson series):

$$|\psi\rangle_I(t) = \left(\hat{1} - \frac{i}{\hbar}\int_{t_0}^{t} dt_1 \hat{V}_I(t_1)\right.$$
$$\left. + \left(\frac{i}{\hbar}\right)^2 \int_{t_0}^{t} dt_1 \int_{t_0}^{t_1} dt_2 \hat{V}_I(t_1)\hat{V}_I(t_2) \mp \ldots \right)|\psi_I(t_0)\rangle \, , \tag{28.15}$$

which may be also symbolically abbreviated as

$$\left(\mathcal{T}e^{-\frac{i}{\hbar}\int_{t_0}^{t} dt_1 \hat{V}_I(t_1)}\right)|\psi_I(t_0)\rangle \, ;$$

the symbol $\mathcal{T}$ is called Dyson's time-ordering operator, because in (28.15)

$$t \geq t_1 \geq t_2 \geq \ldots \, .$$

Apart from the spin, which is a purely quantum mechanical phenomenon, there is a close correspondence between classical and quantum mechanical observables. For example, the classical Hamilton function and the quantum mechanical Hamilton operator usually correspond to each other by the above-mentioned simple replacements. The correspondence is further quantified by the *Ehrenfest theorem*, which states that the expectation values (!) of the

---

[9] This can be easily shown by differentiation.

observables exactly satisfy the *canonical equations of motion* of the classical Hamilton formalism.

The theorem is most simply proved using Heisenberg's representation; i.e., firstly we have

$$\frac{\mathrm{d}\hat{x}_H}{\mathrm{d}t} = \frac{\mathrm{i}}{\hbar}\left[\hat{\mathcal{H}}, \hat{x}_H\right] \quad \text{and} \quad \frac{\mathrm{d}\hat{p}_H}{\mathrm{d}t} = \frac{\mathrm{i}}{\hbar}\left[\hat{\mathcal{H}}, \hat{p}_H\right] .$$

Evaluating the commutator brackets, with

$$\hat{\mathcal{H}} = \frac{\hat{p}^2}{2m} + V(\hat{x}) ,$$

one then obtains the canonical equations

$$\frac{\mathrm{d}\hat{x}_H}{\mathrm{d}t} = \frac{\partial\hat{\mathcal{H}}}{\partial\hat{p}_H} \quad \text{and} \quad \frac{\mathrm{d}\hat{p}_H}{\mathrm{d}t} = -\frac{\partial\hat{\mathcal{H}}}{\partial\hat{x}_H} . \tag{28.16}$$

29 Spin Momentum and the Pauli Principle (Spin-statistics Theorem)

29.1 Spin Momentum; the Hamilton Operator with Spin-orbit Interaction

The Stern-Gerlach experiment (not described here) provided evidence for the half-integral spin of the electron[1]. As a consequence the wavefunction of the electron was assigned an additional quantum number m_s,

$$\psi \equiv \psi(\boldsymbol{r}, m_s) \,,$$

i.e., it is not only described by a position vector $\boldsymbol{r}$ but also by the binary variable

$$m_s = \pm\frac{1}{2} \,,$$

corresponding to the eigenvalues $m_s\hbar$ of the z-component $\hat{S}_z$ of a spin angular momentum. This was suggested by Wolfgang Pauli.

The eigenfunctions of the hydrogen atom are thus given by the following expression, for vanishing magnetic field and neglected spin-orbit interaction (see below), with $s \equiv \frac{1}{2}$:

$$u_{n,l,s,m_l,m_s}(\boldsymbol{r}', m_s') = R_{nl}(r') \cdot Y_{l,m_l}(\theta', \varphi') \cdot \chi_{s,m_s}(m_s') \,. \tag{29.1}$$

Here the two orthogonal spin functions are

a)

$$\chi_{\frac{1}{2},\frac{1}{2}}(m_s') = \delta_{m_s',+\frac{1}{2}} \,,$$

which is identical to the above-mentioned two-spinor

$$\alpha := \uparrow = \begin{pmatrix} 1 \\ 0 \end{pmatrix} \,,$$

b)

$$\chi_{\frac{1}{2},-\frac{1}{2}}(m_s') = \delta_{m_s',-\frac{1}{2}} \,,$$

which is identical with

$$\beta := \downarrow = \begin{pmatrix} 0 \\ 1 \end{pmatrix} \,.$$

---

[1] Apart from the Stern-Gerlach experiment also the earlier Einstein-de Haas experiment appeared in a new light.

Corresponding to the *five "good" quantum numbers* n, l, s, m_l and m_s there are the following *mutually commuting observables*:

$$\hat{\mathcal{H}} \left(\equiv \frac{\hat{p}^2}{2m} - \frac{Z\,e^2}{4\pi\varepsilon_0 r} \right), \quad \hat{L}^2, \quad \hat{S}^2, \quad \hat{L}_z \quad \text{and} \quad \hat{S}_z .$$

In contrast, in a constant magnetic induction

$$\boldsymbol{B} = \boldsymbol{e}_z B_z = \mu_0 \boldsymbol{H} = \mu_0 \boldsymbol{e}_z H_z$$

without the spin-orbit interaction (see below) one obtains

$$\hat{\mathcal{H}} \equiv \frac{(\hat{\boldsymbol{p}} - e\boldsymbol{A})^2}{2m} - \frac{Z\,e^2}{4\pi\varepsilon_0 r} - g\mu_B \boldsymbol{H} \cdot \frac{\hat{\boldsymbol{S}}}{\hbar}$$

(the last expression is the so-called "Pauli term", which is now considered). Here we have without restriction of generality:

$$\boldsymbol{A} = \frac{1}{2}[\boldsymbol{B} \times \boldsymbol{r}] = \frac{B_z}{2}(-y, x, 0) .$$

For electrons *in vacuo* the factor g is found experimentally and theoretically to be almost exactly 2 (more precisely: 2.0023.... Concerning the number 2, this value results from the relativistic quantum theory of Dirac; concerning the correction, 0.0023..., it results from quantum electrodynamics, which are both *not* treated in this volume.) The quantity

$$\mu_B = \frac{\mu_0 e \hbar}{2m}$$

is the Bohr magneton (an elementary magnetic moment);

$$\mu_0 = 4\pi \cdot 10^{-7} \text{ Vs}/(\text{Am})$$

is the vacuum permeability (see Part II). The spin-orbit interaction (see below) has so far been neglected.

Now, including in $\hat{\mathcal{H}}$ all relevant terms (e.g., the kinetic energy, the Pauli term etc.) in a systematic expansion and adding the spin-orbit interaction as well, one obtains the following expression:

$$\hat{\mathcal{H}} = \frac{\hat{p}^2}{2m} - \frac{Z\,e^2}{4\pi\varepsilon_0 r} - k(r)\frac{\hat{\boldsymbol{L}}}{\hbar} \cdot \frac{\hat{\boldsymbol{S}}}{\hbar} - \mu_B \boldsymbol{H} \cdot \frac{\left(g_l\hat{\boldsymbol{L}} + g_s\hat{\boldsymbol{S}} \right)}{\hbar}$$
$$+ \frac{\mu_0^2 e^2}{8m} \boldsymbol{H}^2 \cdot \left(x^2 + y^2 \right) . \tag{29.2}$$

Here the last term, $\propto \boldsymbol{H}^2$, describes the *diamagnetism* (this term is mostly negligible). In contrast, the penultimate term, $\propto \boldsymbol{H}$, describes the *Zeeman effect*, i.e., the influence of an external magnetic field, and the temperature-dependent *paramagnetism*, which has both orbital and spin contributions. Generally this term is important, and through the spin g-factor $g_s = 2$ one recognizes at once the anomalous behavior of the spin momentum $\hat{\boldsymbol{S}}$ compared with the orbital angular momentum $\hat{\boldsymbol{L}}$, where the orbital g-factor g_l is trivial: $g_l = 1$.

The third-from-last term is the spin-orbit interaction (representing the so-called "fine structure" of the atomic spectra, see below):

In fact, an electron orbiting in an *electrostatic* field $\boldsymbol{E}$ partially experiences this electric field, in the co-moving system, as a *magnetic* field $\boldsymbol{H}'$, cf. Part II:

$$\boldsymbol{H}' = -\frac{\boldsymbol{v} \times \boldsymbol{E}}{\mu_0 c^2} \, .$$

With

$$\boldsymbol{E} = \frac{Z|e|}{4\pi\varepsilon_0 r^3}\boldsymbol{r}$$

one obtains a correction to the Hamilton operator of the form

$$-k(r)\frac{\hat{\boldsymbol{L}}}{\hbar} \cdot \frac{\hat{\boldsymbol{S}}}{\hbar} \, ,$$

which has already been taken into account in (29.2). Here $-k(r)$ is positive and $\propto 1/r^3$; the exact value follows again from the relativistic Dirac theory.

In the following sections we shall now assume that we are dealing with an atom with many electrons, where L and S (note: capital letters!) are the quantum numbers of the *total* orbital angular momentum and *total* spin momentum of the electrons of this atom (Russel-Saunders coupling, in contrast to $j-j$ coupling[2]). Furthermore, the azimuthal quantum numbers m_l and m_s are also replaced by symbols with capital letters, M_L and M_S. The Hamilton operator for an electron in the outer electronic shell of this atom is then of the previous form, but with capital-letter symbols and (due to screening of the nucleus by the inner electrons) with $Z \to 1$.

29.2 Rotation of Wave Functions with Spin; Pauli's Exclusion Principle; Bosons and Fermions

Unusual behavior of electronic wave functions, which (as explained above) are half-integral spinor functions, shows up in the behavior with respect to spatial rotations. Firstly, we note that for a usual function $f(\varphi)$ a rotation about the z-axis by an angle α can be described by the operation

$$\left(\hat{D}_\alpha f\right)(\varphi) := f(\varphi - \alpha) \, .$$

---

[2] Here the *total spins* j of the single electrons are coupled to the *total spin* J of the atom.

As a consequence one can simply define the unitary rotation operator $\hat{D}_\alpha$ by the identity

$$\hat{D}_\alpha := e^{-i\alpha\hat{L}_z/\hbar} \; ; \quad \text{i.e., with} \quad \hat{L}_z = \frac{\hbar}{i}\frac{\partial}{\partial\varphi}$$

one obtains

$$\hat{D}_\alpha f(\varphi) = e^{-\alpha\frac{\partial}{\partial\varphi}} f(\varphi) \, ,$$

which is the Taylor series for $f(\varphi - \alpha)$.

As a consequence it is natural to define the rotation of a spinor of degree J about the z-axis, say, by

$$e^{-i\alpha\hat{J}_z/\hbar}|J, M_J\rangle \, .$$

The result,

$$e^{-i\alpha M_J}|J, M_J\rangle \, ,$$

shows that spinors with half-integral J (e.g., the electronic wave functions, if the spin is taken into account) are not reproduced (i.e., multiplied by 1) after a rotation by an angle $\alpha = 2\pi$, but only change sign, because

$$e^{-2\pi i/2} \equiv -1 \, .$$

Only a rotation by $\alpha = 4\pi$ leads to reproduction of the electronic wave function.

As shown below, *Pauli's exclusion principle* is an immediate consequence of this rotation behavior. The principle describes the *permutational behavior* of a wave function for N identical particles (which can be either *elementary* or *compound* particles). Such particles can have half-integral spin (as electrons, or He^3 atoms, which have a nucleus composed of one neutron and two protons, plus a shell of two electrons). In this case the particles are called *fermions*. In contrast, if the particles have *integral* spin (as e.g., pions, and He^4 atoms) they are called *bosons*.

In the position representation one then obtains the N-particle wave function: $\psi(1, 2, \ldots, N)$, where the variables $i = 1, \ldots, N$ are *quadruplets*

$$i = (\boldsymbol{r}'_i, (M'_S)_i)$$

of position and spin variables.

In addition to the square integrability

$$\int d1 \int \ldots \int dN |\psi(1, 2, \ldots, N)|^2 \overset{!}{=} 1 \, , \quad \text{with} \quad \int d1 := \sum_{M_1} \int d^3r_1 \, ,$$

one postulates for the Pauli principle the following *permutational behavior* for the exchange of two particles i and j (i.e., for the simultaneous interchange of the quadruplets describing position *and* spin of the two particles):

(Pauli's relation) $\psi(\ldots, j, i, \ldots) \overset{!}{=} (-1)^{2S} \psi(\ldots, i, j, \ldots)$,

$$\text{i.e., with} \quad (-1)^{2S} = -1 \quad \text{for } \textit{fermions} \text{ and}$$
$$+1 \quad \text{for } \textit{bosons} \text{ .} \qquad (29.3)$$

An immediate consequence is Pauli's exclusion principle for electrons: electrons have $S = \frac{1}{2}$; thus the wavefunctions must be antisymmetric. However, an anti-symmetric product function

$$(u_{i_1}(1) \cdot u_{i_2}(2) \cdot \ldots \cdot u_{i_N}(N))_{\text{asy}}$$

vanishes identically as soon as two of the single-particle states

$$|u_{i_\nu}\rangle (\nu = 1, 2, \ldots, N)$$

are identical (see also below). In other words, in an N-electron quantum mechanical system, no two electrons can be the same single electron state.

Thus, both the *rotational behavior* of the states and the *permutational behavior* of the wave function for N identical particles depend essentially on

$$S \left(= \frac{S}{\hbar} \right) ,$$

or more precisely on $(-1)^{2S}$, i.e., on whether S is integral (bosons) or half-integral (fermions) (the so-called "spin-statistics theorem"). In fact, this is a natural relation, since the permutation of particles i and j can be obtained by a correlated rotation by $180°$, by which, e.g., particle i moves along the upper segment of a circle from r_i to r_j, while particle j moves on the lower segment of the circle, from r_j to r_i. As a result, a rotation by 2π is effected, which yields the above-mentioned sign.

Admissible states of a system of identical fermions are thus linear combinations of the already mentioned anti-symmetrized product functions, i.e., linear combinations[3] of so-called *Slater determinants* of orthonormalized single-particle functions. These *Slater determinants* have the form

$$\psi_{u_1, u_2, \ldots, u_N}^{\text{Slater}}(1, 2, \ldots, N)$$
$$:= \frac{1}{\sqrt{N!}} \sum_{P \binom{i_1 \ \cdots \ i_N}{1 \ldots N}} (-1)^P u_1(i_1) \cdot u_2(i_2) \cdot \ldots \cdot u_N(i_N) \quad (29.4)$$

$$\equiv \frac{1}{\sqrt{N!}} \begin{vmatrix} u_1(1) \ , & u_2(1) \ , & \ldots, & u_N(1) \\ u_1(2) \ , & u_2(2) \ , & \ldots, & u_N(2) \\ \cdots \ , & \cdots \ , & \ldots, & \cdots \\ u_1(N) \ , & u_2(N) \ , & \ldots, & u_N(N) \end{vmatrix} . \qquad (29.5)$$

---

[3] Theoretical chemists call this combination phenomenon "configuration interaction", where a single Slater determinant corresponds to a fixed configuration.

Every single-particle state is "allowed" at most once in such a Slater determinant, since otherwise the determinant would be zero (→ Pauli's exclusion principle, see above).

This principle has enormous consequences not only in atomic, molecular, and condensed matter physics, but also in biology and chemistry (where it determines *inter alia* Mendeleev's periodic table), and through this the way the universe is constructed.

A more technical remark: if one tries to approximate the ground-state energy of the system optimally by means of a single anti-symmetrized product function (Slater determinant), instead of a sum of different Slater determinants, one obtains the so-called *Hartree-Fock approximation*[4].

Remarkably, sodium atoms, even though they consist exclusively of fermions, and both the total spin momentum of the electronic shell and the total spin of the nucleus are half-integral, usually behave as *bosons* at extremely low temperatures, i.e., for T less than $\sim 10^{-7}$ Kelvin, which corresponds to the ultraweak "hyperfine" coupling of the nuclear and electronic spins. However in the following sections, the physics of the electronic shell will again be at the center of our interest although we shall return to the afore-mentioned point in Part IV.

---

[4] This is the simplest way to characterize the Hartree-Fock approximation in a nutshell.

30 Spin-orbit Interaction; Addition of Angular Momenta

30.1 Composition Rules for Angular Momenta

Even if the external magnetic field is zero, M_L and M_S are no longer *good quantum numbers*, if the spin-orbit interaction is taken into account. The reason is that $\hat{L}_z$ and $\hat{S}_z$ no longer commute with $\hat{\mathcal{H}}$. However, $\hat{\boldsymbol{J}}^2$, the square of the total angular momentum

$$\hat{\boldsymbol{J}} := \hat{\boldsymbol{L}} + \hat{\boldsymbol{S}} \,,$$

and the z-component

$$\hat{J}_z := \hat{L}_z + \hat{S}_z \,,$$

do commute with each other and with $\hat{\mathcal{H}}$, and also with $\hat{\boldsymbol{L}}^2$ and $\hat{\boldsymbol{S}}^2$: the five operators $\hat{\mathcal{H}}$, $\hat{\boldsymbol{L}}^2$, $\hat{\boldsymbol{S}}^2$, $\hat{\boldsymbol{J}}^2$ and $\hat{J}_z$ (see below) all commute with each other. In fact we have

$$2\hat{\boldsymbol{L}} \cdot \hat{\boldsymbol{S}} \left(\propto \delta\hat{\mathcal{H}} \right) \equiv \hat{\boldsymbol{J}}^2 - \hat{\boldsymbol{L}}^2 - \hat{\boldsymbol{S}}^2 \,.$$

Thus, by including the spin-orbit interaction, a complete system of mutually commuting operators, including $\hat{\mathcal{H}}$, in a shell model, now consists of $\hat{\mathcal{H}}$, $\hat{\boldsymbol{L}}^2$, $\hat{\boldsymbol{S}}^2$, $\hat{\boldsymbol{J}}^2$ and $\hat{J}_z$. The corresponding "good quantum numbers" are: N, L, S, J and M_J (no longer N, L, S, M_L and M_S).

In the following, the radial functions $R_{N,L}$ can be omitted. Then, by linear combination of the

$$Y_{L,M_L} (\theta', \varphi') \cdot \chi_{S,M_S} (M_S')$$

(see below) one must find abstract states $\mathcal{Y}_{L,S,J,M_J} (\theta', \varphi', M_S')$:

$$\mathcal{Y}_{L,S,J,M_J} (\theta', \varphi', M_S') := \sum_{M_L,M_S} c_{L,S,J,M_J}^{(M_L,M_S)} \cdot Y_{L,M_L} (\theta', \varphi') \chi_{S,M_S} (M_S') \,,$$

(30.1)

which are *eigenstates* of $\hat{\boldsymbol{J}}^2$ and $\hat{J}_z$ with *eigenvalues* $\hbar^2 J(J+1)$ and $\hbar M_J$.[1]

---

[1] Using a consistent formulation a related abstract state $|\psi_{L,S,J,M_J}\rangle$ is defined by $\mathcal{Y}_{L,S,J,M_J}(\theta', \varphi', M_S') = \langle \theta', \varphi', M_J' | \psi_{L,S,J,M_J} \rangle :=$ r.h.s. of (30.1) ..., which is similar to defining an abstract state function $|\psi\rangle$ by the function values $\psi(\boldsymbol{r}) := \langle \boldsymbol{r} | \psi \rangle$.

By straightforward, but rather lengthy methods, or more generally by group theory[2], this can be shown to be exactly possible if J is taken from the following set:

$$J \in \{L + S, L + S - 1, \ldots, |L - S|\} \, .$$

Furthermore, M_J must be one of the $2J + 1$ integral or half-integral values $-J, -J + 1, \ldots, +J$, and $M_J = M_L + M_S$.

The coefficients in (30.1) are called *Clebsch-Gordan coefficients*. (N.B. Unfortunately there are many different conventions in use concerning the formulation of the coefficients.)

30.2 Fine Structure of the p-Levels; Hyperfine Structure

Without the spin-orbit interaction the p-levels ($l = 1$) of the outermost electron of an alkali atom are *sixfold*, since including spin one has $2 \times (2l + 1) = 6$ states. Under the influence of the spin-orbit interaction the level splits into a *fourfold* $p_{\frac{3}{2}}$-level, i.e., with

$$j = l + \frac{1}{2} \, ,$$

with *positive* energy shift ΔE, and a *twofold* $p_{\frac{1}{2}}$-level, i.e., with

$$j = l - \frac{1}{2} \, ,$$

with an energy shift which is twice as large (but *negative*) $-2\Delta E$; the "center of energy" of the two levels is thus conserved.

This so-called fine structure splitting gives rise, for example, to the well-known "sodium D-lines" in the spectrum of a sodium salt.

Analogously to the "fine structure", which is is based on the coupling of $\hat{\boldsymbol{L}}$ and $\hat{\boldsymbol{S}}$ of the electrons,

$$\delta \hat{\mathcal{H}}_{\text{fine}} \propto k(r) \hat{\boldsymbol{L}} \cdot \hat{\boldsymbol{S}} \, ,$$

there is an extremely weak "hyperfine structure" based on the coupling of the total angular momentum of the electronic system, $\hat{\boldsymbol{J}}$, with the nuclear spin, $\hat{\boldsymbol{I}}$, to the *total angular momentum* of the atom,

$$\hat{\boldsymbol{F}} := \hat{\boldsymbol{J}} + \hat{\boldsymbol{I}} \, .$$

We have

$$\delta \hat{\mathcal{H}}_{\text{hyperfine}} = c \cdot \delta(\boldsymbol{r}) \hat{\boldsymbol{J}} \cdot \hat{\boldsymbol{I}} \, ,$$

where the factor c is proportional to the magnetic moment of the nucleus and to the magnetic field produced by the electronic system at the location of the nucleus. In this way one generates, e.g., the well-known "21 cm line" of the neutral hydrogen atom, which is important in radio astronomy.

---

[2] See e.g., *Gruppentheorie und Quantenmechanik*, lecture notes (in German), [2]

30.3 Vector Model of the Quantization of the Angular Momentum

The above-mentioned mathematical rules for the quantization of the angular momenta can be visualized by means of the so-called *vector model*, which also serves heuristic purposes.

Two classical vectors $\boldsymbol{L}$ and $\boldsymbol{S}$ of length

$$\hbar\sqrt{L \cdot (L+1)} \quad \text{and} \quad \hbar\sqrt{S \cdot (S+1)}$$

precess around the *total angular momentum*

$$\boldsymbol{J} = \boldsymbol{L} + \boldsymbol{S} \,,$$

a vector of length

$$\hbar\sqrt{J \cdot (J+1)} \,,$$

with the above constraints for the admitted values of J for given L and S. On the other hand, the vector $\boldsymbol{J}$ itself precesses around the z-direction, where the J_z-component takes the value $\hbar M_J$. (At this point, a diagram is recommended.)

31 Ritz Minimization

The important tool of so-called "Ritz minimization" is based on a theorem, which can easily be proved under general conditions.

For example, let the Hamilton operator $\hat{\mathcal{H}}$ be bounded from below, with *lower spectral limit* E_0 (usually the true energy of the ground state); then one has for all states ψ of the region of definition

$$\mathcal{D}_{\hat{\mathcal{H}}} \quad \text{of} \quad \hat{\mathcal{H}} \,,$$

i.e., *for almost all states ψ of the Hilbert space $(a - \forall \psi)$*:

$$E_0 \leq \frac{\left\langle \psi \middle| \hat{\mathcal{H}} \psi \right\rangle}{\langle \psi | \psi \rangle}_{\text{"}a - \forall \psi\text{"}} . \tag{31.1}$$

If the lower spectral limit corresponds to the point spectrum of $\hat{\mathcal{H}}$, i.e., to a "ground state" ψ_0 and not to the continuous spectrum, the equality sign in (31.1) applies iff

$$\hat{\mathcal{H}} | \psi_0 \rangle = E_0 | \psi_0 \rangle \,.$$

The theorem is the basis of Ritz approximations (e.g., for the ground state), where one tries to optimize a set of variational parameters to obtain a minimum of the r.h.s. of (31.1). As an example we again mention the Hartree-Fock approximation, where one attempts to minimize the energy expectation of states which are represented by a single Slater determinant instead of by a weighted sum of different Slater determinants.

Generally the essential problem of Ritz minimizations is that one does not vary all possible states contained in the region of definition of $\hat{\mathcal{H}}$, but only the states of an *approximation set* $\mathcal{T}$, which may be infinite, but is nonetheless often too small. Usually even in the "infinite" case the set $\mathcal{T}$ does not contain the true ground state ψ_0, and one does not even know the distance of $\mathcal{T}$ from ψ_0.

The main disadvantage of the Ritz approximations is therefore that they are "uncontrolled", i.e., some intuition is needed.

For example, if one wants to obtain the ground states of (i) the *harmonic oscillator*,

$$\psi_0(x) \propto \mathrm{e}^{-\frac{x^2}{2x_0^2}} \,,$$

and (ii) the H-atom,

$$\psi_{n\equiv1}(r) \propto e^{-\frac{r}{a_0}} \ ,$$

by a Ritz minimization, the respective functions must be contained in $\mathcal{T}$;

$$x_0 \left(= \sqrt{\frac{\hbar}{2m\omega_0}}\right) \quad \text{and} \quad a_0 \left(\frac{\hbar^2}{ma_0^2} \overset{!}{=} \frac{e^2}{4\pi\varepsilon_0 \cdot a_0}\right)$$

are the characteristic lengths. For the n-th shell of the hydrogen atom the characteristic radius is $r_n \equiv n \cdot a_0$.

32 Perturbation Theory for Static Problems

32.1 Formalism and Results

Schrödinger's perturbation theory is more systematic than the Ritz method. Let there exist a perturbed Hamilton operator

$$\hat{\mathcal{H}} = \hat{H}_0 + \lambda \cdot \hat{H}_1 \, ,$$

with a real *perturbation parameter* λ.

The unperturbed Hamilton operator $\hat{H}_0$ and the perturbation $\hat{H}_1$ shall be independent of time; furthermore, it is assumed that the spectrum of $\hat{H}_0$ consists of the set of discrete eigenstates $u_n^{(0)}$ with corresponding eigenvalues $E_n^{(0)}$. The following *ansatz* is then made:

$$\begin{aligned}
u_n &= u_n^{(0)} + \lambda \cdot u_n^{(1)} + \lambda^2 \cdot u_n^{(2)} + \dots \, , \quad \text{and analogously :} \\
E_n &= E_n^{(0)} + \lambda \cdot E_n^{(1)} + \lambda^2 \cdot E_n^{(2)} + \dots \, .
\end{aligned} \tag{32.1}$$

Generally, these perturbation series do *not* converge (see below), similarly to the way the Taylor series of a function does *not* converge in general.

In particular, non-convergence occurs if by variation of λ the perturbation changes the spectrum of a Hamilton operator not only *quantitatively* but also *qualitatively*, e.g., if a Hamilton operator which is bounded from below is changed somewhere within the assumed convergence radius R_λ into an operator which is *unbounded*. In fact, the perturbed harmonic oscillator is non-convergent even for apparently "harmless" perturbations of the form

$$\delta\hat{\mathcal{H}} = \lambda \cdot x^4 \quad \text{with} \quad \lambda > 0 \, .$$

The reason is that for $\lambda < 0$ the Hamiltonian would always be unbounded (i.e., $R_\lambda \equiv 0$).

However, even in the case of non-convergence the perturbational results are still useful, i.e., as an asymptotic approximation for small λ, as for a Taylor expansion, whereas the true results would then often contain exponentially small terms that cannot be treated by simple methods, e.g., corrections

$$\propto e^{-\frac{\text{constant}}{\lambda}} \, .$$

With regard to the results of Schrödinger's perturbation theory one must distinguish between the two cases of (i) *non-degeneracy* and (ii) *degeneracy* of the unperturbed energy level.

a) If $u_n^{(0)}$ is *not degenerate*, one has

$$E_n^{(1)} = \left\langle u_n^{(0)} \middle| \hat{H}_1 u_n^{(0)} \right\rangle , \quad \text{and} \quad E_n^{(2)} = - \sum_{m(\neq n)} \frac{\left| \left\langle u_m^{(0)} \middle| \hat{H}_1 u_n^{(0)} \right\rangle \right|^2}{E_m^{(0)} - E_n^{(0)}} ,$$
(32.2)

as well as

$$\left| u_n^{(1)} \right\rangle = - \sum_{m(\neq n)} \left| u_m^{(0)} \right\rangle \frac{\left\langle u_m^{(0)} \middle| \hat{H}_1 u_n^{(0)} \right\rangle}{E_m^{(0)} - E_n^{(0)}} .$$
(32.3)

Hence the 2nd order contribution of the perturbation theory for the energy of the ground state is always negative. The physical reason is that the perturbation leads to the admixture of excited states ($\hat{=}$ polarization) to the unperturbed ground state.

b) In the *degenerate case* the following results apply.
   If (without restriction)

$$E_1^0 = E_2^{(0)} = \ldots = E_f^{(0)} ,$$

then this degeneracy is generally *lifted* by the perturbation, and the values $E_n^{(1)}$, for $n = 1, \ldots f$, are the *eigenvalues* of the Hermitian $f \times f$ perturbation matrix

$$V_{i,k} := \left\langle u_i^{(0)} \middle| \hat{H}_1 u_k^{(0)} \right\rangle , \quad \text{with} \quad i, k = 1, \ldots, f .$$

To calculate the new eigenvalues it is thus again only necessary to know the ground-state eigenfunctions of the *unperturbed* Hamilton operator. Only these functions are needed to calculate the elements of the perturbation matrix. Then, by diagonalization of this matrix, one obtains the so-called *correct linear combinations* of the ground-state functions, which are those linear combinations that diagonalize the perturbation matrix (i.e., they correspond to the eigenvectors). One can often guess these correct linear combinations, e.g., by symmetry arguments, but generally the following systematic procedure must be performed:

– Firstly, the eigenvalues, $E(\equiv E^{(1)})$, are calculated, i.e. by determining the zeroes of the following determinant:

$$P_f(E) := \begin{vmatrix} V_{1,1} - E , & V_{1,2} & , V_{1,3}, \ldots, & V_{1,f} \\ V_{2,1} & , V_{2,2} - E , & V_{2,3}, \ldots, & V_{2,f} \\ \ldots & , \quad \ldots & , \quad \ldots & , \quad \ldots \\ V_{f,1} & , \quad V_{f,2} & , V_{f,3}, \ldots, & V_{f,f} - E \end{vmatrix} .$$
(32.4)

– Secondly, if necessary, a calculation of the corresponding *eigenvector*

$$c^{(1)} := (c_1, c_2, \ldots, c_f)$$

(which amounts to $f - 1$ degrees of freedom, since a complex factor is arbitrary) follows by insertion of the eigenvalue into $(f - 1)$ independent matrix equations, e.g., for $f = 2$ usually into the equation

$$(V_{1,1} - E) \cdot c_1 + V_{1,2} \cdot c_2 = 0 .$$

(The *correct linear combination* corresponding to the eigenvalue $E = E^{(1)}$ is then: $\sum_{i=1}^{f} c_i \cdot u_i^{(0)}$, apart from a complex factor.)

Thus far for the degenerate case we have dealt with first-order perturbation theory. The next-order results (32.2) and (32.3), i.e., for $E_n^{(2)}$ and $u_n^{(1)}$, are also valid in the degenerate case, i.e., for $n = 1, \ldots, f$, if in the sum over m the values $m = 1, \ldots, f$ are excluded.

32.2 Application I: Atoms in an Electric Field; The Stark Effect of the H-Atom

As a first example of perturbation theory with degeneracy we shall consider the Stark effect of the hydrogen atom, where (as we shall see) it is *linear*, whereas for other atoms it would be *quadratic*.

Consider the unperturbed energy level with the principal quantum number $n = 2$. For this value the (orbital) degeneracy is $f = 4$

$$\left(u_i^{(0)} \equiv u_{n,l,m}^{(0)} \propto Y_{00} , \quad \propto Y_{10} , \quad \propto Y_{1,+1} \quad \text{and} \quad \propto Y_{1,-1} \right) .$$

The perturbation is

$$\lambda \hat{H}_1 = -eF \cdot z ,$$

with the electric field strength F.[1]

With $z = r \cdot \cos \theta$ one finds that only the elements

$$V_{21} = V_{12} \propto \langle Y_{00} | \cos \theta Y_{10} \rangle$$

of the 4×4 perturbation matrix do not vanish, so that the perturbation matrix is easily diagonalized, the more so since the states $u_3^{(0)}$ and $u_4^{(0)}$ are not touched at all.

---

[1] It should be noted that the perturbed Hamilton operator, $\hat{\mathcal{H}} = \hat{\mathcal{H}}_0 - \lambda \cdot z$, is not *bounded*, so that in principle also *tunneling through the barrier* to the resulting continuum states should be considered. However the probability for these tunneling processes is exponentially small and, as one calls it, *non-perturbative*.

The relevant eigenstates (correct linear combinations) are

$$u_{\pm}^{(0)} := \frac{1}{\sqrt{2}} \left(u_1^{(0)} \pm u_2^{(0)} \right) \; ;$$

they have an electric dipole moment

$$e \cdot \left\langle u_{\pm}^{(0)} \middle| z u_{\pm}^{(0)} \right\rangle = \pm 3 e \cdot a_0 \; ,$$

which orients parallel or antiparallel to the electric field, where one obtains an induced energy splitting.

Only for H-atoms (i.e., for A/r potentials) can a *linear* Stark effect be obtained, since for other atoms the states with $l = 0$ are no longer degenerate with $l = 1$. For those atoms second-order perturbation theory yields a *quadratic* Stark effect; in this case the electric dipole moment itself is "induced", i.e., $\propto F$. As a consequence the energy splitting is now $\propto F^2$ and the ground state of the atom is always reduced in energy, as mentioned above.

32.3 Application II: Atoms in a Magnetic Field; Zeeman Effect

In the following we shall use an electronic shell model.

Under the influence of a (not too strong) magnetic field the $f(= 2J + 1)$-fold degenerate states

$$|\psi_{N,L,S,J,M_J}\rangle$$

split according to their azimuthal quantum number

$$M_J = -J, -J + 1, \ldots, J \; .$$

In fact, in first-order degenerate perturbation theory the task is to diagonalize (with the correct linear combinations of the states $|M_J\rangle := \left| \psi_{N,L,S,J,M_J}^{(0)} \right\rangle$) the $f \times f$-matrix

$$V_{M_J, M_J'} := \langle M_J | \hat{H}_1 | M_J' \rangle$$

induced by the Zeeman perturbation operator

$$\hat{H}_1 := -\mu_B H_z \cdot \left(\hat{L}_z + 2\hat{S}_z \right) \; .$$

It can be shown that this matrix *is* already diagonal; i.e., with the above-mentioned basis one already *has* the correct linear combinations for the Zeeman effect.

In linear order w.r.t. H_z we obtain

$$E_{N,L,S,J,M_j} = E_{N,L,S,J}^{(0)} - g_J(L,S) \cdot \mu_B \cdot H_z \cdot M_J \; . \tag{32.5}$$

Here $g_J(L, S)$ is the *Landé factor*

$$g_J(L, S) = \frac{3J(J+1) - L(L+1) + S(S+1)}{2J(J+1)} \, . \tag{32.6}$$

In order to verify (32.6) one often uses the so-called Wigner-Eckart theorem, which we only mention. (There is also an elementary "proof" using the vector model for the addition of angular momenta.)

33 Time-dependent Perturbations

33.1 Formalism and Results; Fermi's "Golden Rules"

Now we assume that in Schrödinger's representation the Hamilton operator

$$\hat{\mathcal{H}}_S = \hat{H}_0 + \hat{V}_\omega \exp(-\mathrm{i}\omega t) + \hat{V}_{-\omega} \exp(+\mathrm{i}\omega t) \qquad (33.1)$$

is already explicitly (monochromatically) time-dependent. Due to the "Hermicity" of the Hamilton operator it is also necessary to postulate that

$$\hat{V}_{-\omega} \equiv \hat{V}_\omega^+ .$$

Additionally, it is assumed that the perturbation is "switched-on" at the time $t_0 = 0$ and that the system was at this time in the (Schrödinger) state

$$\left(\psi_S^{(0)}\right)_i (t) := u_i^{(0)} \mathrm{e}^{-\mathrm{i}\omega_i^{(0)} t} .$$

Here and in the following we use the abbreviations

$$\omega_i^{(0)} := E_i^{(0)}/\hbar \quad \text{and} \quad \omega_{fi} := \omega_f^{(0)} - \omega_i^{(0)} .$$

We then expand the function $\psi_S(t)$, which develops out of this initial state ($i =$ "initial state"; $f =$ "final state") in the Schrödinger representation, as follows:

$$\psi_S(t) = \sum_n c_n(t) \cdot \exp\left(-\mathrm{i}\omega_n^{(0)} t\right) \cdot u_n^{(0)}(\boldsymbol{r}) . \qquad (33.2)$$

(In the "interaction representation" (label: I), related to Schrödinger's representation by the transformation $|\psi_I(t)\rangle := \mathrm{e}^{\mathrm{i}\hat{H}_0 t/\hbar}|\psi_S(t)\rangle$, we obtain instead : $\psi_I(t) = \sum_n c_n(t) \cdot u_n^{(0)}(\boldsymbol{r})$, and also the matrix elements of the perturbation operator simplify to $\langle(\psi_I)_n(t)|\hat{V}_I(t)|(\psi_I)_i(t)\rangle = \mathrm{e}^{\mathrm{i}(\omega_n - \omega_i - \omega)t}$. $\left\langle u_n^{(0)}\left|\hat{V}_\omega\right|u_i^{(0)}\right\rangle + \ldots$, where $+ \ldots$ denotes terms, in which ω is replaced by $(-\omega)$.)[1]

---

[1] For Hermitian operators we can write $\langle\psi|\hat{A}|\phi\rangle$ instead of $\langle\psi|\hat{A}\phi\rangle$.

For $f \neq i$ the Schrödinger equation yields the following result:

$$c_f(t) = -\frac{1}{\hbar} V_{fi}^{(0)} \cdot \frac{e^{i(\omega_{fi}-\omega)t} - 1}{\omega_{fi} - \omega} . \tag{33.3}$$

Here only the linear terms in $\hat{V}$ have been considered and the non-resonant terms

$$\propto \hat{V}_\omega^+ e^{+i\omega t} , \quad \text{i.e., with} \quad \omega \to (-\omega) ,$$

have also been neglected; $V_{fi}^{(0)}$ stands for

$$\left\langle u_f^{(0)} \Big| \hat{V}_\omega \Big| u_i^{(0)} \right\rangle .$$

By squaring the above result one obtains

$$|c_f(t)|^2 = \frac{1}{\hbar^2} \left| V_{fi}^{(0)} \right|^2 \cdot \frac{\sin^2 \frac{(\omega_{fi}-\omega)t}{2}}{\left(\frac{\omega_{fi}-\omega2}{2} \right)^2} . \tag{33.4}$$

This corresponds to a periodic increase, followed by a decrease, with the Poincaré repetition time

$$\Delta t = 2\pi / |\omega_{fi} - \omega| ,$$

which is extremely long near a resonance of the denominator.

Thus, with a source of radiation consisting of n uncorrelated "radiators" of (almost) the same frequency $\omega_\alpha \approx \omega$, e.g.,

$$\hat{V}_\omega e^{-i\omega t} \to \sum_{\alpha=1}^{n} \hat{V}_{\omega_\alpha} e^{i(r(\alpha)-\omega_\alpha t)}$$

with random phases $r(\alpha)$, one obtains the n-fold result of (33.4) (if the frequencies are identical). In contrast, if the radiation were *coherent* (e.g., laser radiation), one would obtain the n^2-fold result. However in that case it makes no sense to interrupt the time-dependent perturbation series, as we did, after the lowest order.

In fact, at this point the transition from *coherent and reversible* quantum mechanics to *incoherent and irreversible* behavior occurs, as in statistical physics ($\to$ Part IV).

Thus, if one has a *continuum* of sources of *incoherent* radiation, i.e., with

$$\sum_{\omega_\alpha} \dots \to \int d\omega_\alpha \cdot \varrho_\gamma(\omega_\alpha) \dots ,$$

then one obtains as *transition rate* $W_{i \to f}$ ($\equiv$ transition probability $i \to f$ divided by the time t):

$$W_{i \to f} := \lim_{t \to \infty} \frac{\overline{|c_f(t)|^2}}{t} = \frac{2\pi}{\hbar^2} \overline{\left| V_{fi}^{(0)} \right|^2} \cdot \varrho_\gamma(\omega_{fi}) . \tag{33.5}$$

In the above proof we have used the identity

$$\lim_{t\to\infty}\left(\frac{\sin^2\frac{(\omega_{fi}-\omega)\cdot t}{2}}{\left(\frac{\omega_{fi}-\omega2}{2}\right)^2\cdot t}\right)\equiv 2\pi\delta(\omega_{fi}-\omega)\;.$$

The matrix elements have been incoherently averaged, as expressed by the 'bar' in (33.5).

Equation (33.5) describes transitions from a discrete energetically lower level i to an energetically higher level f by induced absorption of continuous radiation of frequency ω with an incoherent density $\varrho_\gamma(\omega)$. See also Fig. 33.1 below.

Conservation of energy, $\omega = \omega_f - \omega_i$, is explicitly given by the δ-function in the above formal correspondence. By permutation of f and i and the simultaneous replacement $\omega \to (-\omega)$ one obtains almost the same (33.5) for the *induced emission* of radiation. But there is also a *spontaneous emission* of radiation, which has an emission rate $\propto |\omega_{fi}|^3$. This fact makes it hard (since spontaneous emission should be avoided) to obtain the necessary occupation of a high-energy level for X-ray lasers.

A formula similar to (33.5) is also obtained for incoherent transitions from a discrete state i into a continuum $\mathcal{K}$ with the (continuous) density $\varrho_f(E)$ of the final states:

$$W_{i\to\mathcal{K}} = \frac{2\pi}{\hbar}\overline{\left|V_{fi}^{(0)}\right|^2}\varrho_f(E_i+\hbar\omega)\;. \tag{33.6}$$

Such formulae are called Fermi's "golden rules".

An *induced absorption process* is illustrated in Fig. 33.1 by means of a so-called Feynman diagram. The corresponding *induced emission process* would instead have an *outgoing* wiggly line to the right. Concerning *translation invariance* (which does not apply to defective or amorphous solids) the related *momentum conservation* provides an example for the presence (and consequences) of selection rules (see below).

33.2 Selection Rules

Selection rules arise naturally from Fermi's "golden rules". The "selection" refers to the (squared) *matrix elements* appearing in the "golden rules", and refers essentially to their predicted vanishing or nonvanishing due to characteristic symmetry arguments.

To give a simple but typical example we consider a perturbation with so-called σ symmetry, i.e., $\hat{V} \propto z$, thus $\propto \cos\vartheta$, i.e., $\propto Y_{l_2=1,m_l\equiv 0}$, and an isotropic initial state, i.e., without angular dependence, i.e.,

$$|i\rangle \propto Y_{l_i\equiv 0,m_l\equiv 0}\;.$$

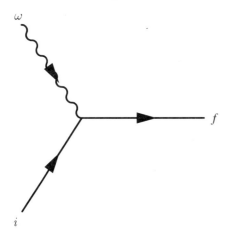

Fig. 33.1. Feynman diagram for an induced absorption process. The solid lines with symbols i and f correspond to an initial state $|\psi_i\rangle$ and a final state $|\psi_f\rangle$, e.g., to an atom with energy levels E_i and $E_f(> E_i)$ and to particles or quasi-particles propagating from the left with momenta $\boldsymbol{p}_i$ and $\boldsymbol{p}_f := \boldsymbol{p}_i + \hbar\boldsymbol{q}$. The wiggly line labelled ω represents the radiation quantum of frequency ω with momentum $\hbar\boldsymbol{q}$, which "pumps the system" from the initial state to the final state. Energy conservation, $E_f = E_i + \hbar \cdot \omega$, is always obeyed

As a consequence the matrix element

$$\langle f| \cos\vartheta |i\rangle$$

is then only nonvanishing, if

$$|f\rangle \propto Y_{l_f \equiv 1, m_l \equiv 0} , \quad \text{i.e., for} \quad l_f - l_i \equiv 1 , \quad m_f \equiv m_i (= 0) .$$

Further selection rules for other cases, e.g., for perturbations with $\pi_{\pm}$-symmetry, i.e.,

$$\hat{V} \propto (x \pm \mathrm{i}y) ,$$

or for multipole radiation beyond the dipole case, are obtained analogously, corresponding to the addition rules for angular momenta.

34 Magnetism: An Essentially Quantum Mechanical Phenomenon

34.1 Heitler and London's Theory of the H_2-Molecule; Singlet and Triplet States; the Heisenberg Model

This chapter also serves as preparation for the subsequent section on the interpretation of quantum mechanics. Firstly we shall treat the hydrogen *molecule* according to the model of Heitler and London, which is a most-important example.

The Hamilton operator for the two electrons is

$$\hat{\mathcal{H}} = \frac{\hat{\boldsymbol{p}}_1^2}{2m} + \frac{\hat{\boldsymbol{p}}_2^2}{2m} +$$
$$\frac{e^2}{4\pi\varepsilon_0} \cdot \left(-\frac{1}{r_{1A}} - \frac{1}{r_{2B}} + \left\{ \frac{1}{r_{12}} + \frac{1}{R_{AB}} - \frac{1}{r_{1B}} - \frac{1}{r_{2A}} \right\} \right) \quad . \quad (34.1)$$

Here r_{12} is the distance between the two electrons; R_{AB} is the separation of the two nuclei, which can be assumed to be at fixed positions, because the mass of the nuclei is ≈ 2000 times larger than that of the electrons; and r_{1B} is the distance of the first electron from nucleus B, etc.. The sum in braces, i.e., the *last* four terms on the r.h.s., can be considered as a perturbation of the *first* four terms.

Since the Hamilton operator does not depend on the spin and is permutationly symmetrical w.r.t. 1 and 2, the eigenfunctions can be written as *products* of position functions and spin functions, and they must have a well-defined *parity* w.r.t. permutations of the position variables $\boldsymbol{r}_1$ and $\boldsymbol{r}_2$.

Furthermore, due to Pauli's exclusion principle, both the position and spin functions must have *complementary permutation behavior*, i.e., a symmetric position function

$$\Phi_+(\boldsymbol{r}_1, \boldsymbol{r}_2)$$

(*symmetric* w.r.t. permutations of $\boldsymbol{r}_1$ and $\boldsymbol{r}_2$) must be multiplied by an *antisymmetric* spin function $\chi_-(1,2)$, and *vice versa*.

This leads to so-called *triplet* products

$$\psi_{tr.} := \Phi_-(\boldsymbol{r}_1, \boldsymbol{r}_2) \cdot \chi_+(1,2) \quad (34.2)$$

and *singlet* products

$$\psi_{si.} := \Phi_+(\boldsymbol{r}_1, \boldsymbol{r}_2) \cdot \chi_-(1,2) \, . \tag{34.3}$$

The names *singlet* and *triplet* are self-explanatory, i.e., there are three orthonormalized *triplet spin functions* $\chi_+(1,2)$:

$$
\begin{aligned}
|S = 1, M = +1\rangle &:= \alpha(1)\alpha(2) & &= \uparrow\uparrow \\
|S = 1, M = 0\rangle &:= \frac{1}{\sqrt{2}}(\alpha(1)\beta(2) + \beta(1)\alpha(2)) &= \frac{1}{\sqrt{2}}(\uparrow\downarrow + \downarrow\uparrow) \\
|S = 1, M = -1\rangle &:= \beta(1)\beta(2) & &= \downarrow\downarrow \, ,
\end{aligned}
\tag{34.4}
$$

but only one orthonormal *singlet spin function* $\chi_-(1,2)$:

$$|S = 0, M = 0\rangle := \frac{1}{\sqrt{2}}(\alpha(1)\beta(2) - \alpha(2)\beta(1)) \quad = \frac{1}{\sqrt{2}}(\uparrow\downarrow - \downarrow\uparrow) \, . \tag{34.5}$$

The functions are simultaneously eigenfunctions of the relevant operators for the total angular momentum, i.e.

$$\left(\hat{\boldsymbol{S}}_1 + \hat{\boldsymbol{S}}_2\right)^2 |S, M\rangle = \hbar^2 S(S+1)|S, M\rangle \quad \text{and}$$

$$\left(\left(\hat{S}_z\right)_1 + \left(\hat{S}_z\right)_2\right) |S, M\rangle = \hbar M |S, M\rangle \, ,$$

with $S = 1$ for the *triplet* states and $S = 0$ for the *singlet* state ($\to$ exercise). (What has been said in Section 30.1 about the addition of angular momenta and the so-called Clebsch-Gordan coefficients can be most easily demonstrated with this problem.)

Approximating the $\Phi_\pm$ by ground state functions of the hydrogen atom:

$$\Phi_\pm(\boldsymbol{r}_1, \boldsymbol{r}_2) := \frac{u_A(\boldsymbol{r}_1)u_B(\boldsymbol{r}_2) \pm u_B(\boldsymbol{r}_1)u_A(\boldsymbol{r}_2)}{\sqrt{2(1 \pm |S_{A,B}|^2)}} \, , \tag{34.6}$$

then, with

a) the so-called *overlap integral*

$$S_{A,B} := \int \mathrm{d}^3 r_1 (u_A(\boldsymbol{r}_1))^* u_B(\boldsymbol{r}_1) \, , \tag{34.7}$$

b) the *Coulomb integral*

$$C_{A,B} := \frac{e^2}{4\pi\varepsilon_0} \iint \mathrm{d}^3 r_1 \, \mathrm{d}^3 r_2 \frac{u_A^*(\boldsymbol{r}_1)u_B^*(\boldsymbol{r}_2)u_B(\boldsymbol{r}_2)u_A(\boldsymbol{r}_1)}{r_{12}} \, , \tag{34.8}$$

and

c) the *exchange integral*

$$J_{A,B} := \frac{e^2}{4\pi\varepsilon_0} \iint d^3 r_1 \, d^3 r_2 \frac{u_A^*(\boldsymbol{r}_1) u_B^*(\boldsymbol{r}_2) u_A(\boldsymbol{r}_2) u_B(\boldsymbol{r}_1)}{r_{12}} \qquad (34.9)$$

apart from minor corrections we obtain the following result:

$$E_{\text{triplet}} \cong \text{constant} \quad + \frac{C_{A,B} - J_{A,B}}{1 - |S_{A,B}|^2},$$
$$E_{\text{singlet}} \cong \text{constant} \quad + \frac{C_{A,B} + J_{A,B}}{1 + |S_{A,B}|^2}. \qquad (34.10)$$

Here not only the numerator but also the denominator is important; *viz*, surprisingly the *triplet* product is energetically higher(!) than the *singlet* product by an amount given by

$$\Delta E \cong 2 \cdot \frac{C_{A,B}|S_{A,B}|^2 - J_{A,B}}{1 - |S_{A,B}|^4}. \qquad (34.11)$$

This is a *positive* quantity, since $C_{A,B}$ is as large as $\mathcal{O}(10)$ eV, while $J_{A,B}$ is only $\mathcal{O}(1)$ eV, such that because of the rather large value of $|S_{A,B}|^2$ the energy difference ΔE is unexpectedly > 0. For the ground state of the hydrogen molecule, the quantity ΔE is in fact of the order of magnitude of $\mathcal{O}(10)$ eV. (*The actual functions $E_{\text{singlet}}(R_{AB})$ and $E_{\text{triplet}}(R_{AB})$ for the hydrogen molecule can be found in all relevant textbooks and should be sketched; approximate values for the equilibrium separation and the dissociation energy of the hydrogen molecule can be ascertained from textbooks of experimental molecular physics.*)

Because of the complementary permutation behavior of position and spin functions these results, which are due to the interplay of the Coulomb interaction and the Pauli principle, can also be obtained by an *equivalent spin operator*

$$\hat{H}^{\text{eff}}\left(\hat{\boldsymbol{S}}_1, \hat{\boldsymbol{S}}_2\right)$$

(see below, (34.13)), as introduced by Dirac. This spin operator, which acts on the spin factor attached to the position function, is just an equivalent description, replacing the genuine effects of Coulomb interaction plus Pauli principle, and contained in the *ansatz* for the two-electron function by complementary products, (34.2) and (34.3), together with the two-electron Hamilton operator $\mathcal{H}$ of (34.1).

The above replacement of

$$\mathcal{H}\left(\hat{\boldsymbol{p}}_1, \hat{\boldsymbol{p}}_2, \boldsymbol{r}_1, \boldsymbol{r}_2\right) \quad \text{by} \quad \hat{H}^{\text{eff}}\left(\hat{\boldsymbol{S}}_1, \hat{\boldsymbol{S}}_2\right)$$

is admittedly quite subtle. As a help towards understanding, see the following sketch (Fig. 34.1).

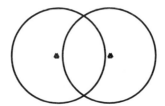

Fig. 34.1. A schematic aid to understanding the Heitler-London theory. The chemical bonding between s-states (spherical symmetry) favors *symmetrical* permutation behavior of the position variables (whereas the Coulomb repulsion acts against it) with a strong overlap between the nuclei (sketched as triangles). The spin function, which (in its permutational behavior) is complementary to the position variables, must be *antisymmetric*, i.e., a *singlet*. Thus one has *diamagnetism* as the usual state of diatomic molecules. *Paramagnetism* (or *ferromagnetism* in a solid) is only obtained if the overlap is strongly reduced (e.g., by non-s-symmetry, by Coulomb repulsion, or by "merging" of the nuclei; see the next section)

Apart from a constant, the Dirac spin operator yields the same energy spectrum as before. The operator is

$$\hat{H}^{\text{eff}}\left(\hat{\boldsymbol{S}}_A, \hat{\boldsymbol{S}}_B\right) = -2 \cdot J_{A,B}^{\text{eff}} \hat{\boldsymbol{S}}_A \cdot \hat{\boldsymbol{S}}_B \ . \tag{34.12}$$

Applied to the spin functions $|S, M\rangle$ it yields the result

$$-J_{A,B}^{\text{eff}} \cdot \left\{ S(S+1) - \frac{3}{2} \right\} |J, M\rangle \ ,$$

since

$$2 \cdot \hat{\boldsymbol{S}}_A \cdot \hat{\boldsymbol{S}}_B \equiv \left(\hat{\boldsymbol{S}}_A + \hat{\boldsymbol{S}}_B \right)^2 - \hat{\boldsymbol{S}}_A^{\,2} - \hat{\boldsymbol{S}}_B^{\,2} \ .$$

With (34.11) we have:

$$J_{A,B}^{\text{eff}} = \frac{J_{A,B} - C_{A,B} \cdot |S_{A,B}|^2}{1 - |S_{A,B}|^4} \ . \tag{34.13}$$

The natural generalization of equation (34.12) to systems of many atoms, i.e., a sum over A and B of terms similar to (34.12), is called the Heisenberg model of magnetism. One can show that it also applies to systems with $S > \frac{1}{2}$.[1]

In particular we have seen above that the effective exchange couplings J_{lm}^{eff} entering this model result from the *interplay of Coulomb interaction and Pauli principle*. Actually the two entities $J_{A,B}^{\text{eff}}$ and $J_{A,B}$ can have *different* sign. For example, if $J_{lm}^{\text{eff}} \leq 0$ (although $J_{A,B}$, as can be shown (e.g., [24],

---

[1] To obtain further insight one should see the following section on *Hund's rule*, which shows that within an atom there can be a strong ferromagnetic coupling between certain intra-atomic spins.

p. 392), is always ≥ 0), one obtains *diamagnetic* molecules (this is the most common case, e.g., the nitrogen molecule) and nonmagnetic or *antiferromagnetic* solids. In contrast, for $J_{lm}^{\text{eff}} \geq 0$ one obtains *paramagnetic* molecules (e.g., the oxygen molecule, which is rather an exceptional case) and nonmagnetic or *ferromagnetic* solids.

At this point quantum mechanics comes fully into play in all details and leads to important phenomena (magnetism, superconductivity,...), which cannot be explained by classical physics. For example, Bohr and van Leeuwen proved many years before the discovery of quantum mechanics that magnetism cannot be explained solely by orbital angular moments (e.g., by Ampère's current loops; see Part II).

34.2 Hund's Rule. Why is the O_2-Molecule Paramagnetic?

The majority of diatomic molecules are *diamagnetic*, in accordance with the theory of Heitler and London for the H_2-molecule, since in most cases the outer electrons of a diatomic molecule only have a single orbital at their disposal. According to the Pauli principle this orbital can at most be occupied by two electrons with opposite spins, e.g.,

$$\psi = u_A(\boldsymbol{r}_1)u_A(\boldsymbol{r}_2)\chi_-(1,2) \ .$$

However some diatomic molecules, e.g., O_2, turn out to be *paramagnetic*, which does not comply with the Heitler and London theory. If two or more orthogonal, energetically degenerate orbitals can be occupied, e.g., two $\pi_\pm$-molecular orbitals in the case of the oxygen molecule or two or more of the five 3d-orbitals in the case of a 3d-ion as Mn^{2+}, then the two electrons can choose between the following three possibilities.

(i)   Both electrons occupy the same orbital (due to the Pauli principle this is only possible with opposite spins); or

(ii) and (iii) they occupy different orbitals. This is possible either in a *singlet* state (case (ii)) or in a *triplet state* (case (iii)).

Hund's rule for the dominance of the configuration with *maximum possible multiplicity* states that of these possibilities case (iii) is favored:

Case (i) is excluded on energy grounds because of the large Coulomb repulsion of the electrons; this is roughly characterized by a Coulomb[2] integral, which involves one and the same orbital:

$$E_{(i)} = U_{AA} = (e^2/(4\pi\varepsilon_0)) \cdot \iint \mathrm{d}^3r_1\mathrm{d}^3r_2|u_A(\boldsymbol{r}_1)|^2 \cdot |u_A(\boldsymbol{r}_2)|^2/r_{12}$$

---

[2] These intra-orbital Coulomb integrals are often called *Hubbard integrals* or *screened Hartree integrals*.

(note A instead of B as the second index!). In contrast we have

$$E_{(ii),(iii)} = C_{A,B} \pm J_{A,B} \,,$$

where in general U_{AA} is significantly larger than $C_{A,B}$, which (on the other hand) is five to ten times larger than $J_{A,B}$. Case (iii), i.e., the *triplet* state, is thus favored over case (i) and case (ii); the latter, because the overlap integral $S_{A,B}$ vanishes *in the present case*, $S_{A,B} = \langle u_A | u_B \rangle \equiv 0$, such that *here* the energy difference between the triplet state (iii) and the singlet state (ii) is given by the direct (here called *Hund's rule*) exchange integral $J_{A,B}$; this integral (as already stated) is always non-negative[3]. Furthermore, as an intra-atomic integral the Hund's rule exchange interaction may be significantly stronger than the inter-atomic exchange integrals appearing in Heitler-London molecules.

---

[3] For the oxygen molecule the "Hund's rule exchange" is $J_{A,B} \approx +0.1$ eV(i.e. > 0), as opposed to $J_{A,B}^{\mathrm{eff}} < 0$ in the preceding section.

35 Cooper Pairs; Superconductors and Superfluids

The so-called *super* effects ("superconductivity" and "superfluidity", and recently (2004) also "supersolids") are typical quantum mechanical phenomena. The following will be now discussed here: (i) conventional superconductivity, (ii) the superfluidity of He^4 and He^3 and (iii) so-called high-temperature superconductivity.

i) Conventional metallic superconductors, e.g., Pb, have critical temperatures T_s below roughly 10 to 20 K; for $T < T_s$ the electric current (a) flows *without* energy losses, and (b) (below a critical value) the magnetic induction is completely *expelled* from the interior of the sample by the action of supercurrents flowing at the boundary (the so-called Meissner effect, see Part II).

It was discovered (not all that long ago) that the characteristic charge for these effects is not $= e$, but $= 2e$, i.e., twice the elementary charge.

In fact, for the temperature range considered, in the electron liquid so-called "Cooper pairs" form, i.e., pairs of electrons which "surround" each other – metaphorically speaking – at a large radius R between ≈ 50 and ≈ 1000 Å $(= 100\,\text{nm})$.[1]

The wave function of such a Cooper pair thus (i) consists of a (spherically) symmetrical position factor $\Phi(|\boldsymbol{r}_1 - \boldsymbol{r}_2|)$, typically an s-function; as a consequence, the remaining spin factor, (ii), must be antisymmetric. The result is so-called *singlet pairing*.

Viewed as compound particles these Cooper pairs thus behave *roughly* as bosons; thus they can condense into a collective state, the so-called *pair condensate*, which moves without resistance through the host system. Energetically the electrons use – in a highly cooperative manner – a small interaction effect, which results from the fact that the host system can be slightly *deformed* (so-called electron-phonon interaction). This has been known since 1957 (i.e., approximately half a century after the experimental discovery(!)); 1957 was the year of the formulation of the so-called BCS theory, named after Bardeen, Cooper and Schrieffer, [25].)

ii) Under normal pressure, He^4-gas becomes a liquid at 4.2 K; on further decrease of the temperature the normal fluid becomes *superfluid* at 2.17 K.

---

[1] A reminder: the characteristic atomic length is the *Bohr radius* $a_0 = 0.529$ Å $(= 0.0529\,\text{nm})$.

This means that below this temperature there is a component of the liquid which flows without energy dissipation.

He^4-atoms (2 protons, 2 neutrons, 2 electrons) as composite particles are *bosons*. Thus, even though there is a strong interaction between the boson particles, they can condense without pairing, and the transition to superfluidity consists in the formation of such a condensate.

The case of He^3 is fundamentally different. This atom consists of two protons, only one neutron, and again two electrons; thus as a composite particle it is a *fermion*. Originally one did not expect superfluidity at all in He^3. All the greater was the surprise when in 1972 Osheroff et al., [26], found that also a He^3-liquid becomes a superfluid, but admittedly at much lower temperatures, i.e. below 2.6 mK. In fact at these low temperatures (not yet ultralow, see Part IV) also in this liquid Cooper pairs of He^3-atoms are formed, where, however, this time the interaction favors *triplet* pairing, i.e. now the position factor of a Cooper pair is antisymmetric, e.g., with $l = 1$, and the spin function is symmetric, i.e., a *triplet function*. As a consequence, in the superfluid state of He^3 both the expectation value of $\hat{\boldsymbol{L}}$ and that of $\hat{\boldsymbol{S}}$ can be different from zero at the same time, which makes the theory very complex. In fact in 2003 a physics Nobel prize was awarded to Anthony Legget,[2] who was involved in the theory of He^3-superfluidity.

iii) Finally, some words on *high-temperature superconductors*. This is a certain class of non-metallic superconductors with many CuO_2-planes, and values of T_s around 120 K, i.e., six times higher than usual. In these systems the electrons also seem to form pairs; the "elementary charge" of the carriers of superconductivity is again $2e$, but the underlying mechanism is not yet clear, although the systems have been under scrutiny since 1986/87. Perhaps one is dealing with Bose-Einstein condensation (see Part IV) of electron-pair aggregates[3]. The "phononic mechanism" of conventional metallic superconductivity (see above) is apparently replaced here by electronic degrees of freedom themselves. Moreover it is also not accidental that the planar radius of the respective Cooper pairs is much smaller than usual.

The electronic degrees of freedom also include antiferromagnetic spin correlations, which seem to play an essential role. Furthermore, the position-space factor seems to correspond to $d_{x^2-y^2}$ and/or d_{xy} pairing, i.e., to attractive and repulsive interactions, respectively, at four alternating planar axes distinguished by $90°$.

---

[2] Nobel prize winners in 2003: Alexej A. Abrikosov, Vitalij L. Ginzburg, Anthony J. Legget, see below.

[3] There may be a relation to the crossover from Bardeen-Cooper-Schrieffer behavior of (weakly coupled) *ultracold* molecules of fermions to Bose-Einstein condensation of (strongly coupled) preformed fermion pairs, when a so-called Feshbach resonance is crossed. See PartIV.

36 On the Interpretation
of Quantum Mechanics
(Reality?, Locality?, Retardation?)

36.1 Einstein-Podolski-Rosen Experiments

In a famous paper published in 1935 [29] Einstein, Podolski and Rosen discussed the implications of quantum mechanics, and correctly pointed out several consequences that do not agree with common sense. For the following two decades in Princeton, Einstein and his group tried to obtain an "acceptable" theory by augmenting quantum mechanics with so-called "hidden variables". Only a few years after Einstein's death, however, John Bell at CERN showed that this is *not* possible (see below).

The phenomena related to the "unacceptable consequences" involve the notion of *entangled states*. In the paper of Einstein, Podolski and Rosen entangled states are constructed from eigenstates of the position and momentum operators $\hat{x}$ and $\hat{p}_x$. However, it is more obvious to form entangled states from spin operators. For example, in the *singlet* state $|S = 0, M = 0\rangle$ of two interacting electrons, see (34.5), the spin states of the single particles are *entangled*, which means that the result *cannot be written as a product of single-particle functions*. For the *triplet* states, in (34.4), two of them, $|S = 1, M = \pm 1\rangle$, are *not* entangled, while the third state, $|S = 1, M = 0\rangle$, is as entangled as the *singlet* state.[1]

Thus, if one has a two-fermion s-wave decay of a system (the "source"), with an original *singlet state* of that system, where two identical fermions leave the source in diametrically opposed directions, and if one finds by measurements on these particles (each performed with a single-particle measuring apparatus at large distances from the source) a well-defined z-component $\hat{S}_z$ of the first particle (e.g., the *positive* value $+\frac{\hbar}{2}$), then one always finds simultaneously(!), with a similar apparatus applied to the second diametrically-opposed particle, the *negative* value $-\frac{\hbar}{2}$, and *vice versa*, if quantum mechanics is correct (which it is, according to all experience; see Fig. 36.1).

However, since the same singlet wave function $|S = 0, M = 0\rangle$ diagonalizes not only the z-component of the total spin,

---

[1] For three particles, there are two different classes of entangled states, *viz* $|\psi\rangle_1 \propto |\uparrow\uparrow\uparrow\rangle + |\downarrow\downarrow\downarrow\rangle$ and $|\psi\rangle_2 \propto \uparrow\uparrow\downarrow + \downarrow\uparrow\uparrow + \uparrow\downarrow\uparrow$, and for four or more particles, there are even more classes, which have not yet been exploited.

$$\mathcal{S}_z := S_z(1) + S_z(2) \, ,$$

but also (as can easily be shown, $\rightarrow$ exercises) the x-component

$$\mathcal{S}_x := S_x(1) + S_x(2) \, ,$$

with the same eigenvalue $\hbar \cdot M \, (= 0)$, identical statements can also be made for a measurement of $\hat{S}_x$ instead of $\hat{S}_z$ (this is true for a singlet state although the operators do not commute).

These consequences of quantum mechanics seemed "unacceptable" to Einstein, since they violate three important postulates, which according to Einstein an "acceptable" theory should always satisfy:

a) the postulate of *reality*, i.e., that even *before* the measurement the state of the system is already determined (i.e., it appears unacceptable that by figuratively switching a button on the measuring equipment from $\hat{S}''_z$ to $\hat{S}''_x$ the observed spins apparently switch from alignment (or anti-alignment) along the z-direction to alignment (or anti-alignment) along the x-direction;

b) the *locality* of the measuring process (i.e., a measurement at position 1 should not determine another property at a different location 2), and

c) *retardation* of the propagation of information (i.e., information from a position 1 should reach a position 2 only after a finite time (as in electrodynamics, see Part II, with a minimum delay of $\Delta t := \frac{|r_1 - r_2|}{c}$).

Fig. 36.1. Singlet decay processes (schematically). A quantum system has decayed from a state without any angular momentum. We assume that the decay products are two particles, which travel diametrically away from each other in opposite directions, and which are always correlated despite their separation. In particular $|\psi\rangle$ always has *even* symmetry w.r.t. exchange of the spatial positions of the two particles, whereas the spin function is a *singlet*; i.e., if a measuring apparatus on the r.h.s. prepares a spin state with a value $\frac{\hbar}{2}$ in a certain direction, then a simultanous measurement on the opposite side, performed with a similar apparatus, prepares the value $-\frac{\hbar}{2}$. (This is stressed by the small line segments below the r.h.s. and above the l.h.s. of the diagram, where we purposely do not present arrows in the z or x direction, since any direction is possible. Purposely also no origin has been drawn to stress that this correlation is always present.)

In contrast, in quantum mechanics the individual state of the system is determined (i.e., *prepared*) just by the measurement[2]. This concerns the problem of reality.

Above all, however, quantum mechanics is *non-local*, i.e., $|\psi\rangle$ already contains all the information for all places, so that no additional information must be transferred from place 1 to 2 or *vice versa*[3]. This simultaneously concerns problems 2 and 3.

Of course, Einstein's objections must be taken very seriously, the more so since they really pinpoint the basis of the formalism. The same applies to the efforts, although inappropriate after all (see below), to modify Schrödinger's theory by adding *hidden variables* to obtain an *acceptable* theory (i.e. *"realistic"*, *"local"* and *"retarded"*). Ultimately, nine years after Einstein's death, it was shown by Bell, [30], a theorist at CERN, that the situation is more complex than Einstein assumed. Bell proved that Einstein's "acceptability" postulates imply certain inequalities for the outcome of certain experiments (the so-called "Bell inequalities" for certain four-site correlations), which significantly differ from the prediction of quantum mechanics. As a consequence, predictions of either quantum mechanics or (alternatively) the quasi-classical "acceptability" postulates of Einstein, Podolski and Rosen, can now be verified in well-defined experiments. Up to the present time many such *Bell experiments* have been performed, and quantum mechanics has always "won" the competition.

36.2 The Aharonov-Bohm Effect; Berry Phases

In the following, both the non-locality of quantum mechanics and wave-particle duality appear explicitly. We shall consider a *magnetized* straight wire along the z-axis. Since we assume that the magnetization in the wire points in the z-direction, the magnetic induction outside the wire vanishes everywhere,

$$\boldsymbol{B}_{\text{outside}} \equiv 0 \ .$$

Thus, an electron moving outside should not take any notice of the wire, as it "passes by" according to the classical equation of motions

$$m \cdot \frac{\mathrm{d}\boldsymbol{v}}{\mathrm{d}t} = e \cdot \boldsymbol{v} \times \boldsymbol{B} \ .$$

---

[2] For example, a singlet two-spin function contains a coherent superposition of both cases $(m_s)(1) = +\hbar/2$ and $(m_s)(1) = -\hbar/2$; they only become mutually exclusive by the (classical) measurement equipment.

[3] It is perhaps not accidental that difficulties are encountered (e.g., one has to perform renormalizations) if one attempts to unify quantum mechanics with *special relativity*, which is a local theory (in contrast to *general relativity*, which is, however, not even renormalizable).

However, quantum mechanically, electrons are represented by a wave function $\psi(\boldsymbol{r}, t)$, i.e., due to this fact they also "see" the vector potential $\boldsymbol{A}(\boldsymbol{r})$ of the magnetic field, and this is different from zero outside the wire, giving rise to a nontrivial magnetic flux Φ_0 (see below), a gauge invariant quantity, which causes a visible interference effect. In detail: For a so-called "symmetrical gauge" we have

$$\boldsymbol{A} \equiv \boldsymbol{e}_\varphi A_\varphi(r_\perp)$$

where the azimuthal component

$$A_\varphi(r_\perp) = \frac{\Phi_0}{2\pi r_\perp} , \quad \text{and} \quad r_\perp := \sqrt{x^2 + y^2} , \quad \text{where} \quad \Phi_0 = B\pi R^2$$

is the magnetic flux through the cross-section of the wire.

As already mentioned, the magnetic field vanishes identically outside of the wire, i.e.,

$$B_z = \frac{\mathrm{d}(r_\perp \cdot A_\varphi)}{r_\perp \mathrm{d}r_\perp} = 0 .$$

However, it is not $\boldsymbol{B}$, but $\boldsymbol{A}$, which enters the equation of motion for the quantum-mechanical probability amplitude

$$i\hbar \frac{\mathrm{d}\psi}{\mathrm{d}t} = \mathcal{H}\psi = \left(\frac{\hbar\nabla}{i} - e\boldsymbol{A}\right)^2 \psi/(2m) .$$

Hence the probability amplitude is influenced by $\boldsymbol{A}$, and finally one observes interferences in the counting rate of electrons, which pass either side of the wire and enter a counter behind the wire.

The results are gauge invariant, taking into account the set of three equations (24.16). The decisive fact is that the closed-loop integral

$$\oint_W \boldsymbol{A} \cdot \mathrm{d}\boldsymbol{r} = \Phi_0(\neq 0) ,$$

which determines the interference is gauge-invariant. The wave function effectively takes the wire fully into account, including its interior, although the integration path W is completely outside.

Aharonov-Bohm interferences have indeed been observed, e.g., by Börsch and coworkers, [27], in 1961, thus verifying that quantum mechanics is nonlocal, in contrast to classical mechanics.

In this connection we should mention the general notion of a so-called *Berry phase*; this is the (position dependent) phase difference[4] that results in an experiment if the wave function not only depends on $\boldsymbol{r}$ but also on parameters $\boldsymbol{\alpha}$ (e.g., on an inhomogeneous magnetic field), such that these parameters change adiabatically slowly along a closed loop. This formulation is consciously rather lax, since the concept of a Berry phase is based on general topological relations (e.g., parallel transport in manifolds), see [28].

---

[4] We remind ourselves that wavefunctions $\psi(\boldsymbol{r})$, which differ only by a global complex factor, describe the same state.

36.3 Quantum Computing

The subject of *quantum computing* is based on the fact that quantum-mechanical wave functions are superposable and can interfere with each other:

$$\psi = \psi_1 + \psi_2, |\psi|^2 = |\psi_1|^2 + |\psi_2|^2 + 2 \cdot \mathcal{R}e(\psi_1^* \cdot \psi_2) \ .^5$$

This is a field of research which has seen great progress in the last few years.

Firstly we should recall that classical computing is based on binary digits; e.g., the decimal number "9" is given by $1 \times 2^3 + 1 \times 2^0$, or the bit sequence 1001; as elementary "bits" one only has zero and unity; hence N-digit binary numbers are the vertices of a 2^N-dimensional hypercube. In contrast, quantum computing is based on the ability to superimpose quantum mechanical wave functions. One considers N-factor product states of the form

$$|\Psi\rangle = \prod_{\nu=1}^{N} \left(c_0^{(\nu)} \left| \psi_0^{(\nu)} \right\rangle + c_1^{(\nu)} \left| \psi_1^{(\nu)} \right\rangle \right) \ ;$$

i.e., as in Heisenberg spin systems with $S = \frac{1}{2}$ one uses the Hilbert space

$$\mathcal{H}^N_{\text{``2-level''}} \ ,$$

a (tensor) product of N "2-level systems", where the *orthonormalized* basis states

$$\left| \psi_0^{(\nu)} \right\rangle \quad \text{and} \quad \left| \psi_1^{(\nu)} \right\rangle$$

are called "quantum bits" or *qubits*. Of course $|\Psi\rangle$ is, as usual, only defined up to a complex factor.

The way that large amounts of computation velocity can be gained is illustrated by the following example.

Assume that (i) the unitary operator $\hat{U}$, i.e., a complex rotation in Hilbert-space, is applied to the initial state $|\Psi_0\rangle$ to generate the intermediate state

$$|\Psi_1\rangle = \hat{U}|\Psi_0\rangle \ ,$$

before (ii) another unitary operator $\hat{V}$ is applied to this intermediate state $|\Psi_1\rangle$ to generate the end state

$$|\Psi_2\rangle = \hat{V}|\Psi_1\rangle \ .$$

Now, the unitary product operator

$$\hat{W} := \hat{V}\hat{U}$$

---

[5] Several comprehensive reviews on quantum computing and quantum cryptography can be found in the November issue 2005 of the German "Physik Journal".

directly transforms the initial state $|\Psi_0\rangle$ into the final state $|\Psi_2\rangle$. In general a classical computation of the product matrix $\hat{W}$ is very complex; it costs many additions, and (above all!) multiplication processes, since

$$W_{i,k} = \sum_j V_{i,j} U_{j,k} \;.$$

In contrast, *sequential execution* by some kind of "experiment",

$$|\Psi_0\rangle \rightarrow |\Psi_1\rangle \rightarrow |\Psi_2\rangle \;,$$

may under certain circumstances be a relatively easy and fast computation for a skilled experimenter.

One could well imagine that in this way in special cases (in particular for $N \rightarrow \infty$) even an exponentially difficult task (i.e., growing exponentially with the number of digits, N, or with the size of the system, L) can be transformed to a much less difficult problem, which does not grow exponentially but only polynomially with increasing N and L.

In fact there are computation which are classically very difficult, e.g., the decomposition of a very large number into prime factors. One can easily see that $15 = 3 \times 5$ and $91 = 7 \times 13$; but even for a medium-sized number, e.g. 437, factorization is not easy, and for very large numbers the task, although systematically solvable, can take days, weeks, or months, even on modern computers. In contrast, the same problem treated by a quantum computer with a special algorithm (the Shor algorithm), tailored for such computers and this problem, would be solved in a much shorter time.

This is by no means without importance in daily life: it touches on the basis of the encoding principles used by present-day personal computers for secure messages on the internet, i.e., so-called PGP encoding (PGP $\hat{=}$ "pretty good privacy").

According to this encoding, every user has two "keys", one of which, the "public key" of the receiver, is used by the sender for *encoding* the message. This "public key" corresponds to the afore-mentioned "large number" $\mathcal{Z}$. But for fast *decoding* of the message the receiver also needs a *private key*, which corresponds to the decomposition of $\mathcal{Z}$ into prime factors, and this key remains known only to the receiver[6].

A "spy", knowing the computer algorithms involved and the "public key" of the receiver, e.g., his (or her) *"large number"*, can thus in principle calculate the corresponding "private key", although this may take weeks or months and would not matter for a lot of short-term transactions by the receiver until the "spy" has finished his computation. But if the "spy" could use a "quantum computer", this would be a different matter.

---

[6] According to this so-called PGP-concept (PGP $\hat{=}$ "Pretty Good Privacy") the *private key* is only "effectively private", see below, similar to a very large integer being factorized by a finite-time computation.

Fortunately, even now (i.e., in our age of classical computers), quantum mechanics offers a totally different way of *encoding*, called quantum cryptography (see below) which, in contrast to the classical encoding schemes is absolutely secure. However, because of the "coherency" requirement[7] of quantum mechanics the method suffers from the problem of range, so that at present it can only be applied over distances smaller than typically ~ 10 to $100\,\text{km}$[8].

After these preliminary remarks on quantum cryptography, we return to quantum computing.

There are other relevant examples where quantum computing would be much more effective than classical computing, e.g., in the task of sorting an extremely large set of data, where on a quantum computer the Grover algorithm would be much faster than comparable algorithms on classical computers.

One may ask oneself why under these circumstances have quantum computers not yet been realized, even though intense work has been going on in this field for many years. The difficulties are hidden in the notion of "a skilled experimentalist", used above, because it is necessary to ensure that, (i), during the preparation of the initial state $|\psi_0\rangle$; then, (ii), during the execution of all operations on this state; and finally, (iii), during all measurements of the results the *coherence* and *superposability* of all signals should be essentially undisturbed. This implies *inter alia* that errors (due to the necessary automatic correction) should only happen with an extremely small probability, e.g., $< 10^{-4}$, and that all experiments performed in the course of a quantum computation should be extremely well-controlled. Additionally, the system should be "scalable" to $N \gg 1$.

Many different suggestions have indeed been made for producing a quantum computer, one of which will be outlined in the next section.

36.4 2d Quantum Dots

Two-dimensional quantum dots can be regarded as artificial atoms with diameters in the region of $\sim 100\text{Å}(= 10\,\text{nm})$ in a two-dimensional electron (or hole) gas ["2DEG" (or "2DHG")]. A two-dimensional electron gas is formed, e.g., at the planar interface between a $GaAs$-semiconductor region (for $z < 0$) and an $Al_{1-x}Ga_xAs$-region (for $z > 0$). At the interface there is a deflexion of the energy bands, and as a consequence an attractive potential trench $V(x, y, z)$ forms, which is, however, attractive only w.r.t. the z-coordinate,

---

[7] The coherency demands do not allow, for quantum mechanical purposes, the usual *amplification* of the signals, which is necessary for the transmission of electromagnetic signals over hundreds or thousands of kilometers.

[8] A quantum cryptographical encoding/decoding software has been commercially available since the winter of 2003/2004.

whereas parallel to the interface (in the x- and y-directions) the potential is constant. Although the electrons are thus in a bound state w.r.t. the z coordinate, they can move freely parallel to the interface, however, with an effective mass m^* that can be much smaller than the electron mass *in vacuo*, m_e (e.g., $m^* = 0.067 m_e$).

Similarly to the case of a transistor, where one can fine tune the current between the source and drain very sensitively via the gate voltage, it is experimentally possible to generate a local depression region of the potential energy $V(\mathbf{r})$. Often the local region of this depression is roughly circular and can be described by a parabolic *confining potential*, as for a 1d-harmonic oscillator, but modified to 2d (the range of the confined region is $x^2 + y^2 \leq R^2$, with $R \approx 50\,\text{Å} (= 5\,\text{nm})$, and the magnitude of the depression is typically $-3.5\,\text{meV}$).

As a consequence, in the resulting two-dimensional potential well one has a finite number of bound states of the "2DEG" (so-called *confined electrons*). For simplicity the confinement potential is (as already mentioned) usually described by a parabolic potential (see below). In this way one obtains a kind of *artificial* 2d-atom, or so-called 2d-quantum dot[9], with a characteristic radius R in the region of 5 nm, which, on the one hand, is microscopically small, but conversely (in spite of the suggestive name "quantum dot") very large compared to a normal atom, which has a diameter in the range of 0.1 nm.

The electrons of the 2d-quantum dot are described by the following single-particle Hamiltonian (where the spin is neglected, which makes sense[10]):

$$\mathcal{H} = \frac{p_{r_\perp}^2}{2m^*} + \frac{(p_\varphi - e \cdot A_\varphi)^2}{2m^*} + \frac{m^* \omega_0^2}{2} r_\perp^2 , \tag{36.1}$$

where

$$r_\perp := \sqrt{x^2 + y^2} \quad \text{and} \quad \varphi = \arctan \frac{y}{x}$$

are the usual planar polar coordinates. The azimuthal quantity A_φ is the only component of our vector potential belonging to the constant magnetic induction $\mathbf{B}$ ($= \text{curl} \mathbf{A}$[11]), which points in the z-direction.

If more electrons are confined to the dot one must write down a sum of such terms, and additionally the Coulomb repulsion of the electrons and the Pauli principle must be taken into account.

---

[9] There are also 3d-quantum dots.

[10] The effective mass m^* is much smaller than the free-electron mass m_e, which enters into the Bohr magneton $\mu_B = \frac{\mu_0 e \hbar}{2m_e}$. Thus in the considered "artificial atoms" the (orbital) second term on the r.h.s. of (36.1) dominates the spin term $-g^* \cdot \mu_B \hat{\mathbf{S}} \cdot (\mathbf{B}/\mu_0)$. Here $g^* = -0.44$ is the effective Landé factor of the system, and the other quantities have their usual meaning (see above).

[11] We remind ourselves that $\mathbf{A}$ is not unique.

An intermediate calculation results in an even simpler expression for the magnetic induction: with the *ansatz*

$$A_\varphi \equiv \frac{B_0}{2} \cdot r_\perp$$

and the formula

$$B_z = \frac{\mathrm{d}(r_\perp \cdot A_\varphi)}{r_\perp \cdot \mathrm{d}r_\perp} \quad (B_{r_\perp} \text{ and } B_\varphi \text{ vanish}) ,$$

one obtains

$$\boldsymbol{B} \equiv \boldsymbol{e}_z \cdot B_0 .$$

Again a product ansatz is helpful:

$$\psi(r_\perp, \varphi) \overset{!}{=} R(r_\perp) \cdot \mathrm{e}^{\mathrm{i}m\varphi} .$$

With this ansatz one obtains an explicit solution of the Schrödinger equation corresponding to (36.1). Due to the 2π-periodicity w.r.t. φ the variable m must of course be an integer. However, for $B_z \neq 0$ one obtains only chiral cylindrical symmetry (i.e., m and $-m$ are not equivalent). This implies that with $\boldsymbol{B} \neq 0$ there is of course a preferred chirality, i.e., clockwise or counter-clockwise.

The equations around (36.1) are known as *Fock-Darwin theory*. This yields one of the few explicit solutions of nontrivial quantum mechanical problems, and in 2001 there was even a set of written examination questions on this theory for teacher students (file 6 of U.K.'s exercises in winter 2003).

Finally, from such artificial atoms (i) one can create *artificial molecules*; (ii) the spin state of these molecules can be calculated in a kind of Heitler-London theory, see above; and (iii) these states can be fine-tuned in a very controlled way as is necessary for quantum computing. In fact, one of the most promising suggestions of quantum computation is based on these systems. (In this context we should mention the theoretical work of Loss and DiVicenzo, [33], plus the experiments of Vandersypen and Kouwenhoven, [34].)

36.5 Interaction-free Quantum Measurement; "Which Path?" Experiments

In principle the Aharonov-Bohm experiment can already be considered as a form of interaction-free quantum measurement, since the electron "measures" the presence of the magnetized wire without coming into direct contact with it. But there is an indirect interaction via the vector potential, which is influenced by the wire.

Interaction-free quantum measurements in another sense (see below) have been performed more recently by Anton Zeilinger's group at Innsbruck[12] in

---

[12] Now at the university of Vienna.

Austria (see e.g., the paper in the journal "Spektrum der Wissenschaften" (1997, issue 1) which has already been mentioned, [32]).

Starting with a laser beam propagating in the x-direction, the photons of this beam are linearly polarized in the vertical ($\pm z$)-direction (polarization $P_0 = \pm 0°$); a *beam splitter* then follows which produces two coherent beams propagating, e.g., in the x-direction and the y-direction, respectively (i.e., along two different paths W_1 and W_2) with complementary diagonal polarizations

$$P_1 = \pm 45° \quad \text{and} \quad P_2 = \pm 135° .$$

Thirdly, the two beams meet again at a crossing point "X", where by interference the original polarization

$$P_0 = \pm 0°$$

is restored. Next, a *detector* follows, which is inserted at the continuation of the path W_1, but counts (per construction) only photons of polarization

$$P_D = \pm 90° .$$

As a consequence, this detector will *never* count any photon of the beam, since all photons leaving X have the orthogonal polarization

$$P_0 = \pm 0° .$$

Thus, with such a detector one cannot state which way a photon emitted by the source reached the crossing point.

Only when the interference is blocked, e.g., by interrupting the path W_2, will the detector start counting photons: a photon emitted by the source reaches the point with probability $\frac{1}{2}$ via the "non-blocked" path, i.e., with polarization

$$P_1 = \pm 45° ,$$

and is counted by the detector

$$P_D = \pm 90° ,$$

again with probability $\frac{1}{2}$.

In this way, i.e., if the detector records a count, one not only has a statement about, (i), *which path* the electron traveled (along W_1), but also one knows, (ii), that the alternative path W_2 is blocked (possibly by a container carrying a bomb; *cf.* the section on Schrödinger's cat). Morever, this fact, i.e., the possible presence of a bomb in path W_2, has been established here via an *interaction-free quantum measurement*, i.e., without making the bomb explode.

Such experiments (and many similar ones) on the non-locality of quantum mechanics (of course without involving any bomb) have been realized by Zeilinger et al.[13]. It is obvious that this has potential applications.

---

[13] In this connection, the recent book [32] should be mentioned.

36.6 Quantum Cryptography

Quantum cryptography is another topical application of quantum mechanics which has been practicable for some years (long before the advent of quantum computing). Quantum cryptography should of course be contrasted to classical cryptography, which is largely based on the PGP concept that we described in some detail in a previous Sect. 36.3 on quantum computing. Whereas classical cryptography is intrinsically *insecure*, since

a) the factorization of large integers, on which it is typically based, is always possible in principle; i.e., if one has enough computing power and patience (the computer may calculate for days, or weeks, or months) one can restore the private key of the receiver by "factorizing" his (or her) public key, and
b) if this does not work in a reasonable time, a spy can find out the *private key* i.e., the way the receiver decodes the contents of a message encoded by his (or her) public key, simply by *eavesdropping*

– in contrast, quantum cryptography is secure (which we stress again: although it is not based on quantum computers).

Within quantum cryptography, a *private key* not only becomes nonessential, rendering eavesdropping obsolete, but (almost incredibly) it is detrimental, because, through the act of eavesdropping, a spy will automatically uncover his or her own presence. The reason is essentially that quantum mechanical measurements usually perturb the measured state, and this can be discovered by cooperation between the sender and receiver.

In the following a well known protocol used in quantum cryptography is described in some detail.

The sender of the message (referred to as Alice) sends signals to the receiver (Bob) with linearly polarized photons with four different polarization directions

$$P = \pm 0°\,, \quad P = \pm 90° \quad P = \pm 45°\,, \quad P = \pm 135°$$

(i.e., "horizontally/vertically/right-diagonally/left-diagonally"). The information assignment A of the signals shall be fixed, e.g.,

$$A \equiv 0 \quad \text{for either} \quad P = \pm 0° \quad \text{or} \quad P = \pm 135°\,, \quad \text{and}$$
$$A \equiv 1 \quad \text{for either} \quad P = \pm 90° \quad \text{or} \quad P = \pm 45°\,.[14]$$

If one has a preference for spin systems, or if the equipment suggests doing so, instead of the optical polarization

$$either \quad P = \pm 90° \quad \text{or} \quad P = \pm 45°$$

---

[14] In contrast to fixed assignments, random assignments are also possible, which the sender and the receiver of the message must agree about before the process, or during it.

one can equivalently use the 2-spinors

$$either \quad \begin{pmatrix} 1 \\ 0 \end{pmatrix} \quad or \quad \frac{1}{\sqrt{2}} \begin{pmatrix} 1 \\ 1 \end{pmatrix} \quad for \quad A \equiv 1 \,;$$

and instead of

$$either \quad P = \pm 0° \quad or \quad P = \pm 135°$$

the spinors

$$either \quad \begin{pmatrix} 0 \\ 1 \end{pmatrix} \quad or \quad \frac{1}{\sqrt{2}} \begin{pmatrix} 1 \\ -1 \end{pmatrix}$$

can be used for $A \equiv 0$. The two classes of spinors named by *either* and *or* yield the eigenvectors of $\hat{S}_z$ and $\hat{S}_x$, with eigenvalues $\pm \frac{\hbar}{2}$.

It is important here that these different eigenvectors with the same eigenvalues of *either* $\hat{S}_z$ *or* $\hat{S}_x$ are *not* orthogonal, whereas according to Einstein, Podolski and Rosen's quasi-classical reasoning one would always tend to associate *orthogonal* states to *alternative* polarizations ($\hat{S}_z$ alternatively to S_x)

Thus far we have described the "public" part of the code. For the "secret" part, Alice and Bob, by some kind of "mutual preparation algorithm" (see the Appendix), come to an agreement, (i) as to which *basis sequence* b_1, b_2, $b_3,\ldots$ they will use in the course of the message, i.e., for the n-th bit either the x-basis (diagonal polarization) or the z-basis (rectilinear polarization), and (ii) which bits n of the sequence of signals contain the message. This part of the agreement (ii) can even be published (e.g., on the internet), whereas the *secret* part of the agreement, ((i), detailing which basis is chosen for the n-th bit), remains unknown to the public.[15]

The details of the "preparation algorithm" are complex, but straighforward, and (as mentioned) are described in the Appendix. But the result (which has also just been mentioned) is simple: the public community, i.e., also a spy ("Adam" or "Eve"), only knows that Alice and Bob use the same basis (z or x) for the bits containing information, but they do *not* know which one. As a consequence, the spy does *not* know whether Alice and Bob interprete a signal as 0 or as 1. If for example the x-basis is used at the n-th digit (which the spy does *not* know), then he (or she) does *not* know whether the signal is in the first component only, or in both components. If he (or she) chooses erroneously (with 50% probability of error) the first alternative and received a signal in the first component (which we can assume to be positive), then the spy does not know whether this component must be completed by a second number. This means that he (or she) does not know whether this second component, which has not been measured, is positive (representing a "1") or negative (representing a "0"); again there is 50% probability of error. Thus, if the spy continues his (or her) activity and sends instead of the

---

[15] The public, including the "spy", only know that sender and receiver use the same basis, but not which one.

original signals, which he (or she) has tried to analyse, "substitute signals" to Bob, these will be wrong with a high probability.

Consequently, Bob, by cooperating with Alice, can uncover the activities of the spy. Thus the spy not only has no chance of obtaining information by eavesdropping, but instead of being successful will necessarily reveal his (or her) presence to Alice and Bob.

For this concept at several places we have used essentially the nonclassical properties of the singlet state

$$|\psi_{\text{singlet}}\rangle \propto |\uparrow\downarrow - \downarrow\uparrow\rangle .$$

A main property of this state is the Einstein-Podolski-Rosen property of "entanglement".

In fact, the above method can even be simplified by using entangled *source* states, e.g., where Bob and Alice receive complementary versions of the signal. In this case *all* digits can be used for the message, i.e., the *mutual preparation* becomes trivial with respect to the *public* part, (ii), but the *secret* part, (i), still applies.

In this case too, the eavesdropping efforts of a spy would be unsuccessful and only reveal his own presence. One may compare the relevant paper in the journal "Physikalische Blätter" 1999, issue 6.

To enable quantum cryptography one must of course ensure *coherence*, i.e., the *ability to interfere*, along the whole length of propagation of the signals. For distances which are larger than typically 10 to 100 km this condition is violated by the present glass fibre technology, since de-coherence by the necessary restoration of the original signal strength can only be avoided up to this sort of distance. With regard to the first commercial realization in the winter of 2003/2004 the reader is referred to a footnote in the section on quantum computing.

37 Quantum Mechanics:
Retrospect and Prospect

In retrospect, let us briefly look again at some of the main differences and similarities between classical and quantum mechanics.

Classical mechanics of an N-particle system takes place in a 6N-dimensional *phase space* of coordinates plus momenta. Observables are arbitrary real functions of the 6N variables. The theory is deterministic and local, e.g., Newton's equations apply with forces which act at the considered moment at the respective position. Measurements can in principle be performed arbitrarily accurately, i.e., they only "state" the real properties of the system[1].

In contrast, in quantum mechanics the state of an N-particle system is described by an equivalence class of vectors ψ of a complex *Hilbert space* $\mathcal{HR}$, where the equivalence relation is given by multiplication with a globally constant complex factor; i.e., one considers "rays" in $\mathcal{HR}$. Already the description of the system is thus more complicated, the more so since the intrinsic angular momentum (*spin*) does not appear in classical physics and has unexpected non-classical properties, e.g., concerning the rotation behavior. The functions ψ must not only be (ia) square-integrable w.r.t. the position variables r_j of N particles $(j = 1, \ldots, N)$, but they must also be (ib) square-summable w.r.t. the respective spin variables $(m_s)_j$, and (ii) for identical particles the Pauli principle, (iia) i.e. permutation-antisymmetry for fermions (half-integer spin quantum number, e.g., electrons or quarks), and (iib) permutation-symmetry for bosons (integer spin quantum number, e.g., pions or gluons), must be considered. The Pauli principle has very important consequences in everyday life, e.g., the Mendeleev (or periodic) table of elements in chemistry.

With regard to (i) one must in any case have

$$\sum_{m_1=\pm 1/2} \cdots \sum_{m_N=\pm 1/2} \int_{r_1} \mathrm{d}^3 r_1 \cdots \int_{r_N} \mathrm{d}^3 r_N |\psi(r_1, m_1; \ldots; r_N, m_N)|^2 \stackrel{!}{=} 1 \;.$$

(37.1)

The integrand in (37.1) has the meaning of a multidimensional probability density.

---

[1] Newtonian mechanics is thus "realistic" in the sense of Einstein, Podolski and Rosen.

Furthermore, quantum mechanics is only *semi-deterministic*:

a) On the one hand, *between two measurements one has a deterministic equation of motion*, e.g., in the Schrödinger representation the equation,

$$i\hbar\dot{\psi} = \mathcal{H}\psi \, ,$$

and in the Heisenberg representation, which is equivalent, the equation of motion for the operators. These equations of motion are determined by the Hamilton operator, which corresponds largely, but not completely, to the classical Hamilton function $H(\boldsymbol{r}, \boldsymbol{p})$, where it is "almost only" necessary to replace the classical variables by operators, e.g., $\boldsymbol{r}$ by a multiplication operator and

$$\boldsymbol{p} \quad \text{by} \quad \frac{\hbar}{\mathrm{i}}\nabla \, ;$$

"almost only", but not completely, due to *spin*, which has no classical analogue.

As a consequence of spin there are additions to the Hamilton operator, in which the spin (vector) operator appears, which, (i), has an anomalous behavior w.r.t. the g-factor (visible in the coupling to a magnetic field). Furthermore, (ii), spin plays a decisive role in the *Pauli principle*, which is in itself most important. Finally, (iii), spin is coupled to the orbital angular momentum through the spin-orbit coupling, which is, e.g., responsible for the spectral fine structure of atoms, molecules and nuclei.

b) On the other hand, for the results of a measurement there are only "probability statements". The reason is essentially that observables are now described by Hermitian operators, which once more correspond largely (but again not completely) to quantities appearing in classical mechanics. In general, operators are not commutable. As a consequence *Heisenberg's uncertainty principle* restricts the product

$$\left(\delta\hat{A}\right)_{\psi}^{2} \cdot \left(\delta\hat{B}\right)_{\psi}^{2}$$

of expectation values, by stating that this product of the variances of two series of measurements for non-commutable Hermitian operators $\hat{A}$ and $\hat{B}$, with

$$\left(\delta\hat{A}\right)_{\psi}^{2} := \langle\psi|\hat{A}^{2}|\psi\rangle - \left(\langle\psi|\hat{A}|\psi\rangle\right)^{2} \, ,$$

should be

$$\geq \frac{1}{4}\left|\langle\psi|\left[\hat{A}, \hat{B}\right]|\psi\rangle\right|^{2} \, , \quad \text{where} \quad \left[\hat{A}, \hat{B}\right] := \hat{A} \cdot \hat{B} - \hat{B} \cdot \hat{A}$$

is the so-called commutator, i.e.,

$$[\hat{p}, \hat{x}] = \frac{\hbar}{\mathrm{i}} \, .$$

Thus, by taking square roots, with appropriate definitions of the uncertainty or *fuzziness* of position and momentum in the state ψ, one obtains the somewhat *unsharp* formulation: *position and momentum cannot simultaneously be precisely determined by a series of measurements,*

$$\delta \hat{x} \cdot \delta \hat{p} \geq \frac{\hbar}{2} .$$

With the definition

$$\delta \hat{k} := \delta \hat{p}/\hbar$$

for the *fuzziness* of the de-Broglie wavenumber one thus obtains

$$\delta \hat{k} \cdot \delta \hat{x} \geq 1/2 ,$$

i.e., a relation, in which the wave properties of matter dominate ($k = 2\pi/\lambda$).

However, the wave aspect of matter is only one side of the coin; *Heisenberg's uncertainty relation* covers both sides.

In fact, in quantum mechanics *wave-particle duality* with all its consequences applies: matter not only possesses *particle properties*, e.g., those described in Newtonian mechanics, but also *wave properties*, e.g., described by wave equations for probability amplitudes, with the essential property of *coherent superposition* and *interference* of these amplitudes. At this place one should also mention the *tunnel effect*.

Quantum mechanical measurements do not "state" properties of a system, but instead they "prepare" properties.

In this respect quantum mechanics is not *"realistic"* but *"preparing"* in the sense of Einstein, Podolski and Rosen. In particular, typically (but not always) a *measurement* changes the state of a system.

The totality of these statements is the so-called *Copenhagen interpretation* of quantum mechanics. After the objections of Einstein, Podolski and Rosen in their paper of 1935 were disproved by Bell experiments, it has now been accepted without controversy for several decades. The Copenhagen interpretation is also in agreement with *non-local* behavior, e.g., Aharonov-Bohm experiments. According to these experiments, the ψ-function, in contrast to the classical particle, does not "see" the local magnetic field and the Lorentz forces exerted by it, cf. Part II (so-called "local action"), but instead it "sees", through the vector potential, its non-local flux ("remote action"). Other consequences of the interpretation based, e.g., on *entanglement, coherence* and *interferences*, have only recently been systematically exploited.

After more than a century (nonrelativistic) quantum mechanics – which has been the main theme of this part of the text – is essentially a closed subject (as are Newtonian mechanics, Part I, and Maxwell's electrodynamics, Part II), but with regard to future prospects, this does not necessarily imply that all consequences of these theories (particularly of quantum theory) have been fully understood or exploited.

For "John Q. Public" some aspects of quantum mechanics, e.g., spin and the Pauli principle, are (unwittingly) of crucial importance in everyday life, e.g., for the simple reason that the periodic table of elements, and therefore the whole of chemistry and essential parts of biology, depend strongly on these properties, which are not at all "classical". Simply for this reason, at school or university, one should thus tolerate a partial lack of understanding of these topics; however, one should be careful about introducing potentially wrong explanations (e.g., spinning-top models for *spin*, which are popular, but essentially inadequate). Instead it may be better to admit that the reasons for the behavior are truly complicated.

In any case, quantum mechanics is a discipline which contains considerable scope for intellectual and philosophical discussion about the nature of matter, radiation, interaction, and beyond.

Finally, in the sections on quantum computation and quantum cryptography we have tried to make clear that quantum mechanics is a field with a promising future and that the next few decades might bring surprises, perhaps not only with regard to novel applications.

38 Appendix: "Mutual Preparation Algorithm" for Quantum Cryptography

Firstly, Alice (the sender of the message) with her *private key*[1] generates a sequence consisting of (+)- or (-)-symbols, which correspond to the instantaneous alignment of her "transmitter", e.g., an optical polarizer. The instantaneous polarization can be *horizontally/vertically* ($0°/90°$, $\hat{=}\hat{S}_z$) or *left/right diagonally* ($135°/45°$, $\hat{=}\hat{S}_x$).

Next she publicly sends to Bob (the receiver) – for their mutual preparation of the encoding of the message to follow – a data set consisting of a long random sequence of (0)- or (1)-bits, a *test message*, where the bits of the (0/1)-sequence are closely correlated to the (+/-)-sequence according to the following rules.

The test message contains at the position n either a "1", e.g., for vertical polarization, $P = 90°$, and for right-diagonal polarization, $P = 45°$; or a "0" for horizontal polarization and for left-diagonal polarization, $P = 0°$ and $P = 135°$.

Thirdly: Bob, using the same rules, receives the *test message* with another (+/-) analyzer sequence, taken from his own private key (which is not known by any other person, not even by Alice!).

Fourthly: Thereafter Bob informs Alice publicly which sequence he actually received, i.e. he sends to Alice the message "?" (e.g., an empty bit), if his analyzer was in the wrong polarization (e.g., if Alice transmitted a 0°-signal at the n-th place, whereas Bob's analyzer was (i) positioned to 90°, so that no signal was received; or (ii) positioned to 135° or 45°, so that the meaning was not clear, since the incoming signal had equally strong components for 1 and 0). In the remaining cases he sends to Alice either a "1" or a "0", according to what he received.

Fifthly: Alice finally compares this message from Bob with her own test message and fixes the numbers n_1, n_2, n_3, etc., where both agree. She informs Bob publicly about these numbers, and that her message, which she will send next, will be completely contained in these "*sensible bits*" and should be interpreted according to the known rules, whereas the remaining bits can be skipped.

Thus it is publicly known, which are the "sensible bits", and that at these bits Bob's analyzer and Alice's polarizer have the same polarization, but

---

[1] if existent; otherwise the sequence is generated *ad hoc*.

whether this is the horizontal/vertical polarization or, instead, the left/right-diagonal polarization is not known to anyone, except to Alice and Bob, inspite of the fact that the communications, including the rules, were completely public.

Although this procedure is somewhat laborious, the final result is simple, as stated above (see Sect. 36.6).

Part IV

Thermodynamics and Statistical Physics

39 Introduction and Overview to Part IV

This is the last course in our compendium on theoretical physics. In Thermodynamics and Statistical Physics we shall make use of a) classical non-relativistic mechanics as well as b) non-relativistic quantum mechanics and c) aspects of special relativity. Whereas in the three above-mentioned subjects one normally deals with just a few degrees of freedom (i.e. the number of atoms N in the system is usually 1 or of the order of magnitude of 1), in thermodynamics and statistical physics N is typically $\approx 10^{23}$; i.e. the number of atoms, and hence degrees of freedom, in a volume of $\approx 1\,\mathrm{cm}^3$ of a gas or liquid under normal conditions is extremely large. *Microscopic* properties are however mostly unimportant with regard to the collective behavior of the system, and for a gas or liquid only a few *macroscopic* properties, such as pressure p, temperature T and density ϱ characterize the behavior.

Quantum mechanics usually also deals with a small number of degrees of freedom, however with *operator* properties which lead to the possibility of discrete energy levels. In addition the *Pauli principle* becomes very important as soon as we are dealing with a large number of identical particles (see below).

In classical mechanics and non-relativistic quantum mechanics we have $v^2 \ll c^2$, with typical atomic velocities of the order of

$$|\langle \psi | \hat{v} \psi \rangle| \approx \frac{c}{100} \ .$$

However statistical physics also includes the behavior of a photon gas, for example, with particles of speed c. Here of course *special relativity* has to be taken into account (see Part I)[1].

In that case too the relevant macroscopic degrees of freedom can be described by a finite number of *thermodynamic potentials*, e.g., for a photon gas by the *internal energy* $U(T,V,N)$ and *entropy* $S(T,V,N)$, or by a single combination of both quantities, the *Helmholtz free energy*

$$F(T,V,N) := U(T,V,N) - T \cdot S(T,V,N) \ ,$$

where T is the thermodynamic temperature of the system in degrees Kelvin (K), V the volume and N the number of particles (number of atoms or

---

[1] In some later chapters even aspects of *general relativity* come into play.

molccules). In *phenomenological thermodynamics* these thermodynamic potentials are subject to measurement and analysis (e.g., using differential calculus), which is basically how a chemist deals with these quantities.

A physicist however is more likely to adopt the corresponding laws of statistical physics, which *inter alia* make predictions about how the above functions have to be calculated, e.g.,

$$U(T, V, N) = \sum_i E_i(V, N) p_i(T, V, N) \, .$$

Here $\{E_i \equiv E_i(V, N)\}$ is the energy spectrum of the system, which we assume is countable, and where the p_i are thermodynamic probabilities, usually so-called Boltzmann probabilities

$$p_i(T, \ldots) = \frac{\exp - \frac{E_i}{k_B T}}{\sum_j \exp - \frac{E_j}{k_B T}} \, ,$$

with the Boltzmann constant $k_B = 1.38 \ldots 10^{-23}$ J/K. We note here that a temperature $T \approx 10^4$ K corresponds to an energy $k_B T$ of ≈ 1 eV.

Using a further quantity $Z(T, V, N)$, the so-called *partition function*, where

$$Z(T, V, N) := \sum_j \exp - \beta E_j \quad \text{and} \quad \beta := \frac{1}{k_B T} \, ,$$

we obtain

$$U(T, V, N) \equiv -\frac{\mathrm{d}}{\mathrm{d}\beta} \ln Z \; ;$$

$$F(T, V, N) \equiv -k_B T \cdot \ln Z \; ;$$

$$S(T, V, N) \equiv -\frac{\partial F}{\partial T} \equiv -k_B \sum_j p_j \ln p_j \, .$$

There are thus three specific ways of expressing the entropy S:

a) $\quad S = \dfrac{U - F}{T} \, ,$

b) $\quad S = -\dfrac{\partial F}{\partial T} \, , \quad$ and

c) $\quad S = -k_B \sum_j p_j \ln p_j \, .$

The forms a) and b) are used in chemistry, while in theoretical physics c) is the usual form. Furthermore entropy plays an important part in information theory (the Shannon entropy), as we shall see below.

40 Phenomenological Thermodynamics: Temperature and Heat

40.1 Temperature

We can subjectively understand what *warmer* or *colder* mean, but it is not easy to make a quantitative, experimentally verifiable equation out of the inequality $T_1 > T_2$. For this purpose one requires a thermometer, e.g., a mercury or gas thermometer (see below), and *fixed points* for a temperature scale, e.g., to set the melting point of ice at normal pressure ($\hat{=}$ 760 mm mercury column) exactly to be $0\,^\circ$ Celsius and the boiling point of water at normal pressure exactly as $100\,^\circ$ Celsius. Subdivision into equidistant intervals between these two fixed points leads only to minor errors of a *classical* thermometer, compared to a *gas* thermometer which is based on the equation for an ideal gas:

$$p \cdot V = N k_B T$$

where p is the pressure ($\hat{=}$ force per unit area), V the volume (e.g., = (height) x (cross-sectional area)) of a gas enclosed in a cylinder of a given height with a given uniform cross-section, and T is the temperature in Kelvin (K), which is related to the temperature in Celsius, Θ, by:

$$T = 273.15 + \Theta$$

(i.e., $0\,^\circ$ Celsius corresponds to 273.15 Kelvin). Other temperature scales, such as, for example, Fahrenheit and Réaumur, are *not* normally used in physics. As we shall see, the Kelvin temperature T plays a particular role.

In the ideal gas equation, N is the number of molecules. The equation is also written replacing $N \cdot k_B$ with $n_{\mathrm{Mol}} \cdot R_0$, where R_0 is the *universal gas constant*, n_{Mol} the number of moles,

$$n_{\mathrm{Mol}} := \frac{N}{L_0}, \quad \text{and} \quad R_0 = L_0 \cdot k_B .$$

Experimentally it is known that chemical reactions occur in constant proportions (Avogadro's law), so that it is sensible to define the quantity *mole* as a specific number of molecules, i.e., the Loschmidt number L_0 given by

$$\approx 6.062(\pm 0.003) \cdot 10^{23} .$$

(Physical chemists tend to use the universal gas constant and the number of moles, writing

$$pV = n_{\text{Mol}} \cdot R_0 T \; ,$$

whereas physicists generally prefer $pV = N \cdot k_B T$.)

In addition to *ideal* gases physicists deal with *ideal paramagnets*, which obey Weiss's law, named after the French physicist Pierre Weiss from the Alsace who worked in Strasbourg before the First World War. Weiss's law states that the magnetic moment m_H of a paramagnetic sample depends on the external field H and the Kelvin temperature T in the following way:

$$m_H = \frac{C}{T} H \; ,$$

where C is a constant, or

$$H = \frac{m_H}{C} \cdot T \; .$$

From measurements of H one can also construct a Kelvin thermometer using this ideal law.

On the other hand an ideal *ferromagnet* obeys the so-called *Curie-Weiss law*

$$H = \frac{m_H}{C} \cdot (T - T_c) \quad \text{where} \quad T_c$$

is the critical temperature or *Curie temperature* of the ideal ferromagnet. (For $T < T_c$ the sample is *spontaneously* magnetized, i.e., the external magnetic field can be set to zero[1].)

Real gases are usually described by the so-called *van der Waals equation of state*

$$p = -\frac{a}{v^2} + \frac{k_B T}{v - b} \; ,$$

which we shall return to later. In this equation

$$v = \frac{V}{N} \; ,$$

and a and b are positive constants. For $a = b = 0$ the equation reverts to the ideal gas law.

A T, p-diagram for H_2O (with T as abscissa and p as ordinate), which is *not* presented, since it can be found in most standard textbooks, would show three phases: solid (top left on the phase diagram), liquid (top right) and gaseous (bottom, from bottom left to top right). The *solid-liquid* phase boundary would be almost vertical with a very large negative slope. The negative slope is one of the anomalies of the H_2O system[2]. One sees how

---

[1] A precise definition is given below; i.e., it turns out that the way one arrives at zero matters, e.g., the sign.

[2] At high pressures ice has at least twelve different phases. For more information, see [36].

steep the boundary is in that it runs from the $0\,^\circ$C fixed point at 760 Torr and 273.15 K almost vertically downwards directly to the *triple point*, which is the meeting point of all three phase boundaries. The triple point lies at a slightly higher temperature,

$$T_{\text{triple}} = 273.16\,\text{K} \,,$$

but considerably lower pressure: $p_{\text{triple}} \approx 5$ Torr. The *liquid-gas* phase boundary starting at the triple point and running from "southwest" to "northeast" ends at the *critical point*, $T_c = 647$ K, $p_c = 317$ at. On approaching this point the density difference

$$\Delta\varrho := \varrho_{\text{liquid}} - \varrho_{\text{gas}}$$

decreases continuously to zero. "Rounding" the critical point one remains topologically in the same phase, because the liquid and gas phases differ only quantitatively but not qualitatively. They are both so-called *fluid* phases.

The H_2O system shows two anomalies that have enormous biological consequences. The first is the negative slope mentioned above. (Icebergs float on water.) We shall return to this in connection with the Clausius-Clapeyron equation. The second anomaly is that the greatest density of water occurs at $4\,^\circ$C, not $0\,^\circ$C. (Ice forms on the surface of a pond, whereas at the bottom of the pond the water has a temperature of $4\,^\circ$C.)

40.2 Heat

Heat is produced by friction, combustion, chemical reactions and radioactive decay, amongst other things. The flow of heat from the Sun to the Earth amounts to approximately 2 cal/(cm²s) (for the unit cal: see below). Frictional heat (or "Joule heat") also occurs in connection with electrical resistance by so-called *Ohmic processes*,

$$\mathrm{d}E = R \cdot I^2 \mathrm{d}t \,,$$

where R is the Ohmic resistance and I the electrical current.

Historically heat has been regarded as a substance in its own right with its own conservation law. The so-called *heat capacity* C_V or C_p was defined as the quotient

$$\frac{\Delta Q_w}{\Delta T} \,,$$

where ΔQ_w is the heat received (at constant volume or constant pressure, respectively) and ΔT is the resulting temperature change. Similarly one may define the *specific heat capacities* c_V and c_p:

$$c_V := \frac{C_V}{m} \quad \text{and} \quad c_p := \frac{C_p}{m} \,,$$

where m is the mass of the system usually given in grammes (i.e., the molar mass for chemists). Physicists prefer to use the corresponding heat capacities $c_V^{(0)}$ and $c_p^{(0)}$ *per atom* (or *per molecule*).

The old-fashioned unit of heat "calorie", *cal*, is defined (as in school physics books) by: 1 *cal* corresponds to the amount of heat required to heat $1\,g$ of water at normal pressure from $14.5\,°C$ to $15.5\,°C$. An equivalent definition is:

$$(c_p)_{|H_2O;\text{normal pressure},15\,°C} \overset{!}{=} 1\text{cal/g} \,.$$

Only later did one come to realize that heat is only a specific form of energy, so that today the electro-mechanical equivalent of heat is defined by the following equation:

$$1\text{cal} = 4.186\,\text{J} = 4.186\,\text{Ws} \,. \qquad (40.1)$$

40.3 Thermal Equilibrium and Diffusion of Heat

If two blocks of material at different temperatures are placed in contact, then an equalization of temperature will take place, $T_j \to T_\infty$, for $j = 1, 2$, by a process of heat flowing from the hotter to the cooler body. If both blocks are insulated from the outside world, then the heat content of the system, $Q_{w|\text{"}1+2\text{"}}$, is conserved, i.e.,

$$\Delta Q_{w|\text{"}1+2\text{"}} = C_1\Delta T_1 + C_2\Delta T_2 \equiv (C_1 + C_2)\cdot \Delta T_\infty \,, \quad \text{giving}$$

$$\Delta T_\infty \equiv \frac{C_1\Delta T_1 + C_2\Delta T_2}{C_1 + C_2} \,.$$

We may generalize this by firstly defining the *heat flux density* $\boldsymbol{j}_w$, which is a vector of physical dimension $[cal/(cm^2 s)]$, and assume that

$$\boldsymbol{j}_w = \lambda \cdot \text{grad} T(\boldsymbol{r}, t) \,. \qquad (40.2)$$

This equation is usually referred to as Fick's first law of heat diffusion. The parameter λ is the *specific heat conductivity*.

Secondly let us define the *heat density* ϱ_w. This is equal to the mass density of the material ϱ_M multiplied by the specific heat c_p and the local temperature $T(\boldsymbol{r}, t)$ at time t, and is analogous to the electrical charge density ϱ_e. For the heat content of a volume ΔV one therefore has

$$Q_w(\Delta V) = \int_{\Delta V} d^3 r \varrho_w \,.^3$$

---

3 In this part, in contrast to Part II, we no longer use more than one integral sign for integrals in two or three dimensions.

Since no heat has been added or removed, a conservation law applies to the total amount of heat $Q_w(\mathcal{R}^3)$. Analogously to electrodynamics, where from the conservation law for total electric charge a continuity law results, *viz*:

$$\frac{\partial \varrho_e}{\partial t} + \mathrm{div}\boldsymbol{j}_e \equiv 0 \;,$$

we have here:

$$\frac{\partial \varrho_w}{\partial t} + \mathrm{div}\boldsymbol{j}_w \equiv 0 \;.$$

If one now inserts (40.2) into the continuity equation, one obtains with

$$\mathrm{div}\,\mathrm{grad} \equiv \nabla^2 = \frac{\partial}{\partial x^2} + \frac{\partial}{\partial y^2} + \frac{\partial}{\partial z^2}$$

the *heat diffusion equation*

$$\frac{\partial T}{\partial t} = D_w \nabla^2 T \;, \tag{40.3}$$

where the heat diffusion constant,

$$D_w := \frac{\lambda}{\varrho_M c_p} \;,$$

has the same physical dimension as all diffusion constants, $[D_W] = [cm^2/s]$. (40.3) is usually referred to as Fick's second law.

40.4 Solutions of the Diffusion Equation

The diffusion equation is a prime example of a parabolic partial differential equation[4]. A first standard task arises from, (i), the *initial value* or *Cauchy problem*. Here the temperature variation $T(\boldsymbol{r}, t = t_0)$ is given over all space, $\forall \boldsymbol{r} \in G$, but only for a single time, $t = t_0$. Required is $T(\boldsymbol{r}, t)$ for all $t \geq t_0$. A second standard task, (ii), arises from the *boundary value problem*. Now $T(\boldsymbol{r}, t)$ is given for all t, but only at the boundary of G, i.e., for $\boldsymbol{r} \in \partial G$. Required is $T(\boldsymbol{r}, t)$ over all space G. For these problems one may show that there is essentially just a single solution. For example, if one calls the difference between two solutions $u(\boldsymbol{r}, t)$, i.e.,

$$u(\boldsymbol{r}, t) := T_1(\boldsymbol{r}, t) - T_2(\boldsymbol{r}, t) \;,$$

---

[4] There is also a formal similarity with quantum mechanics (see Part III). If in (40.3) the time t is multiplied by $i/\hbar$ and D_w and T are replaced by $\hbar^2/(2m)$ and ψ, respectively, one obtains the Schrödinger equation of a free particle of mass m. Here $\hbar$, i and ψ have their usual meaning.

then because of the linearity of the problem it follows from (40.3) that

$$\frac{\partial u}{\partial t} = D_w \nabla^2 u \,,$$

and by multiplication of this differential equation with $u(\mathbf{r}, t)$ and subsequent integration we obtain

$$I(t) := \frac{\mathrm{d}}{2\mathrm{d}t} \int_G \mathrm{d}^3 r u^2(\mathbf{r}, t) = D_w \cdot \int_G \mathrm{d}^3 r u(\mathbf{r}, t) \cdot \nabla^2 u(\mathbf{r}, t)$$

$$= - D_w \int_G \mathrm{d}^3 r \{\nabla u(\mathbf{r}, t)\}^2$$

$$+ D_w \oint_{\partial G} \mathrm{d}^2 S u(\mathbf{r}, t)(\mathbf{n} \cdot \nabla) u(\mathbf{r}, t) \,.$$

To obtain the last equality we have used Green's integral theorem, which is a variant on Gauss's integral theorem. In case (ii), where the values of $T(\mathbf{r}, t)$ are always prescribed only on ∂G, the surface integral

$$\propto \oint_{\partial G} \dots$$

is equal to zero, i.e., $I(t) \leq 0$, so that

$$\int_G \mathrm{d}^3 r \cdot u^2$$

decreases until finally $u \equiv 0$. Since we also have

$$\frac{\partial}{\partial t} \equiv 0 \,,$$

one arrives at a problem, which has an analogy in electrostatics. In case (i), where at t_0 the temperature is fixed everywhere in G, initially one has uniqueness:

$$0 = u(\mathbf{r}, t_0) = I(t_0) \,; \quad \text{thus} \quad u(\mathbf{r}, t) \equiv 0 \quad \text{for all} \quad t \geq t_0 \,.$$

One often obtains solutions by using either 1) *Fourier methods*[5] or 2) so-called *Green's functions*. We shall now treat both cases using examples of one-dimensional standard problems:

---

[5] It is no coincidence that L. Fourier's methods were developed in his tract "Théorie de la *Chaleur*".

1a) (*Equilibration of the temperature for a periodic profile*): At time $t = t_0 = 0$ assume there is a *spatially periodic variation* in temperature

$$T(x,0) = T_\infty + \Delta T(x), \quad \text{with} \quad \Delta T(x+a) = \Delta T(x) \quad \text{for all} \quad x \in G.$$

One may describe this by a Fourier series

$$\Delta T(x) = \sum_{n=-\infty}^{\infty} b_n \exp\left(ink_0 x\right),$$

with

$$b_n \equiv \frac{1}{a} \int_0^a dx \exp\left(-ink_0 x\right) \Delta T(x).$$

Here,

$$k_0 := \frac{2\pi}{\lambda_0},$$

where λ_0 is the fundamental wavelength of the temperature profile. One can easily show that the solution to this problem is

$$T(x,t) = T_\infty + \sum_{n=-\infty,\neq 0}^{\infty} b_n e^{ink_0 x} \cdot e^{-n^2 \frac{t}{\tau_0}}, \quad \text{where} \quad \frac{1}{\tau_0} = D_w k_0^2.$$

According to this expression the characteristic diffusion time is related to the *fundamental wavelength*. The time dependence is one of exponential decay, where the upper harmonics, $n = \pm 2, \pm 3, \ldots$, are attenuated much more quickly,

$$\propto e^{-n^2 \frac{t}{\tau_0}},$$

than the fundamental frequency $n = \pm 1$.

1b) A good example of a "boundary value problem" is the so-called *permafrost problem*, which shall now be treated with the help of Fourier methods. At the Earth's surface $z = 0$ at a particular location in Siberia we assume there is explicitly the following temperature profile:

$$T(z=0,t) = T_\infty + b_1 \cos\omega_1 t + b_2 \cos\omega_2 t.$$

T_∞ is the average temperature at the surface during the year,

$$\omega_1 = \frac{2\pi}{365d}$$

is the annual period and

$$\omega_2 = \frac{2\pi}{1d}$$

is the period of daily temperature fluctuations (i.e., the second term on the r.h.s. describes the seasonal variation of the daytime temperature average, averaged over the 24 hours of a day). We write

$$\cos\omega_1 t = \mathcal{R}e\, e^{-i\omega_1 t}$$

and postpone the evaluation of the real part to the very end (i.e., in effect we omit it). Below the surface $z < 0$ we assume that

$$T(z,t) = T_\infty + b_1 e^{i(k_1 z - \omega_1 t)} + b_2 e^{i(k_2 z - \omega_2 t)} ,$$

with complex (!) wavenumbers k_j, $j = 1, 2$. The heat diffusion equation then leads for a given (real) frequency ω_1 or $\omega_2 = 365\omega_1$ to the following formula for calculating the wavenumbers:

$$i\omega_j = D_w k_j^2 , \quad \text{giving} \quad \sqrt{D_w}k_j = \sqrt{\omega_j}\sqrt{i} .$$

With

$$\sqrt{i} = \frac{1 + i}{\sqrt{2}}$$

one obtains for the real part $k_j^{(1)}$ and imaginary part $k_j^{(2)}$ of the wavenumber k_j in each case the equation

$$k_j^{(1)} = k_j^{(2)} = \frac{\sqrt{\omega_j}}{\sqrt{2D_w}} .$$

The *real part* $k_1^{(1)}$ gives a phase shift of the temperature rise in the ground relative to the surface, as follows: whereas for $z = 0$ the maximum daytime temperature average over the year occurs on the 21st June, below the ground (for $z < 0$) the temperature maximum may occur much later. The *imaginary part* $k_1^{(2)}$ determines the temperature variation below ground; it is much smaller than at the surface, e.g., for the seasonal variation of the daytime average we get instead of $T_\infty \pm b_1$:

$$T_\pm(z) = T_\infty \pm e^{-k_1^{(2)}|z|} \cdot b_1 ;$$

however, the average value over the year, T_∞, does not depend on z. The ground therefore only melts at the surface, whereas below a certain depth $|z|_c$ it remains frozen throughout the year, provided that

$$T_\infty + e^{-k_1^{(2)}|z|_c} \cdot b_1$$

lies below $0\,^\circ$C. It turns out that the seasonal rhythm ω_1 influences the penetration depth, not the daily time period ω_2. From the measured penetration depth (typically a few decimetres) one can determine the diffusion constant D_w.

2) With regard to *Green's functions*, it can be shown by direct differentiation that the function

$$G(x,t) := \frac{e^{-\frac{x^2}{4D_w t}}}{\sqrt{4\pi D_w t}}$$

is a solution to the heat diffusion equation (40.3). It represents a special solution of general importance for this equation, since on the one hand for $t \to \infty$, $G(x,t)$ propagates itself more and more, becoming flatter and broader. On the other hand, for $t \to 0 + \varepsilon$, with positive infinitesimal ε[6], $G(x,t)$ becomes increasingly larger and narrower. In fact, for $t \to 0^+$, $G(x,t)$ tends towards the Dirac delta function,

$$G(x, t \to 0^+) \to \delta(t) , \quad \text{since} \quad \frac{e^{-\frac{x^2}{2\sigma^2}}}{\sqrt{2\pi\sigma^2}} \quad \text{for} \quad t > 0$$

gives a Gaussian curve $g_\sigma(x)$ of width $\sigma^2 = 2D_w t$; i.e., $\sigma \to 0$ for $t \to 0$, but always with unit area (i.e., $\int\limits_{-\infty}^{\infty} dx g_\sigma(x) \equiv 1$ as well as $\int\limits_{-\infty}^{\infty} dx x^2 g_\sigma(x) \equiv \sigma^2$, $\forall \sigma$). Both of these expressions can be obtained by using a couple of integration tricks: the "squaring" trick and the "exponent derivative" trick, which we cannot go into here due to lack of space.

Since the diffusion equation is a *linear* differential equation, the *principle of superposition* holds, i.e., superposition of the functions $f(x') \cdot G(x - x', t)$, with any weighting $f(x')$ and at any positions x', is also a solution of the diffusion equation, and thus one can satisfy the Cauchy problem for the real axis, for given initial condition $T(x, 0) = f(x)$, as follows:

$$T(x, t) = \int\limits_{-\infty}^{\infty} dx' f(x') \cdot G(x - x', t) . \qquad (40.4)$$

For one-dimensional problems on the real axis, with the boundary condition that for $|x| \to \infty$ there remains only a time dependence, the Green's function is identical to the fundamental solution given above. If special boundary conditions at finite x have to be satisfied, one can modify the fundamental solution with a suitable (more or less harmless) perturbation and thus obtain the Green's function for the problem, as in electrostatics.

Similar results apply in three dimensions, with the analogous fundamental solution

$$G_{3d}(\boldsymbol{r}, t) =: \frac{e^{-\frac{r^2}{4D_w t}}}{(4\pi D_w t)^{3/2}} .$$

These conclusions to this chapter have been largely "mathematical". However, one should not forget that diffusion is a very general "physical" process; we shall return to the results from this topic later, when treating the *kinetic theory of gases*.

---

[6] One also writes $t \to 0^+$.

41 The First and Second Laws of Thermodynamics

41.1 Introduction: Work

An increment of heat absorbed by a system is written here as δQ and *not* dQ, in order to emphasize the fact that, in contrast to the variables of state U and S (see below), it is *not* a *total* differential. The same applies to *work* A, where we write its increment as δA, and *not* as dA.

Some formulae:

α) *Compressional work*:

The incremental work done during compression of a *fluid* (gas or liquid), is given by $\delta A = -p\mathrm{d}V$. [Work is given by the scalar product of the applied force and the distance over which it acts, i.e., $\delta A = F \cdot \mathrm{d}z = \left(p\Delta S^{(2)}\right) \cdot \left(-\mathrm{d}V/\Delta S^{(2)}\right) = -p\mathrm{d}V$.]

β) *Magnetic work*:

In order to increase the magnetic dipole moment m_H of a magnetic sample (e.g., a fluid system of paramagnetic molecules) in a magnetic field H we must do an amount of work

$$\delta A = H \cdot \mathrm{d}m_H \ .$$

(For a *proof*: see below.)

Explanation: Magnetic moment is actually a vector quantity like magnetic field $\boldsymbol{H}$. However, we shall not concern ourselves with directional aspects here. Nevertheless the following remarks are in order. A magnetic dipole moment $\boldsymbol{m}_H$ at $\boldsymbol{r}'$ produces a magnetic field $\boldsymbol{H}$:

$$\boldsymbol{H}^{(m)}(\boldsymbol{r}) = -\mathrm{grad}\frac{\boldsymbol{m}_H \cdot (\boldsymbol{r} - \boldsymbol{r}')}{4\pi\mu_0|\boldsymbol{r} - \boldsymbol{r}'|^3} \ ,$$

where μ_0 is the vacuum permeability. On the other hand, in a field $\boldsymbol{H}$ a magnetic dipole experiences forces and torques given by

$$\boldsymbol{F}^{(m)} = (\boldsymbol{m}_H \cdot \nabla)\boldsymbol{H} \quad \text{and} \quad \boldsymbol{\mathcal{D}}^{(m)} = \boldsymbol{m}_H \times \boldsymbol{H} \ .$$

This is already a somewhat complicated situation which becomes even more complicated, when in a magnetically polarizable material we have to take the difference between the *magnetic field* H and the *magnetic induction* B into account. As a reminder, B, not H, is "divergence free", i.e.,

$$B = \mu_0 H + J \,,$$

where J is the *magnetic polarization*, which is related directly to the dipole moment m_H by the expression

$$m_H = \int_V d^3 r J = V \cdot J \,,$$

where V is the volume of our homogeneously magnetized system.

In the literature we often find the term "magnetization" instead of "magnetic polarization" used for J or even for m_H, even though, in the mksA system, the term *magnetization* is reserved for the vector

$$M := J/\mu_0 \,.$$

One should not allow this multiplicity of terminology or different conventions for one and the same quantity dm_H to confuse the issue. Ultimately it "boils down" to the pseudo-problem of μ_0 and the normalization of volume. In any case we shall intentionally write here the precise form

$$\delta A = H \cdot dm_H \quad (\text{and not} = H \cdot dM) \,.$$

In order to verify the first relation, we proceed as follows. The inside of a coil carrying a current is filled with material of interest, and the work done on changing the current is calculated. (This is well known exercise.) The work is divided into two parts, the first of which changes the vacuum field energy density $\frac{\mu_0 H^2}{2}$, while the second causes a change in magnetic moment. This is based on the law of conservation of energy applied to Maxwell's theory. For changes in w, the volume density of the electromagnetic field energy, we have

$$\delta w \equiv E \cdot \delta D + H \cdot \delta B - j \cdot E \cdot \delta t \,,$$

where E is the electric field, j the current density and D the dielectric polarization; i.e., the relevant term is the last-but-one expression,

$$H \cdot \delta B \,, \quad \text{with} \quad \delta B = \mu_0 \delta H + \delta J \,,$$

i.e., here the term $\propto \delta J$ is essential for the material, whereas the term $\propto \delta H$, as mentioned, only enhances the field-energy.

If we introduce the particle number N as a variable, then its conjugated quantity is μ, the *chemical potential*, and for the work done on our material system we have:

$$\delta A = -p dV + H dm_H + \mu dN =: \sum_i f_i dX_I \,, \tag{41.1}$$

where dX_i is the differential of the work variables. The quantities dV, dm_H and dN (i.e., dX_i) are *extensive* variables, *viz* they double when the system size is doubled, while p, H and μ (i.e., f_i) are *intensive* variables.

41.2 First and Second Laws: Equivalent Formulations

a) The First Law of Thermodynamics states that there exists a certain *variable of state* $U(T, V, m_H, N)$, the so-called *internal energy* $U(T, X_i)$, such that (*neither the infinitesimal increment* δQ *of the heat gained nor the infinitesimal increment* δA *of the work done alone, but*) the *sum*, $\delta Q + \delta A$, forms the total differential of the function U, i.e.,

$$\delta Q + \delta A \equiv dU = \frac{\partial U}{\partial T}dT + \frac{\partial U}{\partial V}dV + \frac{\partial U}{\partial m_H}dm_H + \frac{\partial U}{\partial N}dN \,,$$

or more generally

$$= \frac{\partial U}{\partial T}dT + \sum_i \frac{\partial U}{\partial X_i}dX_i \,.$$

(This means that the integrals $\int_W \delta Q$ and $\int_W \delta A$ may indeed depend on the integration path W, but not their sum $\int_W (\delta Q + \delta A) = \int_W dU$. For a *closed* integration path both $\oint \delta Q$ and $\oint \delta A$ may thus be non-zero, but their sum is always zero: $\oint (\delta Q + \delta A) = \oint dU \equiv 0$.)

b) As preparation for the Second Law we shall introduce the term *irreversibility*: Heat can either flow *reversibly* (i.e., *without* frictional heat or any other losses occurring) or *irreversibly* (i.e., *with* frictional heat), and as we shall see immediately this is a very important difference. In contrast, the formula $\delta A = -pdV + \dots$ is valid independently of the type of process leading to a change of state variable.

The Second Law states that a variable of state $S(T, V, m_H, N)$ exists, the so-called *entropy*, generally $S(T, X_i)$, such that

$$dS \geq \frac{\delta Q}{T} \,, \tag{41.2}$$

where the equality sign holds exactly when heat is transferred reversibly.

What is the significance of entropy? As a provisional answer we could say that it is a quantitative measure for the complexity of a system. Disordered systems (such as gases, etc.) are generally more complex than regularly ordered systems (such as crystalline substances) and, therefore, they have a higher entropy.

There is also the following important difference between energy and entropy: The energy of a system is a well defined quantity only apart from an additive constant, whereas the entropy is completely defined. We shall see later that

$$\frac{S}{k_B} \equiv -\sum_j p_j \cdot \ln p_j \,,$$

where p_j are the probabilities for the orthogonal system states, i.e.,

$$p_j \geq 0 \quad \text{and} \quad \sum_j p_j \equiv 1 \,.$$

We shall also mention here the so-called Third Law of Thermodynamics or Nernst's Heat Theorem: The limit of the entropy as T tends to zero, $S(T \to 0, X_i)$, is also zero, except if the ground state is degenerate.[1] As a consequence, which will be explained in more detail later, the absolute zero of temperature $T = 0$ (in degrees Kelvin) cannot be reached in a finite number of steps.

The Third Law follows essentially from the above statistical-physical formula for the entropy including basic quantum mechanics, i.e., an energy gap between the (g_0-fold) ground state and the lowest excited state (g_1-fold). Therefore, it is superfluous in essence. However, one should be aware of the above consequence. At the time Nernst's Heat Theorem was proposed (in 1905) neither the statistical formula nor the above-mentioned consequence was known.

There are important consequences from the first two laws with regard to the coefficients of the associated, so-called Pfaff forms or first order differential forms. We shall write, for example,

$$\delta Q + \delta A = dU = \sum_i a_i(x_1, \ldots, x_f)dx_i , \quad \text{with} \quad x_i = T, V, m_H, N .$$

The differential forms for dU are "total", i.e., they possess a stem function $U(x_1, .., x_f)$, such that, e.g.,

$$a_i = \frac{\partial U}{\partial x_i} .$$

Therefore, similar to the so-called *holonomous* subsidiary conditions in mechanics, see Part I, the following integrability conditions are valid:

$$\frac{\partial a_i}{\partial x_k} = \frac{\partial a_k}{\partial x_i} \quad \text{for all} \quad i, k = 1, \ldots, f .$$

Analogous relations are valid for S.

All this will be treated in more depth in later sections. We shall begin as follows.

41.3 Some Typical Applications: C_V and $\frac{\partial U}{\partial V}$; The Maxwell Relation

We may write

$$dU(T, V) = \frac{\partial U}{\partial T}dT + \frac{\partial U}{\partial V}dV = \delta Q + \delta A .$$

---

[1] If, for example, the ground state of the system is spin-degenerate, which presupposes that $H \equiv 0$ for all atoms, according to the previous formula we would then have $S(T \to 0) = k_B N \ln 2$.

The heat capacity at constant volume ($dV = 0$) and constant N ($dN = 0$), since $\delta A = 0$, is thus given by

$$C_V(T, V, N) = \frac{\partial U(T, V, N)}{\partial T} .$$

Since

$$\frac{\partial}{\partial V}\left(\frac{\partial U}{\partial T}\right) = \frac{\partial}{\partial T}\left(\frac{\partial U}{\partial V}\right) ,$$

we may also write

$$\frac{\partial C_V(T, V, N)}{\partial V} = \frac{\partial}{\partial T}\left(\frac{\partial U}{\partial V}\right) .$$

A well-known experiment by Gay-Lussac, where an ideal gas streams out of a cylinder through a valve, produces no thermal effects, i.e.,

$$\frac{\partial}{\partial T}\left(\frac{\partial U}{\partial V}\right) = 0 .$$

This means that for an ideal gas the internal energy does not depend on the volume, $U(T, V) \equiv U(T)$ and $C_V = C_V(T)$, for fixed N.

According to the Second Law

$$dS = \frac{\delta Q|_{\text{reversible}}}{T} = \frac{dU - \delta A|_{\text{rev}}}{T} = \frac{dU + pdV}{T} = \frac{\partial U}{\partial T}\frac{dT}{T} + \left(\frac{\partial U}{\partial V} + p\right)\frac{dV}{T} ,$$

i.e., $\dfrac{\partial S}{\partial T} = \dfrac{1}{T}\dfrac{\partial U}{\partial T}$ and $\dfrac{\partial S}{\partial V} = \dfrac{1}{T}\left(\dfrac{\partial U}{\partial V} + p\right) .$

By equating mixed derivatives,

$$\frac{\partial^2 S}{\partial V \partial T} = \frac{\partial^2 S}{\partial T \partial V} ,$$

after a short calculation this gives the so-called *Maxwell relation*

$$\frac{\partial U}{\partial V} \equiv T\frac{\partial p}{\partial T} - p . \tag{41.3}$$

As a consequence, the *caloric equation of state*, $U(T, V, N)$, is not required for calculating $\frac{\partial U}{\partial V}$; it is sufficient that the *thermal equation of state*, $p(T, V, N)$, is known.

If we consider the *van der Waals* equation (see below), which is perhaps the most important equation of state for describing the behavior of real gases:

$$p = -\frac{a}{v^2} + \frac{k_B T}{v - b} ,$$

it follows with

$$v := \frac{V}{N} \quad \text{and} \quad u := \frac{U}{N} \quad \text{that} \quad \frac{\partial U}{\partial V} = \frac{\partial u}{\partial v} = T\frac{\partial p}{\partial T} - p = +\frac{a}{v^2} > 0 .$$

The consequence of this is that when a *real* gas streams out of a pressurized cylinder (Gay-Lussac experiment with a real gas) then U tends to increase, whereas in the case of thermal isolation, i.e., for constant U, the temperature T must decrease:

$$\frac{\mathrm{d}T}{\mathrm{d}V}\Big|_U = -\frac{\partial U/\partial V}{\partial U/\partial T} \propto -\frac{a}{v^2} \ .$$

(In order to calculate the temperature change for a volume increase at constant U we have used the following relation:

$$0 \stackrel{!}{=} \mathrm{d}U = \frac{\partial U}{\partial T}\mathrm{d}T + \frac{\partial U}{\partial V}\mathrm{d}V \ ,$$

and therefore

$$\frac{\mathrm{d}T}{\mathrm{d}V}\Big|_U = -\frac{\frac{\partial U}{\partial V}}{\frac{\partial U}{\partial T}} \ . \tag{41.4}$$

The negative sign on the r.h.s. of this equation should not be overlooked.)

41.4 General Maxwell Relations

A set of Maxwell relations is obtained in an analogous way for other extensive variables X_i (reminder: $\delta A = \sum_i f_i \mathrm{d}X_i$). In addition to

$$\frac{\partial U}{\partial V} = T\frac{\partial p}{\partial T} - p \quad \text{we also have} \quad \frac{\partial U}{\partial m_H} = H - T\frac{\partial H}{\partial T} \quad \text{and} \quad \frac{\partial U}{\partial N} = \mu - T\frac{\partial \mu}{\partial T} \ ,$$

and in general

$$\frac{\partial U}{\partial X_i} = f_i - T\frac{\partial f_i}{\partial T} \ . \tag{41.5}$$

41.5 The Heat Capacity Differences $C_p - C_V$ and $C_H - C_m$

In order to calculate the difference $C_p - C_V$, we begin with the three relations

$$C_p = \frac{\delta Q_{|p}}{\mathrm{d}T} = \frac{\mathrm{d}U - \delta A}{\mathrm{d}T} \ , \mathrm{d}U = \frac{\partial U}{\partial T}\mathrm{d}T + \frac{\partial U}{\partial V}\mathrm{d}V \quad \text{and} \quad \delta A = -p\mathrm{d}V \ ,$$

and obtain: $\quad C_p = \frac{\delta Q_{|p}}{\mathrm{d}T} = \frac{\mathrm{d}U - \delta A}{\mathrm{d}T} = \frac{\partial U}{\partial T} + \left(\frac{\partial U}{\partial V} + p\right)\left(\frac{\mathrm{d}V}{\mathrm{d}T}\right)_{|p} \ .$

Therefore,

$$C_p \equiv C_V + \left(\frac{\partial U}{\partial V} + p\right)\left(\frac{\mathrm{d}V}{\mathrm{d}T}\right)_{|p} \ , \quad \text{or} \quad C_p - C_V = \left(\frac{\partial U}{\partial V} + p\right)\left(\frac{\mathrm{d}V}{\mathrm{d}T}\right)_{|p} \ .$$

Using the above Maxwell relation we then obtain

$$C_p - C_v = T\frac{\partial p}{\partial T} \cdot \left(\frac{\mathrm{d}V}{\mathrm{d}T}\right)_{|p} , \quad \text{and finally} \quad C_p - C_V = -T\frac{\partial p}{\partial T} \cdot \frac{\frac{\partial p}{\partial T}}{\frac{\partial p}{\partial V}} .$$

Analogously for $C_H - C_m$ we obtain:

$$C_H - C_m = T\frac{\partial H}{\partial T} \cdot \frac{\frac{\partial H}{\partial T}}{\frac{\partial H}{\partial m_H}} .$$

These are general results from which we can learn several things, for example, that the difference $C_p - C_V$ is proportional to the isothermal *compressibility*

$$\kappa_T := -\frac{1}{V}\frac{\partial V}{\partial p} .$$

For incompressible systems the difference $C_p - C_V$ is therefore zero, and for solids it is generally very small.

In the magnetic case, instead of *compressibility* κ_T, the *magnetic susceptibility* $\chi = \frac{\partial m_H}{\partial H}$ is the equivalent quantity.

For an ideal gas, from the thermal equation of state,

$$p = \frac{N}{V}k_B T ,$$

one obtains the compact result

$$C_p - C_V = N k_B .$$

(Chemists write: $C_p - C_V = n_{\mathrm{Mol}} R_0$). It is left to the reader to obtain the analogous expression for an ideal paramagnetic material.

41.6 Enthalpy and the Joule-Thomson Experiment; Liquefaction of Air

In what follows, instead of the internal energy $U(T, V, \ldots)$ we shall introduce a new variable of state, the *enthalpy* $I(T, p, \ldots)$. This is obtained from the internal energy through a type of Legendre transformation, in a similar way to the Hamilton function in mechanics, which is obtained by a Legendre transform from the Lagrange function. Firstly, we shall write

$$I = U + p \cdot V$$

and then eliminate V using the thermal equation of state

$$p = p(T, V, N, \ldots) ,$$

resulting in $I(T, p, m_H, N, \ldots)$. We can also proceed with other extensive variables, e.g., m_H, where one retains at least one extensive variable, usually N, such that $I \propto N$, and then obtains "enthalpies" in which the *extensive* variables $X_i (= V, m_H$ or $N)$ are replaced wholly (or partially, see above) by the *intensive* variables $f_i = p, H, \mu$ etc., giving the enthalpy

$$I(T, p, H, N, \ldots) = U + pV - m_H H \ ,$$

or generally

$$I(T, f_1, \ldots, f_k, X_{k+1}, \ldots) = U(T, X_1, \ldots, X_k, X_{k+1}, \ldots) - \sum_{i=1}^{k} f_i X_i \ .$$

We can proceed in a similar manner with the *work done*. Instead of *extensive work*

$$\delta A = \sum_i f_i dX_i \ ,$$

we define a quantity the *intensive work*

$$\delta A' := \delta A - d \left(\sum_i f_i X_i \right) \ ,$$

such that

$$\delta A' = - \sum_i X_i df_i$$

is valid. The *intensive work* $\delta A'$ is just as "good" or "bad" as the *extensive work* δA. For example, we can visualize the expression for intensive work

$$\delta A' = +V dp$$

by bringing an additional weight onto the movable piston of a cylinder containing a fluid, letting the pressure rise by moving heavy loads from below onto the piston. As a second example consider the magnetic case, where the expression for intensive magnetic work

$$\delta A' = -m_H dH$$

is the change in energy of a magnetic dipole m_H in a variable magnetic field,

$$H \to H + dH \ ,$$

at constant magnetic moment.

From the above we obtain the following equivalent formulation of the first law.

A variable of state, called *enthalpy* $I(T, p, H, N, \ldots)$, exists whose total differential is equal to the sum

$$\delta Q + \delta A' \ .$$

The second law can also be transformed by expressing the entropy (wholly or partially) as a function of intensive variables: e.g., $S(T, p, H, N, \ldots)$. The Maxwell relations can be extended to cover this case, e.g.,

$$\frac{\partial I}{\partial p} = V - T\frac{\partial V}{\partial T} , \quad \frac{\partial I}{\partial H} = T\frac{\partial H}{\partial T} - H ,$$

or, in general

$$\frac{\partial I}{\partial f_j} = T\frac{\partial X_j}{\partial T} - X_j . \tag{41.6}$$

We now come to the *Joule-Thomson effect*, which deals with the stationary flow of a fluid through a pipe of uniform cross-sectional area $(S)_1$ on the input side of a so-called throttle valve, $\mathcal{V}_{thr}$. On the output side there is also a pipe of uniform cross-section $(S)_2$, where $(S)_2 \neq (S)_1$. We shall see that, as a result, $p_1 \neq p_2$. We shall consider stationary conditions where the temperatures T on each side of the throttle valve $\mathcal{V}_{thr}$ are everywhere the same. Furthermore the pipe is thermally insulated ($\delta Q = 0$). In the region of $\mathcal{V}_{thr}$ itself irreversible processes involving turbulence may occur. However we are not interested in these processes themselves, only in the regions of stationary flow well away from the throttle valve. A schematic diagram is shown in Fig. 41.1.

We shall now show that it is not the internal energy $U(= N \cdot u)$ of the fluid which remains constant in this stationary flow process, but the specific enthalpy,

$$i(x) := \frac{I(T, p(x), N)}{N} ,$$

in contrast to the Gay-Lussac experiment. Here x represents the length coordinate in the pipe, where the region of the throttle valve is not included.

In order to prove the statement for $i(x)$ we must remember that at any time t the same number of fluid particles

$$dN_1 = dN_2$$

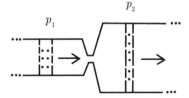

Fig. 41.1. Joule-Thomson process. A fluid (liquid or gas) flowing in a stationary manner in a pipe passes through a throttle valve. The pressure is p_1 on the left and $p_2(< p_1)$ on the right of the valve. There are equal numbers of molecules on average in the shaded volumes left and right of the valve. In contrast to p, the temperatures are everywhere the same on both sides of the valve

pass through the cross-sections S_i at given positions x_i left and right of the throttle valve in a given time interval $\mathrm{d}t$. Under adiabatic conditions the difference in internal energy

$$\delta U := U_2 - U_1$$

for the two shaded sets $S_i \cdot \mathrm{d}x_i$ in Fig. 41.1, containing the same number of particles, is identical to δA, i.e.,

$$\delta A = -p_1 \cdot S_1 \cdot \mathrm{d}x_1 + p_2 \cdot S_2 \cdot \mathrm{d}x_2 \ ,$$

since on the input side (left) the fluid works against the pressure while on the output side (right) work is done on the fluid. Because of the thermal isolation, $\delta Q = 0$, we thus have:

$$\mathrm{d}U = -p_1\mathrm{d}V_1 + p_2\mathrm{d}V_2 \equiv (-p_1v_1 + p_2v_2)\mathrm{d}N \ ;$$

i.e., we have shown that

$$u_2 + p_2v_2 = u_1 + p_1v_1 \ , \quad \text{or} \quad i_2 = i_1 \ ,$$

viz the *specific enthalpy* of the gas is unchanged.[2]

The Joule-Thomson process has very important technical consequences, since it forms the basis for present-day cryotechnology and the Linde gas liquefaction process.

In order to illustrate this, let us make a quantitative calculation of the Joule-Thomson cooling effect (or heating effect (!), as we will see below for an important exceptional case):

$$\left(\frac{\mathrm{d}T}{\mathrm{d}p} \right)_i = -\frac{\frac{\partial i}{\partial p}}{\frac{\partial i}{\partial T}} \ .$$

(This can be described by $\frac{-\frac{\partial i}{\partial p}}{c_p^{(0)}}$, with the molecular heat capacity $c_p^{(0)} := \frac{C_p}{N}$; an analogous relation applies to the Gay-Lussac process.)

Using the *van der Waals equation*

$$p = -\frac{a}{v^2} + \frac{k_B T}{v - b} \ ,$$

see next section, for a real gas together with the Maxwell relation

$$\frac{\partial i}{\partial p} = v - T\frac{\partial v}{\partial T} \quad \text{and} \quad \left(\frac{\partial v}{\partial T} \right)_p = -\frac{\frac{\partial p}{\partial T}}{\frac{\partial p}{\partial v}} \ ,$$

---

[2] If the temperature is different on each side of the throttle valve, we have a situation corresponding to the hotly discussed topic of non-equilibrium transport phenomena (the so-called *Keldysh theory*).

one obtains

$$\left(\frac{\mathrm{d}T}{\mathrm{d}p}\right)_i = \frac{1}{c_p^{(0)}} \frac{\frac{2a}{k_B T} \frac{(v-b)^2}{v^2} - b}{1 - \frac{2a}{v k_B T} \frac{(v-b)^2}{v^2}} . \tag{41.7}$$

The first term of the *van der Waals equation*, $-\frac{a}{v^2}$, represents a negative internal pressure which corresponds to the long-range attractive r^{-6} term of the Lennard-Jones potential (see below). The parameter b in the term $\frac{k_B T}{v-b}$, the so-called co-volume, corresponds to a short-range *hard-core* repulsion $\propto r^{-12}$:

$$b \approx \frac{4\pi\sigma^3}{3} .$$

σ is the characteristic radius of the Lennard-Jones potentials, which are usually written as[3]

$$V_{\text{L.-J.}}(r) = 4\varepsilon \cdot \left(\left(\frac{\sigma}{r}\right)^{12} - \left(\frac{\sigma}{r}\right)^6 \right) . \tag{41.8}$$

In the gaseous state, $v \gg b$, and (41.7) simplifies to

$$\left(\frac{\mathrm{d}T}{\mathrm{d}p}\right)_i \approx \frac{v}{c_p^{(0)}} \left(\frac{\frac{T^{\text{Inv.}}}{T}}{1 - \frac{T^{\text{Inv.}}}{T}} \right) ,$$

where the so-called inversion temperature

$$T^{\text{Inv.}} := \frac{2a}{k_B v} .$$

As long as $T < T^{\text{Inv.}}$, which is normally the case because the inversion temperature usually (but not always!) lies above room temperature, we obtain a cooling effect for a pressure drop across the throttle valve, whereas for $T > T^{\text{Inv.}}$ a heating effect would occur.

In order to improve the efficiency of Joule-Thomson cooling, a countercurrent principle is used. The cooled fluid is fed back over the fluid to be cooled in a heat exchanger and successively cooled until it eventually liquefies. Air contains 78.08 % nitrogen (N_2), 20.95 % oxygen (O_2), 0.93 % argon (Ar) and less than 0.01 % of other noble gases (Ne, He, Kr, Xe) and hydrogen; these are the proportions of permanent gases. A further 0.03 % consists of non-permanent components (H_2O, CO and CO_2, SO_2, CH_4, O_3, etc.). The liquefaction of air sets in at 90 K at normal pressure. For nitrogen the liquefaction temperature is 77 K; for hydrogen 20 K and helium 4.2 K. The inversion temperature for helium is only 20 K; thus in order to liquefy helium using this method, one must first pre-cool it to a temperature

$$T < T^{\text{Inv.}} = 20\,\mathrm{K} .$$

(*He* above 20 K is an example of the "exceptional case" mentioned above.)

---

[3] see (4), Chap. 3, in [37].

41.7 Adiabatic Expansion of an Ideal Gas

The Gay-Lussac and Joule-Thomson effects take place at constant $U(T, V, N)$ and constant

$$i(T, P) := \frac{U + pV}{N} \,,$$

respectively. These subsidiary conditions do not have a special name, but calorific effects which take place under conditions of thermal insulation (with no heat loss, $\delta Q \equiv 0$), i.e., where the entropy $S(T, V, N)$ is constant, are called *adiabatic*.

In the following we shall consider the *adiabatic expansion of an ideal gas*. We remind ourselves firstly that for an *isothermal* change of state of an ideal gas Boyle-Mariotte's law is valid:

$$pV = Nk_B T = \text{constant} \,.$$

On the other hand, for an adiabatic change, we shall show that

$$pV^\kappa = \text{constant} \,, \quad \text{where} \quad \kappa := \frac{C_p}{C_V} \,.$$

We may write

$$\left(\frac{\mathrm{d}T}{\mathrm{d}V} \right)_S = -\frac{\frac{\partial S}{\partial V}}{\frac{\partial S}{\partial T}} = \frac{\frac{\partial U}{\partial V} + p}{C_V} \,,$$

since

$$\mathrm{d}S = \frac{\mathrm{d}U - \delta A}{T} = \frac{C_v \mathrm{d}T}{T} + \frac{\left(\frac{\partial U}{\partial V} + p \right) \mathrm{d}V}{T} \,.$$

Using the Maxwell relation

$$\frac{\partial U}{\partial V} + p = T \frac{\partial p}{\partial T} \,,$$

we obtain

$$\left(\frac{\mathrm{d}T}{\mathrm{d}V} \right)_S = \frac{T \frac{\partial p}{\partial T}}{C_V} \,.$$

In addition, we have for an ideal gas:

$$(C_p - C_V) = \left(\frac{\partial U}{\partial V} + p \right) \left(\frac{\mathrm{d}V}{\mathrm{d}T} \right)_p \equiv p \left(\frac{\mathrm{d}V}{\mathrm{d}T} \right)_p = p \frac{V}{T} = Nk_B \,,$$

so that finally we obtain

$$\left(\frac{\mathrm{d}T}{\mathrm{d}V} \right)_S = -\frac{C_p - C_V}{C_V} \cdot \frac{T}{V} \,.$$

Using the abbreviation

$$\tilde{\kappa} := \frac{C_p - C_V}{C_V} ,$$

we may therefore write

$$\frac{\mathrm{d}T}{T} + \tilde{\kappa}\frac{\mathrm{d}V}{V} = 0 , \quad \text{i.e.,}$$

$$\mathrm{d}\left(\ln\frac{T}{T_0} + \ln\left(\frac{V}{V_0}\right)^{\tilde{\kappa}} \right) = 0 , \quad \text{also}$$

$$\ln\frac{T}{T_0} + \ln\left(\frac{V}{V_0}\right)^{\tilde{\kappa}} = \text{constant} , \quad \text{and}$$

$$T \cdot V^{\tilde{\kappa}} = \text{constant} .$$

With

$$\kappa := \frac{C_p}{C_V} = 1 + \tilde{\kappa} \quad \text{and} \quad p = \frac{Nk_BT}{V}$$

we finally obtain:

$$pV^{\kappa} = \text{constant} .$$

In a V, T diagram one describes the lines

$$\left\{ \left(\frac{\mathrm{d}T}{\mathrm{d}V}\right)_S \equiv \text{constant} \right\}$$

as *adiabatics*. For an ideal gas, since $\tilde{\kappa} > 0$, they are steeper than *isotherms* $T = \text{constant}$. In a V, p diagram the isotherms and adiabatics also form a non-trivial coordinate network with negative slope, where, since $\kappa > 1$, the adiabatics show a more strongly negative slope than the isotherms. This will become important below and gives rise to Fig. 46.2.

One can also treat the adiabatic expansion of a photon gas in a similar way.

42 Phase Changes, van der Waals Theory and Related Topics

As we have seen through the above example of the Joule-Thomson effect, the van der Waals equation provides a useful means of introducing the topic of liquefaction. This equation will now be discussed systematically in its own right.

42.1 Van der Waals Theory

In textbooks on chemistry the van der Waals equation is usually written as

$$\left(p + \frac{A}{V^2}\right) \cdot (V - B) = n_{\mathrm{Mol}} R_0 T \ .$$

In physics however one favors the equivalent form:

$$p = -\frac{a}{v^2} + \frac{k_B T}{v - b} \ , \tag{42.1}$$

with

$$a := \frac{A}{N} \ , \quad v := \frac{V}{N} \ , \quad N k_B = n_{\mathrm{Mol}} R_0 \ , \quad b := \frac{B}{N} \ .$$

The meaning of the terms have already been explained in connection with Lennard-Jones potentials (see (41.8)).

Consider the following p, v-diagram with three typical isotherms (i.e., p versus v at constant T, see Fig. 42.1):

a) At high temperatures, well above T_c, the behavior approximates that of an ideal gas,

$$p \approx \frac{k_B T}{v} \ ,$$

i.e. with negative first derivative

$$\frac{\mathrm{d}p}{\mathrm{d}v} \ ;$$

but in the second derivative,

$$\frac{\mathrm{d}^2 p}{\mathrm{d}v^2} \ ,$$

there is already a slight depression in the region of $v \approx v_c$, which corresponds later to the critical density. This becomes more pronounced as the temperature is lowered.

b) At exactly T_c, for the so-called critical isotherm, which is defined by both derivatives being simultaneously zero:

$$\frac{dp}{dv}_{|T_c, p_c} \equiv \frac{d^2 p}{dv^2}_{|T_c, p_c} \equiv 0 .$$

In addition the critical atomic density $\varrho_c = v_c^{-1}$ is also determined by T_c and p_c.

c) An isotherm in the so-called coexistence region,

$$T \equiv T_3 < T_c .$$

This region is characterized such that in an interval

$$v_1^{(3)} < v < v_2^{(3)}$$

the slope

$$\frac{dp}{dv}$$

of the formal solution $p(T_3, v)$ of (42.1) is no longer negative, but becomes positive. In this region the solution is thermodynamically unstable, since an increase in pressure would increase the volume, not cause a decrease. The bounding points $v_i^{(3)}$ of the region of instability for $i = 1, 2$ (e.g., the lowest point on the third curve in Fig. 42.1) define the so-called "spinodal line", a fictitious curve along which the above-mentioned isothermal compressibility diverges.

On the other hand, more important are the coexistence lines, which are given by $p_i(v)$ for the liquid and vapor states, respectively ($i = 1, 2$). These can be found in every relevant textbook.

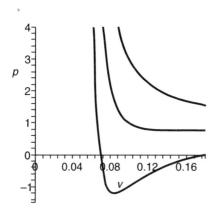

Fig. 42.1. Three typical solutions to van der Waals' equation are shown, from top to bottom: $p = -0.05/v^2 + T/(v - 0.05)$ for, (i), $T = 4$; (ii) $T \equiv T_c = 3$; and, (iii), $T = 2$

The solution for $p(T_3, v)$ is stable outside the spinodal. At sufficiently small v, or sufficiently large v, respectively, i.e., for

$$v < v_1^{\text{liquid}} \quad \text{or} \quad v > v_2^{\text{vapor}} ,$$

the system is in the liquid or vapor state, respectively. One obtains the bounding points of the transition,

$$v_1^{\text{liquid}} \quad \text{and} \quad v_2^{\text{vapor}} ,$$

by means of a so-called *Maxwell construction*, defining the *coexistence region*. Here one joins both bounding points of the curve (42.1) by a straight line which defines the saturation vapor pressure $p_s(T_3)$ for the temperature $T = T_3$:

$$p(T_3, v_1^{\text{liquid}}) = p(T_3, v_2^{\text{vapor}}) = p_s(T_3) .$$

This straight line section is divided approximately in the middle by the solution curve in such a way that (this is the definition of the Maxwell construction) both parts *above* the right part and *below* the left part of the straight line up to the curve are *exactly equal in area*. The construction is described in Fig. 42.2.

This means that the integral of the work done,

$$\oint v dp \quad \text{or} \quad -\oint p dv ,$$

for the closed path from

$$(v_1^{\text{liquid}}, p_s) ,$$

firstly along the straight line towards

$$(v_2^{\text{vapor}}, p_s)$$

and then *back* along the solution curve $p(T_3, v)$ to the van der Waals' equation (42.1), is exactly zero, which implies, as we shall see later, that our fluid

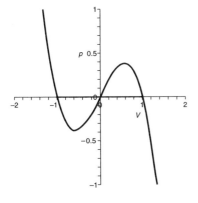

Fig. 42.2. Maxwell construction. The figure shows the fictitious function $p(V) = -(V + 1) \cdot V \cdot (V - 1)$, which corresponds qualitatively to the Maxwell theory; i.e., the l.h.s. and r.h.s. represent the marginal values of the liquid and vapor phases, respectively. The saturation-pressure line is represented by the "Maxwell segment", i.e., the straight line $p \equiv 0$ between $V = -1$ and $V = +1$, where according to *Maxwell's construction* the areas above and below the straight line on the r.h.s. and the l.h.s. are identical

system must possess the same value of the chemical potential μ in both the liquid and vapor states:

$$\mu_1(p_s, T) = \mu_2(p_s, T) \ .$$

If v lies between both bounding values, the value of v gives the ratio of the fluid system that is either in the liquid or the vapor state:

$$v = x \cdot v_1^{\text{liquid}} + (1 - x) \cdot v_2^{\text{vapor}} \ , \quad \text{where} \quad 0 \le x \le 1 \ .$$

Only this straight line (= genuine "Maxwell part") of the coexistence curve and the condition $\mu_1 = \mu_2$ then still have a meaning, even though the *van der Waals curve* itself loses its validity. This is due to the increasing influence of thermal density fluctuations which occur in particular on approaching the critical point

$$T = T_c, p = p_c, v = v_c \ .$$

Near this point according to the *van der Waals theory* the following Taylor expansion would be valid (with terms that can be neglected written as $+ \ldots$):

$$-(p - p_c) = A \cdot (T - T_c) \cdot (v - v_c) + B \cdot (v - v_c)^3 + \ldots \ , \qquad (42.2)$$

since for $T \equiv T_c$ there is a saddle-point with negative slope, and for $T > T_c$ the slope is always negative, whereas for $p = p_s < p_c$ the equation $p \equiv p_s$ leads to three real solutions. The coefficients A and B of the above Taylor expansion are therefore positive.

42.2 Magnetic Phase Changes; The Arrott Equation

In magnetism the so-called Arrott equation,

$$H = A \cdot (T - T_0) \cdot m_h + B \cdot (m_h)^3 \qquad (42.3)$$

has analogous properties to the *van der Waals equation* in the neighborhood of the critical point, (42.2).

T_0 is the Curie temperature. As in the van der Waals equation, the coefficients A and B are positive. For $T \equiv T_0$, i.e., on the *critical isotherm*,

$$H = B \cdot (m_h)^\delta \ ,$$

with a *critical exponent* $\delta = 3$. For $T \gg T_0$ we have

$$H = A \cdot (T - T_0) \cdot m_H \ ,$$

i.e., so-called *Curie-Weiss* behavior. Finally, for $T < T_0$ and positive ΔH we have a linear increase in H, if m_H increases linearly from the so-called spontaneous boundary value $m_H^{(0)}$ that arises for $H = 0^+$, together with a discontinuous transition for negative ΔH. This corresponds to a *coexistence region of positively and negatively magnetized domains*, respectively, along a straight line section on the x-axis, which corresponds to the Maxwell straight line, i.e., for

$$-m_H^{(0)} < m_H < m_H^{(0)} \ .$$

Therefore, one can also perform a Maxwell construction here, but the result, the axis $H \equiv 0$, is trivial, due to reasons of symmetry. However, the fact that the Maxwell line corresponds to a domain structure with $H \equiv 0$, and thus a corresponding droplet structure in the case of a fluid, is by no means trivial.

The *spontaneous magnetization* $m_H^{(0)}$ follows from (42.3) with $H \equiv 0$ for $T < T_0$, and behaves as

$$m_H^{(0)} \propto (T_0 - T)^\beta \ ,$$

with a *critical exponent*

$$\beta = \frac{1}{2} \ .$$

Finally, one can also show that the *magnetic susceptibility*

$$\chi = \frac{\mathrm{d}m_H}{\mathrm{d}H} \left(= \left(\frac{\mathrm{d}H}{\mathrm{d}m_H} \right)^{-1} \right) \ ,$$

is divergent, *viz* for $T > T_0$ as

$$\chi = \frac{1}{A \cdot (T - T_0)} \ .$$

On the other hand, for $T < T_0$ it behaves as

$$\chi = \frac{1}{2A \cdot (T_0 - T)} \ ,$$

i.e., generally as

$$\chi \propto |T - T_0|^\gamma \ , \quad \text{with} \quad \gamma = 1$$

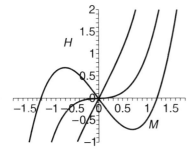

Fig. 42.3. The Arrott equation. The figure shows three typical solutions to the Arrott equation, which are discussed in the text. The equation is $H = A \cdot (T - T_0) \cdot M + B \cdot M^3$ and is plotted for three cases: (i), $T = 3$, $T_0 = 1$; (ii), $T = T_0 = 1$, and (iii), $T = -0.5$, $T_0 = 1$. In all cases we have assumed $A \equiv 1$

and different coefficients for $T < T_c$ and $T > T_c$. The values given above for

$$\beta \left(= \frac{1}{2}\right), \quad \gamma (= 1) \quad \text{and} \quad \delta (= 3)$$

are so-called *molecular field exponents*. Near the critical point itself they must be modified (see below).

42.3 Critical Behavior; Ising Model; Magnetism and Lattice Gas

Molecular field theory (and thus the van der Waals expressions) are no longer applicable in the neighborhood of the critical point, because thermal fluctuations become increasingly important, such that they may no longer be neglected. The spontaneous magnetization is then given by

$$m_H \propto (T_0 - T)^\beta \quad \text{with} \quad \beta \approx \frac{1}{3} \quad \left(not: \quad \frac{1}{2}\right)$$

for a three-dimensional system, and

$$\beta \equiv \frac{1}{8}$$

in a two-dimensional space. As a consequence of thermal fluctuations, instead of the Arrott equation, for $T \geq T_0$ the following *critical equation of state* holds[1]:

$$H = a \cdot |T - T_0|^\gamma m_H + b \cdot |m_H|^\delta + \dots , \qquad (42.4)$$

where

$$\gamma \approx \frac{4}{3} \quad \text{and} \quad \delta \approx 5 \quad \text{for} \quad d = 3 , \quad \text{and}$$

$$\gamma \equiv \frac{7}{4} \quad \text{and} \quad \delta \equiv 15 \quad \text{for} \quad d = 2 .$$

However, independent of the dimensionality the following *scaling law* applies:

$$\delta = \frac{\gamma + \beta}{\beta} ,$$

i.e., the critical exponents are indeed non-trivial and assume values dependent on d, but – independently of d – all of them can be traced back to two values, e.g., β and γ.[2]

---

[1] Here the interested reader might consider the exercises (file 8, problem 2) of winter 1997, [2],

[2] The "scaling behavior" rests on the fact that in the neighborhood of T_c there is only a single dominating length scale in the system, the so-called *thermal correlation length* $\xi(T)$ (see e.g. Chap. 6 in [38]).

The (not only qualitative) similarity between magnetism and phase changes on liquefaction is particularly apparent when it comes to the *Ising model*. In this model one considers a lattice with sites l and m, on which a discrete degree of freedom s_l, which can only assume one of the values $s_l = \pm 1$, is located. The Hamilton function (energy function) of the system is written:

$$\mathcal{H} = -\sum_{l,m} J_{l,m} s_l s_m - \sum_l h_l s_l \quad \text{(Ising model)} . \tag{42.5}$$

The standard interpretation of this system is the *magnetic interpretation*, where the s_l are atomic magnetic moments: $s_l = \pm 1$ correspond to the spin angular momentum

$$s_z = \pm \frac{\hbar}{2} .$$

The energy parameters $J_{l,m}$ are "exchange integrals" of the order of 0.1 eV, corresponding to temperatures around 1000 Kelvin. The parameters h_l correspond to external or internal magnetic fields at the position $\boldsymbol{r}_l$.

However, a *lattice-gas interpretation* is also possible, in which case l and m are lattice sites with the following property: $s_l = \pm 1$ means that the site l is either occupied or unoccupied, and $J_{l,m}$ is the energy with which atoms on the sites $\boldsymbol{r}_l$ and $\boldsymbol{r}_m$ attract ($J_{l,m} > 0$) or repel ($J_{l,m} < 0$) each other.

The similarity between these diverse phenomena is therefore not only qualitative.

One obtains the above-mentioned molecular field approximation by replacing the Hamilton function in the Ising model, (42.5), by the following expression for an (optimized[3]) single-spin approximation:

$$\mathcal{H} \xrightarrow{\approx} \mathcal{H}^{MF} := -\sum_l \left(\sum_m 2J_{l,m} \langle s_m \rangle_T + h \right) \cdot s_l$$
$$+ \sum_{l,m} J_{l,m} \cdot \langle s_l \rangle_T \langle s_l \rangle_T . \tag{42.6}$$

Here, $\langle s_m \rangle_T$ is a thermodynamic expectation value which must be determined in a self-consistent fashion. In the transition from (42.5) to (42.6), which looks more complicated than it is, one has dismantled the cumbersome expression $s_l s_m$, as follows:

$$s_l s_m \equiv \langle s_l \rangle_T s_m + s_l \langle s_m \rangle_T - \langle s_l \rangle_T \langle s_m \rangle_T + (s_l - \langle s_l \rangle_T) \cdot (s_m - \langle s_m \rangle_T) ,$$

where we neglect the last term in this identity, i.e., the "fluctuations", so that from the complicated *non-linear* expression a comparatively simple self-

---

[3] Here the optimization property, which involves a certain *Bogoliubov inequality*, is not treated.

consistent *linear* approximation is obtained:

$$h_l \rightarrow h_l + \sum_m 2J_{l,m}\langle s_m\rangle_T \ .$$

Here, s_l is formally an operator, whereas $\langle s_l\rangle_T$ is a real number.

Analogous molecular field approximations can also be introduced into other problems. A good review of applications in the theory of phase transitions is found in [41].

43 The Kinetic Theory of Gases

43.1 Aim

The aim of this section is to achieve a deeper understanding on a microscopic level of both the thermal equation of state $p(T, V, N)$ and the calorific equation of state $U(T, V, N)$ for a fluid system, not only for *classical* fluids, but also for *relativistic* fluids by additionally taking Bose and Fermi statistics (i.e., intrinsic *quantum mechanical* effects) into account.

Firstly, we shall make a few remarks about the history of the ideal gas equation. The Boyle-Mariotte law in the form $p \cdot V = f(T)$ had already been proposed in 1660, but essential modifications were made much later. For example, the quantity "mole" was only introduced in 1811 by Avogadro. Dalton showed that the total pressure of a gas mixture is made up additively from the partial pressures of the individual gases, and the law

$$\frac{V(\Theta_C)}{V_{0\,\circ\text{C}}} = \frac{273.15 + \Theta_C}{273.15},$$

which forms the basis for the concept of absolute (or Kelvin) temperature, was founded much later after careful measurements by Gay-Lussac († 1850).

43.2 The General Bernoulli Pressure Formula

The Bernoulli formula applies to non-interacting (i.e., "ideal") relativistic and non-relativistic gases, and is not only valid for ideal *Maxwell-Boltzmann gases*, but also for ideal *Fermi* and *Bose* gases. It is written (as will be shown later):

$$p = \left(\frac{N}{V}\right) \cdot \frac{1}{3} \langle m(v)v^2 \rangle_T . \tag{43.1}$$

$$m(v) = \frac{m_0}{\sqrt{1 - \frac{v^2}{c^2}}}$$

is the relativistic mass, with m_0 as rest mass, c the velocity of light and v the particle velocity. The thermal average $\langle \ldots \rangle_T$ is defined from the distribution function $F(\boldsymbol{r}, \boldsymbol{v})$ which still has to be determined:

$$\langle A(\boldsymbol{r}, \boldsymbol{v})\rangle_T := \frac{\iint \mathrm{d}^3 r \mathrm{d}^3 v A(\boldsymbol{r}, \boldsymbol{v}) \cdot F(\boldsymbol{r}, \boldsymbol{v})}{\iint \mathrm{d}^3 r \mathrm{d}^3 v F(\boldsymbol{r}, \boldsymbol{v})} \,. \tag{43.2}$$

$F(\boldsymbol{r}, \boldsymbol{v}) \Delta^3 r \Delta^3 v$ is the number of gas molecules averaged over time with $\boldsymbol{r} \in \Delta^3 r$ and $\boldsymbol{v} \in \Delta^3 v$, provided that the volumes $\Delta^3 r$ and $\Delta^3 v$ are sufficiently small, but *not too small*, so that a continuum approximation is still possible.

In order to verify (43.1) consider an element of area $\Delta^2 S$ of a plane wall with normal $\boldsymbol{n} = \boldsymbol{e}_x$. Let ΔN be the number of gas molecules with directions $\in [\vartheta, \vartheta + \Delta\vartheta)$ and velocities $\in [v, v + \Delta v)$ which collide with $\Delta^2 S$ in a time Δt, i.e., we consider an inclined cylinder of base $\Delta^2 S \boldsymbol{n}$ and height $v_x \Delta t$ containing molecules which impact against $\Delta^2 S$ at an angle of incidence $\in [\vartheta, \vartheta + \Delta\vartheta)$ in a time Δt, where

$$\cos^2 \vartheta = \frac{v_x^2}{v_x^2 + v_y^2 + v_z^2} \,, \quad \text{for} \quad v_x > 0$$

(note that $\boldsymbol{v}$ is parallel to the side faces of the cylinder). These molecules are elastically reflected from the wall and transfer momentum ΔP to it. The pressure p depends on ΔP, as follows:

$$p = \frac{\Delta P / \Delta t}{\Delta^2 S} \,, \quad \text{and} \quad \Delta P = \sum_{"i \in \Delta t"} 2 m_i (v_x)_i \,, \quad \text{i.e,} \quad \Delta P \propto \Delta^2 S \cdot \Delta t \,,$$

with a well-defined proportionality factor. In this way one obtains explicitly:

$$p = \int_{v_x = 0}^{\infty} \mathrm{d} v_x \int_{v_y = -\infty}^{\infty} \mathrm{d} v_y \int_{v_z = -\infty}^{\infty} \mathrm{d} v_z 2 m v_x^2 F(\boldsymbol{r}, \boldsymbol{v}) \,, \quad \text{or}$$

$$p = \int_{\mathcal{R}^3(\boldsymbol{v})} \mathrm{d}^3 v F(\boldsymbol{r}, \boldsymbol{v}) m v_x^2 \,.$$

Now we replace v_x^2 by $v^2/3$, and for a spatially constant potential energy we consider the spatial dependence of the distribution function,

$$F(\boldsymbol{r}, \boldsymbol{v}) = \left(\frac{N}{V}\right) \cdot g(\boldsymbol{v}) \,, \quad \text{with} \quad \int_{\mathcal{R}^3(\boldsymbol{v})} \mathrm{d}^3 v g(\boldsymbol{v}) = 1 \,;$$

this gives (43.1).

a) We shall now discuss the consequences of the Bernoulli pressure formula. For a non-relativistic ideal gas

$$\langle m v^2 \rangle_T = \langle m_0 v^2 \rangle_T =: 2 \cdot \langle \varepsilon_{\text{kin., Transl.}} \rangle_T \,,$$

with the translational part of the kinetic energy. Thus, we have

$$p = \frac{2}{3} n_V \cdot \langle \varepsilon_{\text{kin., Transl.}} \rangle_T \ , \quad \text{where} \quad n_V := \frac{N}{V} \ ;$$

$$\text{i.e.,} \quad p \equiv \frac{2 U_{\text{kin.,Transl.}}}{3V} \ .$$

On the other hand, for a *classical* ideal gas:

$$p = n_V k_B T \ .$$

Therefore, we have

$$\langle \varepsilon_{\text{kin., Transl.}} \rangle_T = \frac{3}{2} k_B T \ .$$

This identity is even valid for interacting particles, as can be shown with some effort. Furthermore it is remarkable that the distribution function $F(\boldsymbol{r}, \boldsymbol{v})$ is not required at this point. We shall see later in the framework of classical physics that

$$F(\boldsymbol{r}, \boldsymbol{v}) \propto n_V \cdot \exp\left(-\beta \frac{mv^2}{2}\right)$$

is valid, with the important abbreviation

$$\beta := \frac{1}{k_B T} \ .$$

This is the Maxwell-Boltzmann velocity distribution, which is in agreement with a more general *canonical* Boltzmann-Gibbs expression for the thermodynamic probabilities p_j for the states of a comparatively *small* quantum mechanical system with discrete energy levels E_j, embedded in a *very large* so-called microcanonical ensemble of molecules (see below) which interact weakly with the *small* system, such that through this interaction only the Kelvin temperature T is prescribed (i.e., these molecules only function as a "thermostat"). This Boltzmann-Gibbs formula states:

$$p_j \propto \exp(-\beta E_j) \ .$$

The expression

$$p \equiv \frac{2 U_{\text{kin.,Transl.}}}{3V}$$

is (as mentioned) not only valid for a *classical* ideal gas (*ideal gases* are so dilute that they can be regarded as interaction-free), but also for a non-relativistic ideal *Bose* and *Fermi* gas. These are gases of *indistinguishable* quantum-mechanical particles obeying Bose-Einstein and Fermi statistics, respectively (see Part III). In Bose-Einstein statistics any number of particles can be in the same single-particle state, whereas in Fermi statistics a maximum of one particle can be in the same single-particle state.

In the following we shall give some examples. Most importantly we should mention that *photons* are bosons, whereas *electrons* are fermions. Other examples are *pions* and *nucleons* (i.e., protons or neutrons) in nuclear physics; *gluons* and *quarks* in high-energy physics and as an example from condensed-matter physics, He^4 atoms (two protons and two neutrons in the atomic nucleus plus two electrons in the electron shell) and He^3 atoms (two protons but only one neutron in the nucleus plus two electrons). In each case the first of these particles is a boson, whereas the second is a fermion.

In the nonrelativistic case, the distribution function is now given by:

$$F(\boldsymbol{r}, \boldsymbol{v}) = \frac{1}{\exp\left[\beta\left(\frac{mv^2}{2} - \mu(T)\right)\right] \mp 1} , \tag{43.3}$$

with -1 for bosons and $+1$ for fermions. The *chemical potential* $\mu(T)$, which is positive at low enough temperatures both for fermions and in the Maxwell-Boltzmann case (where ∓ 1 can be replaced by 0, since the exponential term dominates). In contrast, μ is always ≤ 0 for bosons. Usually, μ can be found from the following condition for the particle density:

$$\int F(\boldsymbol{r}, \boldsymbol{v}) \mathrm{d}^3 v \equiv n_V(\boldsymbol{r}) .$$

Here for bosons we consider for the time being only the *normal case* $\mu(T) < 0$, such that the integration is unproblematical.

b) We now come to *ultrarelativistic behavior and photon gases*. The Bernoulli equation, (43.1), is valid even for a relativistic dependence of mass, i.e.,

$$p = \frac{n_V}{3}\left\langle \frac{m_0 v^2}{\sqrt{1 - \frac{v^2}{c^2}}} \right\rangle_T .$$

Ultrarelativistic behavior occurs if one can replace v^2 in the numerator by c^2, i.e., if the particles almost possess the speed of light. One then obtains

$$p \approx \frac{n_V}{3}\left\langle \frac{m_0 c^2}{\sqrt{1 - \frac{v^2}{c^2}}} \right\rangle_T \quad \text{and}$$

$$p \approx \frac{U(T, V)}{3V} , \quad \text{with} \quad U(T, V) = \left\langle m_0 c^2 \sum_{j=1}^{N} \frac{1}{\sqrt{1 - \frac{v_j^2}{c^2}}} \right\rangle_T ,$$

where the summation is carried out over all N particles. Now one performs the simultaneous limit of $m_0 \to 0$ and $v_j \to c$, and thus obtains for a *photon gas* in a cavity of volume V the result

$$p \equiv \frac{U}{3V} \ .$$

Photons travel with the velocity of light $v \equiv c$, have zero rest mass and behave as relativistic bosons with (due to $m_0 \to 0$) vanishing chemical potential $\mu(T)$.

A single photon of frequency ν has an energy $h \cdot \nu$. At a temperature T the contribution to the internal energy from photons with frequencies in the interval $[\nu, \nu + d\nu)$ for a gas volume V (*photon gas $\to$ black-body (cavity) radiation*) is:

$$dU(T, V) = \frac{V \cdot 8\pi\nu^2 d\nu \cdot h\nu}{c^3 \cdot \left[\exp\left(\frac{h\nu}{k_B T}\right) - 1\right]} \ ,$$

which is Planck's radiation formula (see Part III), which we state here without proof (see also below).

By integrating all frequencies from 0 to ∞ we obtain the Stefan-Boltzmann law

$$U = V\sigma T^4 \ ,$$

where σ is a universal constant. The *radiation pressure* of a photon gas is thus given by

$$p = \frac{U}{3V} = \frac{\sigma T^4}{3} \ .$$

c) We shall now discuss the internal energy $U(T, V, N)$ for a classical ideal gas. Previously we have indeed only treated the translational part of the kinetic energy, but for diatomic or multi-atomic ideal gases there are additional contributions to the energy. The potential energy of the interaction between molecules is, however, still zero, except during direct collisions, whose probability we shall neglect. So far we have only the Maxwell relation

$$\frac{\partial U}{\partial V} = T\frac{\partial p}{\partial T} - p = 0 \ .$$

This means that although we know

$$p = \frac{N}{V}k_B T \quad \text{and thus} \quad \partial U/\partial V = 0 \ ,$$

it is not yet fully clear how the internal energy depends on T. For a *monatomic* ideal gas, however,

$$U \equiv U_{\text{kin., transl.}} \cdot$$

From the Bernoulli formula it follows that

$$U \equiv \frac{3}{2}Nk_B T \ .$$

However, for an ideal gas consisting of *diatomic* molecules the following is valid for an individual molecule

$$\varepsilon \simeq \frac{m_0}{2} \langle v_s^2 \rangle_T + \varepsilon_{\text{rot.}} + \varepsilon_{\text{vibr.}} \, ,$$

where v_s is the velocity of the center of mass, $\varepsilon_{\text{rot.}}$ is the rotational part of the kinetic energy and $\varepsilon_{\text{vibr.}}$ the vibrational part of the energy of the molecule. (The atoms of a diatomic molecule can vibrate relative to the center of mass.) The rotational energy is given by

$$\varepsilon_{\text{rot.}} = \langle L_\perp^2 \rangle_T / (2\Theta_\perp) \, .[1]$$

A classical statistical mechanics calculation gives

$$\langle \varepsilon_{\text{rot.}} \rangle_T = k_B T \, ,$$

which we quote here without proof ($\rightarrow$ exercises). There are thus two extra degrees of freedom due to the two independent transverse rotational modes about axes perpendicular to the line joining the two atoms. Furthermore, vibrations of the atom at a frequency

$$\omega = \sqrt{\frac{k}{m_{\text{red.}}}} \, ,$$

where k is the "spring constant" and

$$m_{\text{red.}} = \frac{m_1 m_2}{m_1 + m_2}$$

the reduced mass (i.e. in this case $m_{\text{red}} = \frac{m_{\text{atom}}}{2}$), give two further degrees of freedom corresponding to vibrational kinetic and potential energy contributions, respectively. In total we therefore expect the relation

$$U(T) = N k_B T \cdot \left(\frac{3 + f}{2} \right)$$

to hold, with $f = 4$. At room temperature, however, it turns out that this result holds with $f = 2$. This is due to effects which can only be understood in terms of quantum mechanics: For both rotation and vibration, there is a discrete energy gap between the ground state and the excited state,

$$(\Delta E)_{\text{rot.}} \quad \text{and} \quad (\Delta E)_{\text{vibr.}} \, .$$

Quantitatively it is usually such that at room temperature the vibrational degrees of freedom are still "*frozen-in*", because at room temperature

---

[1] The analogous longitudinal contribution $\langle L_\|^2 \rangle_T / (2\Theta_\|)$ would be ∞ (i.e., it is frozen-in, see below), because $\Theta_\| = 0$.

$$k_B T \ll (\Delta E)_{\text{vibr.}} ,$$

whereas the rotational degrees of freedom are already fully " *activated* ", since

$$k_B T \gg (\Delta E)_{\text{rot.}} .$$

In the region of room temperature, for a monatomic ideal gas the internal energy is given by

$$U(T) = \frac{3}{2} N k_B T ,$$

whereas for a diatomic molecular gas

$$U(T) = \frac{5}{2} N k_B T .^2$$

This will be discussed further in the chapter on Statistical Physics.

43.3 Formula for Pressure in an Interacting System

At this point we shall simply mention a formula for the pressure in a classical fluid system where there are *interactions* between the monatomic particles. This is frequently used in computer simulations and originates from Robert Clausius. Proof is based on the so-called virial theorem. The formula is mentioned here without proof:

$$p = \frac{2}{3} \frac{U_{\text{kin. transl.}}}{V} + \frac{1}{6} \frac{\left\langle \sum_{i,j=1}^{N} \boldsymbol{F}_{i,j} \cdot (\boldsymbol{r}_i - \boldsymbol{r}_J) \right\rangle_T}{V} .$$

In this expression $\boldsymbol{F}_{i,j}$ is the internal interaction force exerted on particle i due to particle j. This force is exerted in the direction of the line joining the two particles, in a similar way to Coulomb and gravitational forces.

---

^{2} Only for $T \gg 1000K$ for a diatomic molecular gas would the vibrational degrees of freedom also come into play, such that $U(T) = \frac{7}{2} N k_B T$.

44 Statistical Physics

44.1 Introduction; Boltzmann-Gibbs Probabilities

Consider a quantum mechanical system (confined in a large volume), for which all energy values are discrete. Then

$$\hat{H}\psi_j = E_j\psi_j \, ,$$

where $\hat{H}$ is the Hamilton operator and ψ_l the complete set of orthogonal, normalized eigenfunctions of $\hat{H}$. The observables of the system are represented by Hermitian[1] operators $\hat{A}$, i.e., with real eigenvalues. For example, the spatial representation of the operator $\hat{p}$ is given by the differential operator

$$\hat{p} = \frac{\hbar}{i}\nabla$$

and the space operator $\hat{r}$ by the corresponding multiplication operator. $\hbar = h/(2\pi)$ is the reduced Planck's constant, and i the square root of minus one (the imaginary unit). The eigenvalues of $\hat{A}$ are real, as well as the expectation values

$$\langle\psi_j|\hat{A}\ \psi_j\rangle \, ,$$

which are the averages of the results of an extremely comprehensive series of measurements of $\hat{A}$ in the state ψ_j. One can calculate these expectation values, i.e., primary *experimental* quantities, theoretically via the scalar product given above, for example, in the one-particle spatial representation, without spin, as follows:

$$\langle\psi_j|\hat{A}\ \psi_j\rangle = \int \mathrm{d}^3r\psi_j^*(\boldsymbol{r},t)\hat{A}(\hat{\boldsymbol{p}},\hat{\boldsymbol{r}})\psi_j(\boldsymbol{r},t) \, , \quad \text{where}$$

$$\langle\psi_j|\psi_j\rangle = \int \mathrm{d}^3r\psi_j^*(\boldsymbol{r},t)\psi_j(\boldsymbol{r},t) \equiv 1 \, .$$

The thermal expectation value at a temperature T is then

$$\left\langle\hat{A}\right\rangle_T = \sum_j p_j \cdot \langle\psi_j|\hat{A}\psi_j\rangle \, , \tag{44.1}$$

---

[1] more precisely: by self-adjoint operators, which are a) Hermitian and b) possess a *complete* system of eigenvectors.

with the *Boltzmann-Gibbs probabilities*

$$p_j = \frac{e^{-\frac{E_j}{k_B T}}}{\sum_l e^{-\frac{E_l}{k_B T}}} \; .$$

Proof of the above expression will be deferred, since it relies on a so-called "microcanonical ensemble", and the entropy of an ideal gas in this ensemble will first be calculated.

In contrast to quantum mechanics, where *probability amplitudes* are added (i.e., $\psi = c_1\psi_1 + c_2\psi_2 + \ldots$), the *thermal average is incoherent*. This follows explicitly from (44.1). This equation has the following interpretation: the system is with probability p_j in the quantum mechanical *pure state* $\psi_j{}^2$, and the related quantum mechanical expectation values for $j = 1, 2, \ldots$, which are bilinear expressions in ψ_j, are then added like *intensities* as in incoherent optics, and not like *amplitudes* as in quantum mechanics. It is therefore important to note that *quantum mechanical coherence is destroyed by thermalization*. This limits the possibilities of "quantum computing" discussed in Part III (Quantum Mechanics).

The sum in the denominator of (44.1),

$$Z(T) := \sum_l e^{-\frac{E_l}{k_B T}} \; ,$$

is the *partition function*. This is an important function, as shown in the following section.

44.2 The Harmonic Oscillator and Planck's Formula

The partition function $Z(T)$ can in fact be used to calculate "almost anything"! Firstly, we shall consider a quantum mechanical harmonic oscillator. The Hamilton operator is

$$\hat{H} = \frac{\hat{p}^2}{2m} + \frac{m\omega_0^2}{2}\hat{x}^2$$

with energy eigenvalues

$$E_n = \left(n + \frac{1}{2}\right) \cdot \hbar\omega_0 \; , \quad \text{with} \quad n = 0, 1, 2, \ldots \; .$$

Therefore

$$Z(T) = \sum_{n_0}^{\infty} e^{-\beta \cdot (n+\frac{1}{2}) \cdot \hbar\omega_0} \; , \quad \text{with} \quad \beta = \frac{1}{k_B T} \; .$$

---

[2] or the corresponding equivalence class obtained by multiplication with a complex number

Thus

$$Z(T) = e^{\frac{\beta\hbar\omega_0}{2}} \cdot \tilde{Z}(\beta) , \quad \text{with} \quad \tilde{Z}(\beta) := \sum_{n=0}^{\infty} e^{-n \cdot \beta\hbar\omega_0} .$$

From $Z(T)$ we obtain, for example:

$$U(T) = \langle \hat{H} \rangle_T = \sum_{n=0}^{\infty} p_n \cdot E_n = \sum_{n=0}^{\infty} \left(n + \frac{1}{2} \right) \hbar\omega_0 \cdot p_n$$

$$= \hbar\omega_0 \cdot \left(\left\{ \frac{\sum_{n=0}^{\infty} n e^{-(n+\frac{1}{2})\beta\hbar\omega_0}}{\sum_{n=0}^{\infty} e^{-(n+\frac{1}{2})\beta\hbar\omega_0}} \right\} + \frac{1}{2} \right)$$

$$= \hbar\omega_0 \times \left(\left\{ \frac{\sum_{n=0}^{\infty} n e^{-n\beta\hbar\omega_0}}{\sum_{n=0}^{\infty} e^{-n\beta\hbar\omega_0}} \right\} + \frac{1}{2} \right)$$

$$= -\frac{\mathrm{d}}{\mathrm{d}\beta} \ln \tilde{Z}(\beta) + \frac{\hbar\omega_0}{2} .$$

But $\tilde{Z}(\beta)$, which is required at this point, is very easy to calculate as it is an infinite geometric series:

$$\tilde{Z}(\beta) = \sum_{n=0}^{\infty} e^{-n\beta\hbar\omega_0} \equiv \frac{1}{1 - e^{-\beta\hbar\omega_0}} .$$

Finally we obtain

$$U(T) = \hbar\omega_0 \cdot \left(\frac{e^{-\beta\hbar\omega_0}}{1 - e^{-\beta\hbar\omega_0}} + \frac{1}{2} \right) = \hbar\omega_0 \cdot \left(\frac{1}{e^{\beta\hbar\omega_0} - 1} + \frac{1}{2} \right)$$

$$= \tilde{U}(T) + \frac{\hbar\omega_0}{2} , \quad \text{with} \quad \tilde{U}(T) = \frac{\hbar\omega_0}{e^{\beta\hbar\omega_0} - 1} .$$

This expression corresponds to the Planck radiation formula mentioned previously.

As we have already outlined in Part III, Planck (1900) proceeded from the opposite point of view. Experiments had shown that at high temperatures $\tilde{U} = k_B T$ appeared to be correct (Rayleigh-Jeans) while at low temperatures

$$\tilde{U}(T) = \hbar\omega_0 e^{-\beta\hbar\omega_0}$$

(Wien). Planck firstly showed that the expression

$$\tilde{U}(T) = \frac{\hbar\omega_0}{e^{\beta\hbar\omega_0} - 1}$$

(or more precisely: the related expression for entropy) not only interpolated between those limits, but also reproduced all relevant experiments inbetween. He then hit upon energy quantization and the sequence

$$E_n = n\hbar\omega_0 \,,$$

at that time without the *zero-point energy*

$$\frac{\hbar\omega_0}{2} \,.$$

Indeed, the expression

$$\tilde{U}(T) = \frac{\hbar\omega_0}{e^{\beta\hbar\omega_0} - 1}$$

describes the experimentally observed behavior at both high and low temperatures, for example, for $k_B T \gg \hbar\omega_0$ (high T) we have:

$$\left(e^{\frac{\hbar\omega_0}{k_B T}} - 1\right) \approx \frac{\hbar\omega_0}{k_B T} \,.$$

The expression

$$\tilde{U}(T) = \frac{\hbar\omega_0}{e^{\beta\hbar\omega_0} - 1}$$

is valid for a single harmonic oscillator of frequency

$$\nu = \frac{\omega_0}{2\pi} \,.$$

One can regard $\hbar\omega_0$ as the excitation energy of an individual quantum of this *electrodynamic oscillation mode* ($\rightarrow$ photons). In solids other vibrational quanta of this kind exist (phonons, magnons, plasmons etc., see below), which are regarded as "quasi-particles". They have zero *chemical potential* μ, such that the factor

$$\langle n \rangle_T := \left(e^{\beta\hbar\omega_0} - 1\right)^{-1}$$

can be interpreted as the thermal expectation value for the number of quasi-particles n. Thus

$$U(T) = \hbar\omega_0 \cdot \left(\langle n \rangle_T + \frac{1}{2}\right) \,.$$

(Cf. one of the last sections in Part II, Sect. 21.1, dealing with dispersion, or Sect. 53.5 below.)

To summarize, plane electromagnetic waves of wavenumber $k = 2\pi/\lambda$ travelling with the speed of light c in a cavity of volume V, together with the relation

$$\lambda = c/\nu = 2\pi c/\omega_0$$

between wavelength λ and frequency ν (or angular frequency ω_0) can be interpreted as the vibrational modes of a radiation field. Formally this field has the character of an ensemble of oscillators, the *particles* (or *quasi-particles*) of the field, called *photons* (or the like). Therefore from $\tilde{U}(T)$ one obtains Planck's radiation formula by multiplying the result for a single oscillator with

$$V \cdot 2 \cdot \mathrm{d}^3 k/(2\pi)^3 = V \cdot 8\pi\nu^2 \mathrm{d}\nu/c^3 \ .$$

(The factor 2 arises because every electromagnetic wave, since it is transverse, must have two linearly independent polarization directions.)

If there is a small hole (or *aperture*) of area $\Delta^2 S$ in the outer boundary of the cavity, it can be shown that in a time Δt the amount of energy that escapes from the cavity as radiation is given by

$$\Delta I_E := \int_0^\infty \frac{c}{4} \cdot u_T(\omega_0) \mathrm{d}\omega_0 \cdot \Delta^2 S \Delta t \ .$$

One can thus define a related spectral energy flux density $j_E(\omega_0)$ of the cavity radiation

$$j_E(\omega_0) = \frac{c}{4} \cdot u_T(\omega_0) \ . \tag{44.2}$$

(The density $u_T(\omega_0)$ corresponds to the quantity $\tilde{U}(T)$.) This expression is utilized in *bolometry*, where one tries to determine the energy flux and temperature from a hole in the cavity wall[3].

---

[3] A corresponding exercise can be found on the internet, see [2], winter 1997, file 8.

45 The Transition
to Classical Statistical Physics

45.1 The Integral over Phase Space; Identical Particles in Classical Statistical Physics

The transition to classical statistical physics is obtained by replacing sums of the form

$$\sum_l p_l \cdot g(E_l)$$

by integrals over the phase space of the system. For f degrees of freedom we obtain:

$$\int\limits_{p_1\ x_1}\int \cdots \int\limits_{p_f\ x_f}\int \frac{\mathrm{d}p_1\mathrm{d}x_1 \cdot \ldots \cdot \mathrm{d}p_f\mathrm{d}x_f}{h^f} \cdot g(H(p_1,\ldots,p_f,x_1,\ldots,x_f)) \,,$$

where h is Planck's constant. Here the quasi-classical dimensionless measure of integration

$$\frac{\mathrm{d}p_1\mathrm{d}x_1 \cdot \ldots \cdot \mathrm{d}p_f\mathrm{d}x_f}{h^f}$$

is, so to speak, quantized in units of h^f; one can show that this result is invariant for Hamilton motion and corresponds exactly to the factor

$$V \cdot \mathrm{d}^3k/(2\pi)^3$$

(see below) in the Planck formula. It is also obtained with $f = 3$ and the *de Broglie relation*

$$\boldsymbol{p} = \hbar\boldsymbol{k} \,, \quad \text{as well as} \quad V = \Delta^3x \,.$$

With regard to phase-space quantization it is appropriate to mention here Heisenberg's uncertainty principle in the form

$$\Delta x \Delta p_x \sim h \,.$$

The factor

$$V \cdot \mathrm{d}^3k/(2\pi)^3$$

indeed follows by taking a wave-field approach *à la de Broglie*:

$$\psi(x) \propto \mathrm{e}^{\mathrm{i}\boldsymbol{k}\cdot\boldsymbol{r}} \,,$$

and replacing the system volume by a box of equal size with *periodic boundary conditions*, which practically does not alter the volume properties, but makes counting the eigenmodes easier. For a one-dimensional system of size L, due to

$$\psi(x + L) \equiv \psi(x)\,, \quad \text{we have:} \quad k \equiv \frac{2\pi n}{L}\,,$$

where n is integer. Thus

$$\int_{-\infty}^{\infty} \mathrm{d}n \cdot \ldots = \frac{L}{2\pi} \cdot \int_{-\infty}^{\infty} \mathrm{d}k \cdot \ldots\,,$$

and in three dimensions

$$\int \mathrm{d}^3 n \cdot \ldots = \frac{V}{(2\pi)^3} \cdot \int \mathrm{d}^3 k \cdot \ldots\,.$$

To summarize, the transition from a statistical treatment of quantum systems to classical statistical physics corresponds completely with the wave picture of matter of Louis de Broglie, which was indeed fundamental for the introduction of Schrödinger's equation ($\to$ Quantum Mechanics). The factors h^{-f} cancel each other out when calculating expectation values,

$$\left\langle \hat{A}(\boldsymbol{p}, \boldsymbol{x}) \right\rangle_T = \frac{\iint \mathrm{d}p \mathrm{d}x \hat{A}(\hat{p}, \hat{x})}{\iint \mathrm{d}p \mathrm{d}x}\,;$$

but for calculating the *entropy* (see below) they give non-trivial contributions.

There is a further (perhaps even more important) "legacy" from quantum mechanics: the effective *indistinguishability of identical particles*. For N identical particles, whether fermions or bosons, one must introduce a permutation factor $1/N!$ in front of the integral in phase space (and of course put $f = 3N$). Sackur and Tetrode (1913) pointed out long before quantum mechanics had been introduced that the basic property of additivity of entropy, which had been required since Boltzmann, is only obtained by including this permutation factor, even for the case of an ideal gas (see below). The fact that even from the viewpoint of classical physics the molecules of an ideal gas are indistinguishable is an astonishing suggestion, because one could indeed have the idea of painting them in N different colors to make them distinguishable. This could involve a negligible amount of work, but it would significantly change the entropy of the system (see again later).

45.2 The Rotational Energy of a Diatomic Molecule

The rotational behavior of diatomic molecules will now be discussed in more detail. This section refers in general to *elongated molecules*, i.e. where the longitudinal moment of inertia of the molecule $\Theta_{||}$ is negligible compared to the transverse moment of inertia $\Theta_\perp$. In addition the ellipsoid describing the anisotropy of the moment of inertia is assumed here to be a rotational ellipsoid ($\to$ Part I).

The rotational energy of such a molecule, on rotation about an axis perpendicular to the molecular axis, is found from the Hamiltonian

$$\hat{H} = \frac{\hat{L}^2}{2\Theta_\perp} \ .$$

Due to *quantization of the angular momentum* the partition function is given by

$$Z(T) = \sum_{l=0}^{\infty} (2l+1) \cdot e^{-\beta \frac{\hbar^2 l \cdot (l+1)}{2\Theta_\perp}} \ .$$

The factor $(2l+1)$ describes the degeneracy of the angular momentum, i.e. the states

$$Y_{l,m_l}(\vartheta, \varphi) , \quad \text{for} \quad m_l = -l, -l+1, \ldots, +l ,$$

have the same eigenvalues,

$$\hbar^2 l \cdot (l+1) ,$$

of the operator $\hat{L}^2$.

At very low temperatures,

$$k_B T \ll \frac{\hbar^2 \cdot}{2\Theta_\perp} ,$$

rotation is quenched due to the finite energy gap

$$\Delta := \frac{\hbar^2}{\Theta_\perp}$$

between the ground state ($l=0$) and the (triplet) first excited state ($l=1$). The characteristic *freezing* temperature Δ/k_B lies typically at approximately 50 K.

At high temperatures, e.g. room temperature, rotations with $l \gg 1$ are almost equally strongly excited, so that we can approximate the sum

$$Z(T) = \sum_{l=0}^{\infty} (2l+1) \cdot e^{-\beta \frac{\hbar^2 l \cdot (l+1)}{2\Theta_\perp}}$$

by the integral

$$\int_{l=0}^{\infty} dl \cdot 2l \cdot e^{-\beta \frac{\hat{L}^2}{2\Theta_\perp}} \ .$$

Substituting (still for $l \gg 1$) $l \to \frac{L}{\hbar}$, it follows that

$$2l \cdot dl \approx \frac{1}{\pi} \frac{d^2 L}{\hbar^2} \ .$$

We thus arrive at exactly the classical result with two additional degrees of freedom per molecule and an *energy* $\frac{k_BT}{2}$ *per degree of freedom*. This last result is called the *law of equipartition of energy* and is valid for canonical variables p_i and q_i which occur as *bilinear* terms in the classical Hamilton function.

The *vibrations* of diatomic molecules also involve *two* additional degrees of freedom per molecule (kinetic plus potential energy), but at room temperature they are generally still frozen-in, in contrast to the *rotational* degrees of freedom.

The low-temperature behavior of diatomic molecules is thus determined by *quantum mechanics*. For example,

$$ U(T) = [E_0 g_0 e^{-\beta E_0} + E_1 g_1 e^{-\beta E_1} + \ldots]/[g_0 e^{-\beta E_0} + g_1 e^{-\beta E_1} + \ldots] \, , $$

with degeneracy factors g_k, $k = 0, 1, 2 \ldots$, where we write $E_0 = 0$. The high-temperature behavior follows *classical* statistical physics, for example,

$$ U(T) = f \cdot \frac{k_BT}{2} \, . $$

The transition region satisfies the relation

$$ k_BT \approx (E_1 - E_0) \, ; $$

the transition itself can be calculated numerically using the Euler-MacLaurin summation formula.[1] (Remarkably, it is non-monotonic.)

With the above integral expression for classical statistical physics we also automatically obtain as special cases:

a) *the Maxwell-Boltzmann velocity distribution*:

$$ F(\boldsymbol{v})\mathrm{d}^3v \propto e^{-\beta \frac{mv^2}{2}} 4\pi v^2 \mathrm{d}v \, . $$

(The proportionality constant is obtained from the identity

$\int_{\mathcal{R}^3} \mathrm{d}^3 x \frac{e^{-\frac{x^2}{2\sigma^2}}}{(2\pi\sigma^2)^{\frac{3}{2}}} = 1$.)

b) *the barometric pressure equation*:

$$ \frac{n(z)}{n(0)} = \frac{p(z)}{p(0)} = \exp(-\beta mgz) \, . $$

Alternative derivations for both the Maxwell-Boltzmann velocity distribution and the barometric formula can be found in many textbooks, which provide plausible verification of the more general Boltzmann-Gibbs distribution. This means that one can arrive at these results by avoiding the transition from a microcanonical to a canonical or grand canonical ensemble, for example, if one wishes to introduce the Boltzmann-Gibbs distribution into the school curriculum.

---

[1] More details can be found in the "Göschen" booklet on Theoretical Physics by Döring, [39], Band V, Paragraph 16 (in German).

46 Advanced Discussion of the Second Law

46.1 Free Energy

We next define the quantity *Helmholtz Free Energy*

$$F(T, V, m_H, N) := U(T, V, m_H, N) - T \cdot S(T, V, m_H, N) .$$

The meaning of this thermodynamic variable is "available energy", i.e., taking into account the heat loss, given by the entropy S multiplied by the absolute temperature T, which is subtracted from the *internal energy*. Indeed one arrives at this interpretation by forming the differential of F and then using both the first and second laws:

$$\mathrm{d}F = \mathrm{d}U - T \cdot \mathrm{d}S - S \cdot \mathrm{d}T = \delta A + \delta Q - T \cdot \mathrm{d}S - S \cdot \mathrm{d}T \le \delta A - S \cdot \mathrm{d}T .$$

This inequality,

$$\mathrm{d}F \le \delta A - S \cdot \mathrm{d}T ,$$

is actually a way of stating the first and second laws simultaneously. In particular, under isothermal conditions we have

$$\mathrm{d}F \le \delta A_{\mathrm{T}} .$$

For a reversible process the equality holds. Then

$$\mathrm{d}F = -p\mathrm{d}V + H\mathrm{d}m_H + \mu\mathrm{d}N - S\mathrm{d}T , \quad \text{i.e.,}$$

$$\frac{\partial F}{\partial V} = -p ; \quad \frac{\partial F}{\partial m_H} = H ; \quad \frac{\partial F}{\partial N} = \mu \quad \text{and} \quad \frac{\partial F}{\partial T} = -S .$$

By equating the mixed second derivatives

$$\frac{\partial^2 F}{\partial x_1 \partial x_2} = \frac{\partial^2 F}{\partial x_2 \partial x_1}$$

one obtains further Maxwell relations:

$$\frac{\partial S}{\partial V} = \frac{\partial p}{\partial T} , \quad \frac{\partial p}{\partial m_H} = -\frac{\partial H}{\partial V} \quad \text{and} \quad \frac{\partial \mu}{\partial T} = -\frac{\partial S}{\partial N} .$$

By transforming from an *extensive* expression for the work done, for example,

$$\delta A = \ldots + H \mathrm{d}m_H + \ldots \, ,$$

to the corresponding *intensive* quantity, e.g.,

$$\delta A' := \delta A - \mathrm{d}(H m_H) = \ldots - m_H \mathrm{d}H + \ldots \, ,$$

one obtains, instead of the Helmholtz free energy $F(T, V, m_H, N)$, the *Gibbs free energy*

$$F_g(T, V, H, N) := F(T, V, m_H(T, V, H, N), N) - m_H H \, ,^1$$

with

$$\mathrm{d}F_g = -p\mathrm{d}V + \mu \mathrm{d}N - m_H \mathrm{d}H - S\mathrm{d}T$$

and the corresponding relations

$$\frac{\partial F_g}{\partial H} = -m_H \, , \quad \frac{\partial F_g}{\partial T} = -S \, ,$$

etc. The second law is now

$$\mathrm{d}F_g \le \delta A' - S\mathrm{d}T \, .$$

The *Gibbs free energy* $F_g(T, V, H, N)$, and not Helmholtz free energy $F(T, V, m_H, N)$, is appropriate for Hamiltonians with a Zeeman term, e.g., the Ising Hamiltonian (42.5), because the magnetic field is already taken into account in the energy values. Whereas for non-magnetic systems one can show that

$$F(T, V, N) \equiv -k_B T \cdot \ln Z(T, V, N) \, ,$$

for magnetic systems

$$F_g(T, V, H, N) \equiv -k_B T \cdot \ln Z'(T, V, H, N) \, , \quad \text{with}$$
$$Z'(T, V, H, N) \equiv \sum_l \mathrm{e}^{-\beta E_l(V, H, N)} \, .^2$$

46.2 On the Impossibility of Perpetual Motion of the Second Kind

The following six statements can be regarded as equivalent formulations of the second law of thermodynamics:

---

[1] Similarly, for magnetic systems one may also distinguish, e.g., between a *Helmholtz enthalpy* $I_{\mathrm{Helmholtz}}(T, p, m_H, N)$ and a *Gibbs enthalpy* $I_{\mathrm{Gibbs}}(T, p, H, N)$.

[2] Here the prime is usually omitted.

1) It is not possible to construct an *ideal heat pump* in which heat can flow from a colder body to a warmer body without any work being done to accomplish this flow. This formulation is due to Robert Clausius (1857). In other words, energy will not flow spontaneously from a low temperature object to a higher temperature object.
2) An alternative statement due to William Thomson, later Lord Kelvin, is essentially that you cannot create an *ideal heat engine* which extracts heat and converts it all to useful work.

Thus a perpetual motion machine of the second kind, which is a hypothetical device undergoing a cyclic process that does nothing more than convert heat into mechanical (or other) work, clearly does not exist, since it would contradict the second law of thermodynamics. On the other hand, as we shall see later, a *cyclically operating* (or reciprocating) *real* heat pump is perfectly feasible, where an amount of heat Q_2 is absorbed at a low temperature T_2 and a *greater* amount of heat $Q_1 = Q_2 + \Delta Q$ is given off at a higher temperature T_1 ($> T_2$), but – to agree with the first law of thermodynamics – the difference ΔQ must be provided by mechanical work done on the system (hence the term "heat *pump*").

3a) All Carnot heat engines[3] (irrespective of their operating substance) have the same *maximum efficiency*

$$\eta := \frac{\Delta A}{Q_1} = \frac{Q_1 - Q_2}{Q_1} \, , \quad \text{where} \quad \eta \le \frac{T_1 - T_2}{T_1}$$

is also valid. The equal sign holds for a reversible process.

The following (apparently reciprocal) statement is also valid:
3b) All Carnot heat pumps have the same *maximum efficiency*

$$\eta' := \frac{Q_1}{\Delta A} = \frac{Q_1}{Q_1 - Q_2} \, , \quad \text{with} \quad \eta' \le \frac{T_1}{T_1 - T_2} \, .$$

The equal sign again applies for a reversible process.

The efficiencies η and η' are defined in a reciprocal way, and in the optimum case also give reciprocal values. However the inequalities are not reciprocal; the irreversibility of statistical physics expresses itself here, since in both cases we have the inequality sign $\le$, due to, for example, frictional losses.

Figure 46.1 shows the (T, V)-diagram for a Carnot process. Depending on whether we are considering a Carnot heat engine or a Carnot heat pump the cycle (which comprises two *isotherms* and two *adiabatics*) runs either *clockwise* or *anticlockwise*.

---

[3] Usually more practical than *Carnot machines* are *Stirling machines*, for which the *adiabatics* of the Carnot cycle (see below) are replaced by *isochores*, i.e., $V = $ constant.

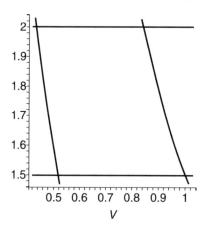

Fig. 46.1. Carnot process in a (V, T) representation. Carnot cycle (running clockwise: $\rightarrow$ heat engine; running anticlockwise: $\rightarrow$ heat pump) in a (V, T)-diagram with isotherms at $T_1 = 2$ and $T_2 = 1.5$ (arbitrary units) and adiabatics $T \cdot V^{5/3} = 1.5$ and 0.5, as for an ideal gas. Further explanation is given in the text

We shall now calculate the value of the optimal efficiency for a Carnot cycle

$$\eta_{\text{opt.}} = 1/\eta'_{\text{opt.}}.$$

by assuming that the working substance is an ideal gas. Firstly, there is an *isothermal* expansion from the upper-left point 1 to the upper-right point $1'$ in Fig. 46.1 at constant temperature T_1. The heat absorbed during this expansion is Q_1. Then we have a further expansion from $1'$ down to $2'$, i.e., from the upper-right to the lower-right (i.e., from $T_1 = 2$ to $T_2 = 1.5$), which occurs *adiabatically* according to

$$pV^\kappa = TV^{\kappa-1} = \text{const.} , \quad \text{where} \quad \kappa = \frac{C_p}{C_v} = \frac{5}{3} ,$$

down to point $2'$ at the lower temperature T_2. Subsequently there is an *isothermal* compression to the left, i.e., from $2' \rightarrow 2$ at T_2, with heat released Q_2. Then the closed cycle W is completed with an *adiabatic* compression leading from 2 up to the initial point 1. It follows that

$$\oint_W dU \equiv 0 ;$$

on the other hand we have

$$\oint_W dU = Q_1 - Q_2 - \Delta A ,$$

where the last term is the *work done by the system*.

The efficiency

$$\eta := \frac{\Delta A}{Q_1} \equiv \frac{Q_1 - Q_2}{Q_1}$$

can now be obtained by calculating the work done during the isothermal expansion

$$A_{1\to 1'} = \int_1^{1'} pdV \equiv Q_1 \;,$$

since for an ideal gas

$$\Delta U_{1\to 1'} = 0 \;.$$

Furthermore

$$\int_1^{1'} pdV = Nk_BT_1 \int_1^{1'} \frac{dV}{V} \;, \quad \text{i.e.} \quad Q_1 = Nk_BT_1 \ln \frac{V_{1'}}{V_1} \;.$$

Similarly one can show that

$$Q_2 = Nk_BT_2 \ln \frac{V_{2'}}{V_2} \;.$$

For the adiabatic sections of the process

$$(1', T_1) \to (2', T_2) \quad \text{and} \quad (2, T_2) \to (1, T_1)$$

we have

$$T_1 V_{1'}^{\kappa'} = T_2 V_{2'}^{\kappa'} \quad \text{and} \quad T_1 V_1^{\kappa'} = T_2 V_2^{\kappa'} \;,$$

where

$$\kappa' := \frac{C_P}{C_V} - 1 = \frac{2}{3} \;.$$

We then obtain

$$\frac{V_{1'}}{V_1} = \frac{V_{2'}}{V_2} \quad \text{and} \quad \eta = 1 - \frac{T_2 \ln \frac{V_{2'}}{V_2}}{T_1 \ln \frac{V_{1'}}{V_1}} \;,$$

finally giving

$$\eta = 1 - \frac{T_2}{T_1} \;,$$

as stated.

We shall now return to the concept of *entropy* by discussing a fourth equivalent version of the second law:

4) The quantity

$$S := \int_1^2 \frac{\delta Q^{\text{rev}}}{T}$$

is a variable of state (where "rev" stands for "reversible"), i.e., the integral does not depend on the path. Using this definition of the state variable *entropy* we may write:

$$\mathrm{d}S \geq \frac{\delta Q}{T} \; ,$$

with the equality symbol for reversible heat flow.

We now wish to show the equivalence between 4) and 3) by considering a Carnot heat engine. According to 4) we have

$$\oint \frac{\delta Q}{T} = \frac{Q_1}{T_1} - \frac{Q_2}{T_2} \; ,$$

but from 3) for a reversible process

$$\eta = 1 - \frac{Q_1}{Q_2} \equiv 1 - \frac{T_2}{T_1} \; , \quad \text{i.e.,} \quad \oint \frac{\delta Q^{\mathrm{rev}}}{T} \equiv 0 \; .$$

For an irreversible process the amount of heat given out at the lower temperature T_2 is greater than in the reversible case. Thus

$$\mathrm{d}S > \frac{\delta Q^{\mathrm{rev}}}{T} \; .$$

One may approximate general *reciprocating* (or *cyclic*) processes by introducing *Carnot coordinates* as indicated in Fig. 46.2 below:

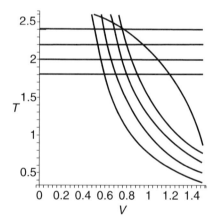

Fig. 46.2. Carnot coordinates. This (V, T) diagram shows a single curved line segment in the upper-right hand part of the diagram and two sets of four horizontal *isotherms* and four non-vertically curved *adiabatics*, respectively, forming a grid $\mathcal{N}$ of so-called Carnot coordinates. For an asymptotic improvement of the grid the single curve is extended to the bounding line ∂G of a two-dimensional region G, and since the internal contributions cancel each other in a pairwise manner, one may show analogously to Stokes's integral theorem that the following is valid: $\oint_{\partial G} \frac{\delta Q^{\mathrm{rev}}}{T} \equiv \sum_{\text{Carnot processes} \in \mathcal{N}} \frac{\delta Q^{\mathrm{rev}}}{T}$. Finally, instead of $\sum_{\dots \in \mathcal{N}}$ one may also write the integral $\int_G$

47 Shannon's Information Entropy

What actually is *entropy* ? Many answers seem to be rather *vague*: Entropy is a measure of *disorder*, for example. An increase in entropy indicates *loss of information*, perhaps. However the fact that entropy is a *quantitative* measure of information, which is very important in both physics and chemistry, becomes clear when addressing the following questions:

– How large is the number $\mathcal{N}$ of *microstates* or *configurations* of a fluid consisting of N molecules ($\nu = 1, \ldots, N$, e.g., $N \sim 10^{23}$), where each molecule is found with a probability p_j in one of f different orthogonal states ψ_j ($j = 1, \ldots, f$)?
– How large is the number $\mathcal{N}$ of texts of length N (= number of digits per text) which can be transmitted through a cable, where each digit originates from an alphabet $\{A_j\}_{j=1,\ldots,f}$ of length f and the various A_j occur with the probabilities (= relative frequencies) p_j?
– How many times $\mathcal{N}$ must a language student guess the next letter of a foreign-language text of length N, if he has no previous knowledge of the language and only knows that it has f letters A_j with probabilities p_j, $j = 1, \ldots, f$?

These three questions are of course identical in principle, and the answer is (as shown below):

$$\mathcal{N} = 2^{N \cdot I} = e^{\frac{N \cdot S}{k_B}} , \quad \text{where} \quad I = \frac{1}{N} \mathrm{ld} \mathcal{N}$$

is the so-called *Shannon information entropy*. Apart from a normalization factor, which arose historically, this expression is identical to the entropy used by physicists. Whereas Shannon and other information theorists used *binary logarithms* ld (logarithms to the base 2) physicists adopt *natural* logarithms (to the base $e = \exp(1) = 2.781\ldots$, $\ln x = (0.693\ldots) \cdot \mathrm{ld} x$). Then we have

$$\frac{S}{k_B} = \frac{1}{N} \ln \mathcal{N} ,$$

which apart from unessential factors is the same as I. Proof of the above relation between $\mathcal{N}$ and I (or $\frac{S}{k_B}$) is obtained by using some "permutation gymnastics" together with the so-called *Stirling approximation*, i.e., as follows.

The number of configurations is

$$\mathcal{N} = \frac{N!}{N_1! N_2! \cdot \ldots \cdot N_f!}$$

because, for a length of text N there are $N!$ permutations where exchange gives in general a new text, except when the digits are exchanged amongst each other. We then use the Stirling approximation: for

$$N \gg 1, \quad N! \cong \sqrt{2\pi N} \left(\frac{N}{e}\right)^N \cdot \left(1 + \mathcal{O}\left(\frac{1}{N}\right)\right) ;$$

i.e., by neglecting terms which do not increase exponentially with N, we may write

$$N! \approx \left(\frac{N}{e}\right)^N .$$

Thus

$$\mathcal{N} \cong \frac{\left(\frac{N}{e}\right)^{(N_1+N_2+\ldots+N_f)}}{\left(\frac{N_1}{e}\right)^{N_1} \cdot \ldots \cdot \left(\frac{N_f}{e}\right)^{N_f}} , \quad \text{or}$$

$$\ln \mathcal{N} \cong -\sum_{j=1}^{f} N_j \cdot \ln\left(\frac{N_j}{N}\right) \equiv -N \sum_{j=1}^{f} p_j \cdot \ln p_j ,$$

which gives

$$\mathcal{N} \cong e^{N \cdot \frac{S}{k_B}}, \text{with} \quad \frac{S}{k_B} \equiv -\sum_{j=1}^{f} p_j \cdot \ln p_j . \tag{47.1}$$

Using the same basic formula for S one can also calculate the thermodynamic entropy $S(T)$, e.g., with the Boltzmann-Gibbs probabilities

$$p_j = \frac{e^{-\beta E_j}}{Z(T)} ,$$

in agreement with the expressions

$$Z(T) = \sum_j e^{-\beta E_j} , \quad U(T) = -\frac{d \ln Z}{d\beta} , \quad F(T) = -k_B T \cdot \ln Z(T) ,$$

$$\text{with} \quad F(T) = U(T) - T \cdot S(T) , \quad \text{i.e.} \quad S(T) \equiv -\frac{\partial F}{\partial T} .$$

The relative error made in this calculation is

$$\mathcal{O}\left(\frac{\ln \sqrt{2\pi N}}{N}\right) ,$$

which for $N \gg 1$ is negligible.

To end this section we shall mention two further, particularly neat formulations of the second law:

5) A spontaneously running process in a closed system can only be reversed by doing work on the system.

This is equivalent to stating:

6) Heat only flows spontaneously from a higher to a lower temperature. This last formulation due to Max Planck goes back to Robert Clausius (see above).[1] At the same time he recognized that it is not easy to prove the second law in a statistical-physical way. This is possible, but only using stochastic methods.[2]

Work done on a closed system, $\delta A > 0$, always leads to the release of heat from the system, $\delta Q < 0$, since

$$dU (\equiv \delta A + \delta Q) = 0 .$$

The entropy must therefore have increased along the "first leg of the cycle", since $\oint dS = 0$. The requirement that we are dealing with a closed system, i.e., $dU = 0$, is thus unnecessarily special, since one can always modify an open system, e.g., with heat input, $\delta Q > 0$, to be a closed one by including the heat source. In this sense, closed modified systems are the "most general" type of system, and we shall see in the following sections in what ways they may play an important part.

---

[1] We should like to thank Rainer Höllinger for pointing this out.
[2] see, e.g., the script *Quantenstatistik* by U.K.

48 Canonical Ensembles in Phenomenological Thermodynamics

48.1 Closed Systems and Microcanonical Ensembles

The starting point for the following considerations is a closed system, corresponding to the so-called *microcanonical ensemble*. The appropriate function of state is $S(U, V, N)$. The relevant extremum principle is $S \overset{!}{=} max.$, so that $\Delta S > 0$ until equilibrium is reached. Furthermore

$$dS = \frac{\delta Q^{\mathrm{rev}}}{T} = \frac{dU - \delta A}{T} = \frac{dU + pdV - \mu dN}{T} \; ; \quad \text{thus}$$

$$\frac{1}{T} = \left(\frac{\partial S}{\partial U}\right)_{V,N} , \quad \frac{p}{T} = \left(\frac{\partial S}{\partial V}\right)_{U,N} \quad \text{and} \quad \frac{\mu}{T} = -\left(\frac{\partial S}{\partial N}\right)_{U,V} .$$

The probabilities p_j are given by

$$p_j = \frac{1}{\mathcal{N}} , \quad \text{if} \quad U - \delta U < E_j \leq U$$

otherwise $p_j = 0$. Here, δU is, e.g., an instrumental uncertainty, and

$$\mathcal{N} \, (= \mathcal{N}(U - \delta U, U))$$

is the total number of states with

$$U - \delta U < E_j \leq U .$$

We shall see below that the value of $\frac{\delta U}{U}$ is not significant unless it is extremely small in magnitude.

48.2 The Entropy of an Ideal Gas from the Microcanonical Ensemble

Inserting the relation

$$p_j \equiv \frac{1}{\mathcal{N}} \quad \text{for} \quad U - \delta U < E_j \leq U ,$$

otherwise $\equiv 0$, into the relation for S gives

$$S = k_B \cdot \ln \mathcal{N} \,, \quad \text{with}$$

$$\mathcal{N} = \mathcal{N}(U - \delta U, U) = \frac{V^N}{h^{3N} N!} \cdot \left\{ \int\limits_{U - \delta U < \frac{p_1^2 + \ldots + p_N^2}{2m} \le U} d^{3N} p \right\} \,.$$

The braced multidimensional integral in $\boldsymbol{p}$-space is the difference in volume between two spherical shells in $\mathcal{R}^{3N}(p)$ with radii

$$R_1(U) := \sqrt{2mU} \quad \text{and} \quad R_2(U - \delta U) := \sqrt{2m \cdot (U - \delta U)} \,.$$

This gives

$$\mathcal{N} = \frac{V^N}{h^{3N} N!} \Omega(3N)(2m)^{3N/2} \cdot \left\{ U^{3N/2} - (U - \delta U)^{3N/2} \right\} \,,$$

where $\Omega(d)$ is the volume of a unit sphere in d-dimensional space.

Here it is important to note that the whole second term, $(U - \delta U)^{3N/2}$, can be neglected compared to the first one, $U^{3N/2}$, except for extremely small δU, since normally we have

$$\left(\frac{U - \delta U}{U} \right)^{3N/2} \ll 1 \,,$$

because N is so large[1], i.e., one is dealing with an exponentially small ratio.

We therefore have

$$\mathcal{N} \equiv \mathcal{N}(U) = \frac{V^N \Omega(3N)(2mU)^{3N/2}}{h^{3N} N!} \quad \text{and}$$

$$S = k_B \ln \mathcal{N} = k_B N \cdot \left(\ln \frac{V}{V_0} + \frac{3}{2} \ln \frac{U}{U_0} + \ln \left[s_0 V_0 U_0^{\frac{3}{2}} \right] \right) \,.$$

Here, V_0 and U_0 are volume and energy units (the exact value is not significant; so they can be arbitrarily chosen). The additional factor appearing in the formula, the "atomic entropy constant"

$$s_0' := \ln \left[s_0 V_0 U_0^{\frac{3}{2}} \right]$$

is obtained from the requirement that (for $N \gg 1$)

$$N \cdot \ln[s_0 V_0 U_0^{\frac{3}{2}}] \overset{!}{=} \ln \left[\frac{(2m)^{3N/2} \Omega(3N)}{h^{3N} N!} V_0^N U_0^{\frac{3N}{2}} \right] \,.$$

---

[1] E.g., one should replace U by 1 and $(U - \delta U)$ by 0.9 und study the sequence $(U - \delta U)^N$ for $N = 1, 2, 3, 4, \ldots$.

The Stirling approximation was used again, since

$$\Omega(3N) \approx \left(\frac{2e\pi}{3N}\right)^{\frac{3N}{2}} .$$

Using

$$N! \cong \left(\frac{N}{e}\right)^N$$

one thus obtains:

$$s_0' \equiv \ln\left[\left(\frac{4\pi m}{h^2}\right)^{\frac{3}{2}} \cdot e^{\frac{5}{2}} \cdot V_0 U_0^{\frac{3}{2}}\right] .$$

The entropy constant therefore only depends on the type of gas via the logarithm of the particle mass m.

The above result is reasonable because, with the factor N, it explicitly expresses the *additivity* for the entropy of an ideal gas, which (as already mentioned) depends crucially on the permutation factor $\frac{1}{N!}$ for identical particles. Furthermore the law

$$\frac{1}{T} = \frac{\partial S}{\partial U}$$

yields the relation between U and T:

$$U = \frac{3}{2}Nk_BT .$$

Similarly

$$\frac{p}{T} = \frac{\partial S}{\partial V}$$

yields $pV = Nk_BT$. Only the expression for the chemical potential μ is somewhat less apparent, but we shall require it later: From

$$\frac{\mu}{T} = -\frac{\partial S}{\partial N}$$

it follows that

$$\mu = \frac{5}{2}k_BT - T \cdot \frac{S}{N} \equiv \frac{U + pV - T \cdot S}{N} .$$

Thus $\mu \equiv \frac{G}{N}$, i.e. the chemical potential μ is identical to the free enthalpy

$$G = U + pV - T \cdot S$$

per particle. This identity is generally valid for a fluid system.

48.3 Systems in a Heat Bath: Canonical and Grand Canonical Distributions

In the previous section we treated closed systems. Now we shall concentrate on closed systems that consist of a *large* system, a so-called heat bath, to which a *small* partial system (which is actually the system of interest) is weakly coupled. For a *canonical ensemble* the coupling refers only to energy, and only serves to fix the temperature. For a so-called *grand canonical ensemble* on the other hand not only exchange of energy takes place with the large system but also particles are exchanged. In this case not only the temperature T of the small system is regulated by the heat bath but also the chemical potential μ.

In the next section we shall consider how the transition from a microcanonical ensemble to a canonical and grand canonical ensemble is achieved mathematically.

For a *canonical ensemble* the appropriate variable of state is the Helmholtz free energy $F(T, V, N)$. For a magnetic system it is the Gibbs free energy $F_g(T, V, H, N)$. They can be obtained from the corresponding *partition functions*

$$Z(T, V, N) := \sum_j e^{-\beta E_j(V, N)} \quad \text{and} \quad Z'(T, V, H, N) := \sum_j e^{-\beta E_j(V, H, N)} \; ,$$

where

$$F(T, V, N) = -k_B T \cdot \ln Z(T, V, N) \quad \text{and}$$
$$F_g(T, V, H, N) = -k_B T \cdot \ln Z'(T, V, H, N) \; .$$

In a grand canonical ensemble not only the energy fluctuates but also the number of particles of the "small system"

$$\hat{H} \psi_j^{(k)} = E_j^{(k)} \psi_j^{(k)} \quad \text{and} \quad \hat{N} \psi_j^{(k)} = N_k \psi_j^{(k)} \; .$$

One therefore has, in addition to the energy index j, a particle-number index k. Thus, in addition to the reciprocal temperature

$$\beta = \frac{1}{k_B T}$$

the *chemical potential* μ appears as a further distribution parameter. Both parameters control the expectation values, i.e.,

$$\langle \hat{N} \rangle_{\beta, \mu} = \left\langle \sum_{j,k} N_k p_j^{(k)} \right\rangle_{\beta, \mu} \quad \text{and}$$

$$U = \langle \hat{H} \rangle_{\beta,\mu} := \left\langle \sum_{j,k} E_j^{(k)} p_j^{(k)} \right\rangle_{\beta,\mu} \quad , \quad \text{where}$$

$$p_j^{(k)}(\beta,\mu) = \frac{e^{-\beta\left(E_j^{(k)}-\mu N_k\right)}}{\mathcal{Z}(\beta,\mu)} ,$$

with the *grand canonical partition function*

$$\mathcal{Z}(\beta,\mu) := \sum_{j,k} e^{-\beta\left(E_j^{(k)}-\mu N_k\right)} .$$

The grand canonical Boltzmann-Gibbs distribution $p_j^{(k)}$ is therefore very similar to the canonical Boltzmann-Gibbs distribution. In particular the grand canonical partition function $\mathcal{Z}$ is related to the grand canonical thermodynamic potential Φ in a similar way as the free energy $F(T,V,N)$ is related to the usual partition function $Z(T,V,N)$:

$$\Phi(T,V,\mu) = -k_B T \cdot \ln \mathcal{Z}(T,V,\mu) .$$

The quantity Φ is the *Gibbs grand canonical potential*; phenomenologically it is formed from the free energy by a Legendre transformation with respect to N:

$$\Phi(T,V,\mu) = F(T,V,N(T,V,\mu)) - \mu N , \quad \text{and} \quad \mathrm{d}\Phi = -p\mathrm{d}V - N\mathrm{d}\mu - S\mathrm{d}T .$$

48.4 From Microcanonical to Canonical and Grand Canonical Ensembles

For an ergodic[2] system one can calculate the results for observables $\hat{A}_1$, which only involve the degrees of freedom of the *small* system "1", according to the microcanonical distribution for $E_I \approx U - \varepsilon_i$.[3]

---

[2] A classical system in a given energy range $U-\mathrm{d}U < E \leq U$ is called *ergodic* if for almost all conformations in this energy region and almost all observables $A(\boldsymbol{p},\boldsymbol{q})$ the *time average* $\frac{1}{t_0} \int_0^{t_0} \mathrm{d}t\, A(\boldsymbol{p}(t),\boldsymbol{q}(t))$ for $t_0 \to \infty$ is almost identical with the so-called *ensemble average*: $\langle A(\boldsymbol{p},\boldsymbol{q}) \rangle = \dfrac{\int_{U-\mathrm{d}U<H(\boldsymbol{p},\boldsymbol{q})\leq U} \mathrm{d}^f p\, \mathrm{d}^f q\, A(\boldsymbol{p},\boldsymbol{q})}{\int_{U-\mathrm{d}U<H(\boldsymbol{p},\boldsymbol{q})\leq U} \mathrm{d}^f p\, \mathrm{d}^f q}$. Most fluid systems are ergodic, but important non-ergodic systems also exist, for example, glasses and polymers, which at sufficiently low temperature often show unusual behavior, e.g. *ageing phenomena* after weeks, months, years, decades or even centuries, because the investigated conformations only pass through untypical parts of phase space, so that for these systems application of the principles of statistical physics becomes questionable.

[3] Here and in the following we shall systematically use *small letters* for the small system and *large letters* for the large system (or heat bath). For example, $U \approx E_I + \varepsilon_i$.

$$\langle A_1 \rangle = \frac{\displaystyle\sum_{U-\mathrm{d}U-\varepsilon_i < E_I \le U-\varepsilon_i} \langle \psi_i | \hat{A}_1 \psi_i \rangle \mathcal{N}^{(2)}(U-\varepsilon_i)}{\displaystyle\sum_{U-\mathrm{d}U-\varepsilon_i < E_I \le U-\varepsilon_i} \mathcal{N}^{(2)}(U-\varepsilon_i)} \,. \tag{48.1}$$

In the following we shall omit the indices $_1$ and $^{(2)}$.

We now introduce a Taylor expansion for the exponent of

$$\mathcal{N}(U-\varepsilon_i)\,,$$

viz:

$$\mathcal{N}(U-\varepsilon_i) \equiv \mathrm{e}^{\frac{S(U-\varepsilon_i)}{k_B}} = \mathrm{e}^{\frac{S(U)}{k_B}} \cdot \mathrm{e}^{-\frac{\varepsilon_i}{k_B}\cdot\frac{\mathrm{d}S}{\mathrm{d}U}} \cdot \mathrm{e}^{-\frac{\varepsilon_i^2 \mathrm{d}^2 S}{2k_B \mathrm{d}U^2}} \cdot \dots \,.$$

The first term is a non-trivial factor, the second term on the right-hand-side of this equation, gives $\mathrm{e}^{-\frac{\varepsilon_i}{k_B T}}$; one can already neglect the next and following terms, i.e., replace the factors by 1, as one sees, for example, for $N \to \infty$ in the term

$$\frac{\varepsilon_i^2}{k_B}\frac{\mathrm{d}^2 S}{\mathrm{d}U^2}\,, \quad \text{with} \quad U \approx \frac{3}{2}Nk_B T\,, \quad \text{thus obtaining} \quad \mathrm{e}^{-\frac{\varepsilon_i^2 \mathrm{d}}{3Nk_B \mathrm{d}T}\left(\frac{1}{k_B T}\right)} \to 1\,,$$

i.e., if one uses a monatomic ideal gas as heat bath and replaces factors

$$\mathrm{e}^{-\frac{\text{const.}}{N}}$$

by unity. Inserting this Taylor expansion into the above formula one obtains the Boltzmann-Gibbs distribution for a canonical system. One may proceed similarly for the case of a grand canonical ensemble.

49 The Clausius-Clapeyron Equation

We shall now calculate the saturation vapor pressure $p_s(T)$, which has already been discussed in the context of van der Waals' theory; $p_s(T)$ is the equilibrium vapor pressure at the interface between the liquid and vapor phase of a fluid. We require here the quantity

$$\frac{\mathrm{d}p_s}{\mathrm{d}T} \; .$$

There are two ways of achieving this.

a) The first method is very simple but rather formal. Inversely to what is usually done it consists of replacing differential quotients where necessary by quotients involving differences, i.e. the non-differentiable transition function from gas to liquid state from the Maxwell straight line is approximated by a gently rounded, almost constant transition function in such a way that for the flat parts of a curve one may equate quotients of differences with corresponding differential quotients.

Having dealt with *mathematical* aspects, we shall now consider the *physics* of the situation: We have $V = N_1 v_1 + N_2 v_2$, where v_1 and v_2 are the atomic specific volumes in liquid and vapor phase respectively, and N_1 and N_2 are the corresponding numbers of molecules. We therefore have $\Delta V = \Delta N \cdot (v_2 - v_1)$, since $N = N_1 + N_2$ is constant. For $N_2 \rightarrow N_2 + \Delta N$ we also have $N_1 \rightarrow N_1 - \Delta N$, and thus from the first law:

$$\Delta U = \delta Q + \delta A = l \cdot \Delta N - p \Delta V \; ,$$

with the (molecular) *specific latent heat of vaporization* $l(T)$ ($\approx 530 \, cal/g_{H_2O}$). Using the Maxwell relation, which is essentially a consequence of the second law, we obtain

$$\left(\frac{\partial U}{\partial V} \right)_T = \frac{\Delta U}{\Delta V} = T \frac{\partial p}{\partial T} - p = T \frac{\mathrm{d}p_S}{\mathrm{d}T} - p_s \; .$$

Thus $\frac{\partial p}{\partial T} = \ldots$, or

$$\frac{\mathrm{d}p_s(T)}{\mathrm{d}T} = \frac{l(T)}{T \cdot (v_2 - v_1)} \; . \tag{49.1}$$

This is known as the *Clausius-Clapeyron* equation. We shall now derive some consequences from this equation, and in doing so we must take the sign into account. In the case of water *boiling*, everything is "normal" provided one is not in the close vicinity of the critical point: v_2 (vapor) $\gg v_1$ (liquid), and thus from (49.1) we obtain

$$\frac{dp_s}{dT} \approx \frac{l}{T \cdot v_2} , \quad \text{and with} \quad v_2 \approx \frac{k_B T}{p_s} :$$

$$\frac{dp_s}{dT} \approx \frac{l \cdot p_s}{k_B T^2} , \quad \text{i.e.,} \quad p_s(T) \approx p_0 \cdot e^{-\frac{l}{k_B T}} ,$$

with constant p_0. *As a result there is a very fast drop in saturation vapor pressure with increasing temperature.*

There is no peculiarity here with regard to the sign; however, the Clausius-Clapeyron equation (49.1) is valid not only for a boiling transition but also for melting. For water-ice transitions, v_2, the atomic specific volume of the liquid, is 10% smaller than v_1, the atomic specific volume of the ice phase[1]. As a result of this,

$$\frac{dp_s}{dT} = \frac{l}{T \cdot (v_2 - v_1)}$$

is now *negative*, in agreement with the anomalous behavior of the phase diagram of H_2O mentioned earlier.

We now come to a second derivation of the Clausius-Clapeyron equation:

b) The method is based on an ideal *infinitesimal Carnot process*, which one obtains by choosing for a given liquid or gas segment in the equation of state the line $p_s(T)$ corresponding to the Maxwell construction as the *lower* Carnot path (i.e. $T_2 \equiv T$), whereas one chooses the saturation pressure line $p_s(T + \Delta T)$ as the *upper* Carnot path (i.e. $T_1 \equiv T + \Delta T$). We then find

$$\Delta A = \left(\int_1 - \int_2 \right) p_s dV = p_s \cdot (v_2 - v_1)\Delta N \overset{!}{=} \frac{\Delta T}{T} \cdot Q_1 ,$$

since

$$\eta = \frac{\Delta A}{Q_1} = \frac{\Delta T}{T} .$$

Using

$$Q_1 = l \cdot \Delta N \quad \text{and} \quad \Delta A = \Delta p_s \cdot (v_2 - v_1)\Delta N = \frac{\Delta T}{T} \cdot Q_1 ,$$

the Clausius-Clapeyron equation is obtained: (49.1).

These derivations imply – as we already know – that the Maxwell relations, the second law, and the statement on the efficiency of a Carnot process are all equivalent, and that in the coexistence region the straight-line Maxwell section, e.g., $p_s(T)$, is essential.

---

[1] Chemists would again prefer to use the specific molar volume.

50 Production of Low and Ultralow Temperatures; Third Law of Thermodynamics

Low temperatures are usually obtained by a process called *adiabatic demagnetization*. *Ultralow* temperatures are achieved (in Spring 2004 the record was $T_{\min} = 0.45 \times 10^{-9}$ Kelvin) in multistage processes, e.g., firstly by adiabatic demagnetization of electron-spin systems, then by adiabatic demagnetization of nuclear spins, thirdly by *laser cooling*, and finally by *evaporation* methods. (Many small steps prove to be effective.) Evaporation cooling is carried out on atomic and molecular gas systems, mainly gases of alkali atoms, that are held in an electromagnetic "trap". The phenomenon of Bose-Einstein condensation is currently being investigated on such systems at *extreme* temperatures ($\lesssim 10^{-7}$ K and lower powers of ten). This will be discussed later. In 2001 the Nobel Prize was awarded for investigations of the Bose-Einstein condensation of ultracold gases of alkali atoms (see below). These investigations could only be performed after it had been discovered how to obtain ultralow temperatures in a reproducible and controllable manner.[1]

Next we shall consider the production of low temperatures in general. The techniques usually depend on "x-caloric effects", e.g., the *magnetocaloric effect*. We shall consider the following examples:

a) *Gay-Lussac's* experiment on the free expansion of a gas from a container (see above). This occurs at a constant internal energy, such that

$$\left(\frac{\mathrm{d}T}{\mathrm{d}V}\right)_U = -\frac{\frac{\partial U}{\partial V}}{\frac{\partial U}{\partial T}} \, .$$

With the Maxwell relation

$$\frac{\partial U}{\partial V} = T\frac{\partial p}{\partial T} - p$$

and *van der Waals'* equation of state

$$p = -\frac{a}{v^2} + \frac{k_B T}{v - b} \quad \text{we obtain} \quad \left(\frac{\mathrm{d}T}{\mathrm{d}V}\right)_U = -\frac{a}{c_v^{(0)} v^2} \, ,$$

---

[1] Nobel Prize winners: Cornell, Ketterle, Wiemann.

i.e. the desired negative value (see above). In this connection we should recall that the exact differential

$$dU = \frac{\partial U}{\partial T} dT + \frac{\partial U}{\partial V} dV \overset{!}{=} 0 \,.$$

b) The *Joule-Thomson* effect. has also been discussed above. This involves a *pressure drop* at constant internal enthalpy *per particle*. One obtains the expression

$$\left(\frac{dT}{dp}\right)_{I/N} = -\frac{\frac{\partial I/N}{\partial p}}{\frac{\partial I/N}{\partial T}} = \ldots = \frac{\frac{2a}{k_B T} - b}{c_p^{(0)} \cdot (1 - \ldots)} \,,$$

i.e., giving a *negative* value above and a *positive* value below the so-called *inversion temperature* $T_{\text{Inv.}}$. Thus, for $T < T_{\text{Inv.}}$ (this is the normal case) a drop in temperature occurs for a reduction in pressure.

c) Thirdly we shall consider the *magnetocaloric effect* or the phenomenon of temperature reduction by *adiabatic demagnetization*, i.e., $dH < 0$ for constant entropy $S(T, H)$. Here one obtains as above:

$$\left(\frac{dT}{dH}\right)_S = -\frac{\frac{\partial S}{\partial H}}{\frac{\partial S}{\partial T}} \,,$$

so that one might imagine reducing the temperature indefinitely, if the entropy $S(T, H)$ behaved in such a way that for $T = 0$ at finite H also $S(0, H)$ were *finite*, with

$$\frac{\partial S}{\partial H} < 0 \,.$$

One would then only need to magnetize the magnetic sample in a first stage (step 1) *isothermally* (e.g., $H \to 2H$) and subsequently (step 2) to demagnetize it *adiabatically*, i.e. at constant S, in order to reach absolute zero $T = 0$ immediately in this second step. This supposed behavior of $S(T, H)$ is suggested by the *high temperature behavior*:

$$S(T, H) \propto a(T) - \frac{b}{H^2} \,.$$

However, it would be wrong to extrapolate this behavior to low temperatures.

In fact, about 100 years ago the *third law of thermodynamics* was proposed by the physico-chemist Walter Nernst. This is known as Nernst's heat theorem, which can be formulated, as follows:

Let $S(T, X)$ be the entropy of a thermodynamic system, where X represents one or more of the variables of state, e.g., $X = V, p$, m_j or H. Then, for $X > 0$, the limit as $S(T \to 0, X)$ is zero; and the convergence to zero is such that the absolute zero of temperature in Kelvin, $T = 0$, *is unattainable*

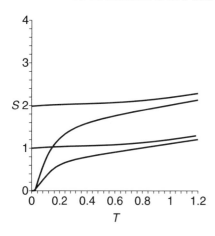

Fig. 50.1. Third Law of Thermodynamics (schematically). The low temperature
behavior of the entropy S between 0 and
4 units is presented vs. the absolute temperature T (here between 0 and 1.2 units);
only the 2nd and 4th curve from above
(i.e., with $S(T = 0) \equiv 0$) are realistic, whereas the 1st and 3rd lines represent false extrapolations suggested by the
high-T asymptotes

in a finite number of steps. In the case of, for example, *adiabatic demagnetization* this results in a countably-infinite number of increasingly small steps
(see Fig. 50.1):

Figure 50.1 shows the qualitative behavior of the entropy $S(T, H)$ of
a paramagnetic system as a function of T for two magnetic field strengths.
The first and third curves from the top correspond to extrapolations suggested by high-temperature behavior; but they do *not* give the *true* behavior
for $T \to 0$. This is instead represented by the second and fourth curves, from
which, for the same high-temperature behavior, we have for all $H \neq 0$, as
postulated by Nernst:

$$S(T = 0, H) \equiv 0 \; .$$

The *third* law, unattainability of absolute zero in a finite number of steps
– which is *not* a consequence of the *second* law – can be relatively easily
proved using *statistical physics* and basic quantum mechanics, as follows.

Consider the general case with degeneracy, where, without loss of generality, $E_0 = 0$. Let the ground state of the system be g_0-fold, and the first
excited state g_1-fold; let the energy difference ($= E_1 - E_0$) be $\Delta(X)$. Then
we obtain for the *x-caloric effect*:

$$\left(\frac{\mathrm{d}T}{\mathrm{d}X}\right)_S = -\frac{\frac{\partial S}{\partial X}}{\frac{\partial S}{\partial T}} \; , \quad \text{where} \quad S = -\frac{\partial F}{\partial T} \quad \text{and} \quad F = -k_B T \cdot \ln Z \; ,$$

with the following result for the partition function:

$$Z = g_0 + g_1 \cdot \mathrm{e}^{-\beta \Delta} + \dots \; .$$

Elementary calculation gives

$$\frac{S(T, X)}{k_B} = \ln g_0 + \frac{g_1}{g_0} \cdot \frac{\Delta}{k_B T} \cdot \mathrm{e}^{-\frac{\Delta}{k_B T}} + \dots \; ,$$

where the dots describe terms which for $k_B T \ll \Delta$ can be neglected. If one
assumes that only Δ, but not the degeneracy factors g_0 and g_1, depend on

X, it follows strictly that

$$\left(\frac{dT}{dX}\right)_S \equiv T \cdot \frac{\frac{\partial \Delta}{\partial X}}{\Delta} + \dots \; ,$$

since the exponentially small factors

$$\propto e^{-\beta \Delta}$$

in the numerator and denominator of this expression cancel each other out. In any case, for $T \to 0$ we arrive at the assertion of unattainability of absolute zero. Furthermore, we find that the assumption, $S \to 0$, in Nernst's heat theorem is unnecessary. In fact, with $g_0 \equiv 2$ for spin degeneracy of the ground state, one obtains:

$$S(T = 0, H \equiv 0) = k_B \ln 2 (\neq 0) \; .$$

In spite of this exception for $H \equiv 0$, the principle of unattainability of absolute zero still holds, since one always starts from $H \neq 0$, where $S(0, H) = 0$.

   In this respect one needs to be clear how the ultralow temperatures mentioned in connection with Bose-Einstein condensation of an alkali atom gas are achieved in a reasonable number of steps. The deciding factor here is that ultimately only the *translational kinetic energy* of the atoms is involved, and not energetically much higher degrees of freedom. Since we have

$$\frac{M \langle v^2 \rangle_T}{2} = \frac{3 k_B T}{2} \; ,$$

the relevant temperature is defined by the mean square velocity of the atoms,

$$k_B T = \frac{M \cdot \langle v^2 \rangle_T}{3} \; ,$$

where we must additionally take into account that the relevant mass M is not that of an electron, but that of a Na atom, which is of the order of 0.5×10^5 larger. One can compare this behavior with that of He^4, where at normal pressure superfluidity (which can be considered as some type of Bose-Einstein condensation for strong interaction) sets in at 2.17 K, i.e., $\mathcal{O}(1)$ K. The mass of a Na atom is an order of magnitude larger than that of a He^4 atom, and the interparticle distance δr in the Na gas considered is three to four orders of magnitude larger than in the He^4 liquid, so that from the formula

$$k_B T_c \approx \frac{\hbar^2}{2M(\delta r)^2}$$

one expects a factor of $\sim 10^{-7}$ to $\sim 10^{-9}$, i.e. temperatures of

$$\mathcal{O}\left(10^{-7}\right) \quad \text{to} \quad \mathcal{O}\left(10^{-9}\right) \text{ K}$$

are accessible.

At ultralow temperatures of this order of magnitude the phenomenon of Bose-Einstein condensation comes into play (see Sect. 53.3).

In this case the third law only plays a part at even lower temperatures where quantization of the translational energy would become noticeable, i.e., at temperatures

$$k_B T \lesssim \frac{\hbar^2}{2M_{\text{eff}} R^2} \;,$$

where R describes the size scale of the sample and M_{eff} may be the mass of a Na atom in the normal state above the Bose condensation (or in the condensed phase: an effective mass). Anyhow, we should remind ourselves that at these low temperatures and the corresponding low particle numbers (some 10^4 to 10^5 instead of 10^{23}) one should work with the microcanonical ensemble, not the canonical or grand canonical ones. However one should also remember that the basic temperature definition

$$M \cdot \langle v_s^2 \rangle = 3kT$$

is also valid for microcanonical ensembles.

51 General Statistical Physics (Formal Completion; the Statistical Operator; Trace Formalism)

Is it really necessary to diagonalize the Hamilton operator $\hat{H}$ of the system ($\hat{H} \to E_j$), if one "just" wants to calculate the partition function

$$Z(T) = \sum_j e^{-\frac{E_j}{k_B T}}$$

of the system and obtain the thermodynamic potentials or thermal expectation values

$$\left\langle \hat{A} \right\rangle_T = \sum_j \frac{e^{-\frac{E_j}{k_B T}}}{Z(T)} \langle \psi_j | \hat{A} \psi_j \rangle \; ?$$

The answer to this *rhetorical question* is of course *negative*. Instead of diagonalizing the Hamiltonian we can make use of the so-called *trace formalism*. This approach is based on the definition:

$$\text{trace } \hat{A} := \sum_j \langle \psi_j | \hat{A} \psi_j \rangle \; ,$$

which is valid for every complete orthonormal[1] basis. It is easy to show that the expression on the r.h.s. of this equation, the sum of the diagonal elements of the matrix

$$A_{i,j} := \langle \psi_i | \hat{A} \psi_j \rangle \; ,$$

is invariant with respect to a base change. It therefore follows, for example, that

$$Z(T) = \text{trace } e^{-\beta \hat{H}} \; ,$$

where

$$e^{-\beta \hat{H}}$$

is the operator, which has in the base of the eigenfunctions of $\hat{H}$ a matrix representation with diagonal elements

$$e^{-\beta E_j} \; .$$

(In another base it can also be defined by the power series $\sum_{n=0}^{\infty} \frac{(-\beta)^n}{n!} \hat{H}^n$.)

---

[1] Orthonormality is not even necessary. On the other hand, operators for which the trace exists, belong to a class of their own (trace class).

In addition there is a Hermitian *density operator* $\hat{\varrho}$, also called *state operator* or *statistical operator*, whose eigenvalues are just the probabilities p_j (ψ_j are the corresponding eigenstates). For example, one can write the relation

$$\frac{S}{k_B} = -\sum_j p_j \ln p_j$$

abstractly as

$$\frac{S}{k_B} = -\text{trace}\{\hat{\varrho} \ln \hat{\varrho}\} \ ,$$

or even more abstractly:

$$\frac{S}{k_B} = -\langle \ln \hat{\varrho} \rangle_{\hat{\varrho}} \ ,$$

just as one also uses, instead of the formula

$$\langle \hat{A} \rangle := \sum_j p_j \langle \psi_j | \hat{A} \psi_j \rangle \ ,$$

the more abstract formula

$$\langle \hat{A} \rangle_{\hat{\varrho}} := \text{trace} \hat{\varrho} \hat{A} \ .$$

However, one must realize that for this additionally gained freedom of avoiding diagonalizing the Hamiltonian there is the penalty of more complicated calculations[2]. For example, it is easy to calculate the partition function if one has already diagonalized $\hat{H}$, whereas without diagonalization of it, calculation of

$$\hat{\varrho} := \frac{\text{e}^{-\beta \hat{H}}}{\text{trace} \, \text{e}^{-\beta \hat{H}}}$$

becomes very difficult. Indeed, the trace of a matrix product involves a double sum, e.g.,

$$\text{trace} \hat{\varrho} \hat{A} = \sum_{j,k} \varrho_{j,k} A_{k,j} \ .$$

---

[2] A type of "*conservation law for effort*" holds here.

52 Ideal Bose and Fermi Gases

In the following section we shall consider identical particles, such as elementary particles and compound particles, as well as *quasi-particles* which are similar to light quanta (*photons*, electromagnetic waves): e.g., sound quanta (*phonons*, elastic waves) and spin-wave quanta (*magnons*). These particles or quasi-particles are either

a) *fermions* (particles or quasi-particles with spin $s = 1/2, 3/2, \ldots$ in units of $\hbar$), such as electrons, protons, neutrons and He^3, as well as *quarks*, from which nucleons are formed (nucleons are compound particles made up of three quarks), or

b) *bosons* (particles or quasi-particles with spin $s = 0, 1, 2, \ldots$ in units of $\hbar$), such as, for example, the *pion*, which is an elementary particle of rest mass $m_0 \approx 273$ MeV/c^2 consisting of two quarks; or a He^4 particle; or one of the above-named quasi-particles which all possess zero rest mass and, as a result, vanishing chemical potential.

Fermionic quasi-particles also exist in solids. For example, in polar semiconductors there are the so-called *polarons*, which are electrons accompanied by an attached *phonon cloud*. This fermionic quasi-particle possesses a non-negligible rest mass.

If these particles or quasi-particles do not interact with each other (or only interact weakly), the energy levels are given by:

$$E_{n_1, n_2, \ldots} = \sum_{n_1=0}^{n_{\max}} \ldots \sum_{n_f=0}^{n_{\max}} \ldots (n_1 \varepsilon_1 + n_2 \varepsilon_2 + \ldots),$$

and the number of particles is:

$$N_{n_1, n_2, \ldots} = \sum_{n_1=0}^{n_{\max}} \ldots \sum_{n_f=0}^{n_{\max}} \ldots (n_1 + n_2 + \ldots).$$

There are thus n_1 particles or quasi-particles in single-particle states of energy ε_1, etc. For fermions, $n_{\max} = 1$, while for bosons, $n_{\max} = \infty$. These statements are fundamental to quantum mechanics (*viz* Pauli's exclusion principle). In addition, if there is no particle interaction (i.e., in a dilute

Bose or Fermi gas), the partition function can be factorized as follows:

$$\mathcal{Z}_{\text{tot}}(\beta,\mu) = \mathcal{Z}_1(\beta,\mu) \cdot \mathcal{Z}_2(\beta,\mu) \cdot \ldots \cdot \mathcal{Z}_f(\beta,\mu) \cdot \ldots \,,$$

where β and μ are *reciprocal temperature* and *chemical potential* which determine the mean values of energy and particle number respectively.

It is therefore sufficient to calculate the partition function for a single factor, i.e., a single one-particle level. For example,

$$\mathcal{Z}_f(\beta,\mu) := \sum_{n_f=0}^{n_{\max}} e^{-\beta \cdot (\varepsilon_f - \mu)n_f} \,.$$

For fermions the sum consists of only two terms ($n_{\max} = 1$); a convergence problem does not arise here. For bosons on the other hand, since $n_{\max}$ is without an upper limit[1], there is an infinite geometric series, which converges if $\mu < \varepsilon_f$[2]. In both cases it then follows that

$$\mathcal{Z}_f(\beta,\mu) = \left(1 \pm e^{-\beta(\varepsilon_f - \mu)}\right)^{\pm 1} \,. \tag{52.1}$$

For the $\pm$-terms we have a plus sign for fermions and a minus sign for bosons. Only a simple calculation is now required to determine $\ln \mathcal{Z}_f$ and the expectation value

$$\langle n_f \rangle_T := \frac{\mathrm{d}}{\mathrm{d}(\beta\mu)} \ln \mathcal{Z}_f \,.$$

One then obtains the fundamental expression

$$\langle n_f \rangle_{T,\mu} = \frac{1}{e^{\beta(\varepsilon_f - \mu)} \pm 1} \,, \tag{52.2}$$

where the plus and minus signs refer to fermions and bosons respectively. For a given temperature

$$T = \frac{1}{k_B \beta}$$

and average particle number N, the chemical potential μ is determined from the auxiliary condition:

$$N = \sum_j \langle n_j \rangle_{T,\mu} \,,$$

---

[1] An analogy from everyday life: With regard to the problem of buying a dress, French women are essentially *fermionically* inclined, because no two French women would buy the same dress, irrespective of the cost. On the other hand, German women are *bosonically* inclined, since they would all buy the same dress, provided it is the least expensive.

[2] The limiting case $\mu \to 0^-$ is treated below in the subsection on Bose-Einstein condensation.

as long as for all j we have $\varepsilon_j < \mu$. (In the boson case, generally $\mu \leq 0$, if the lowest single-particle energy is zero. As already mentioned, the case $\mu = 0$ is treated below in the chapter on Bose-Einstein condensation.)

The classical result of *Boltzmann statistics* is obtained in (52.2) when the exponential term dominates the denominator, i.e., formally by replacing the term ± 1 by zero. It is often stated that different statistics are required for fermions as opposed to bosons or classical particles, but this is not really the case, since the derivation of (52.2) is made entirely within the framework of the grand canonical Boltzmann-Gibbs statistics, and everything is derived together until the difference between fermions and bosons is finally expressed by the value of n_{max} prescribed by the Pauli principle (see above), which depends on the fact that the spin angular momentum (in units of $\hbar$) is an *integer* for bosons ($\to n_{\mathrm{max}} = \infty$) and a *half-integer* for fermions ($\to n_{\mathrm{max}} = 1$). In this respect the (non-classical) property of *spin* is crucial. (We have already seen in quantum mechanics how the Pauli principle is responsible for atomic structure and the periodic table of elements.) It is important to make this clear in school and undergraduate university physics and not to disguise the difficulties in the theory.[3]

---

[3] e.g., one should mention that spin with all its unusual properties is a consequence of *relativistic* quantum theory and that one does not even expect a graduate physicist to be able to understand it fully.

53 Applications I: Fermions, Bosons, Condensation Phenomena

In the following sections we shall consider several applications of *phenomenological thermodynamics* and *statistical physics*; firstly the Sommerfeld theory of electrons in metals as an important application of the Fermi gas formalism, see [40]. Actually, we are not dealing here with a dilute Fermi gas, as prescribed by the above introduction, but at best with a *Fermi liquid*, since the particle separations are as small as in a typical *liquid metal*. However, the essential aspect of the formalism of the previous chapter – which is that interactions between particles can be neglected – is still valid to a good approximation, because electrons avoid each other due to the Pauli principle. As a result, Coulomb interactions are normally relatively unimportant, as long as the possibility of avoidance is not prevented, e.g., in a transverse direction or in $d = 1$ dimension or by a magnetic field.

53.1 Electrons in Metals (Sommerfeld Formalism)

a) The internal energy $U(T, V, N)$ of such an electron system can be written

$$U(T, V, N) = \int_0^\infty d\varepsilon \cdot \varepsilon \cdot g(\varepsilon) \cdot \langle n(\varepsilon) \rangle_{T,\mu} , \qquad (53.1)$$

where $g(\varepsilon)$ is the single-particle density; furthermore, $d\varepsilon g(\varepsilon)$ is equal to the number of single-particle energies ε_f with values in the interval $d\varepsilon$ (i.e., this quantity is $\propto V$).

b) Similarly, for the number of particles N:

$$N = \int_0^\infty d\varepsilon g(\varepsilon) \cdot \langle n(\varepsilon) \rangle_{T,\mu} . \qquad (53.2)$$

The value of the chemical potential at $T = 0$ is usually referred to as the *Fermi energy* ε_F, i.e. $\mu(T = 0) = \varepsilon_F$. Depending on whether we are dealing at $T = 0$ with a *non-relativistic* electron gas or an *ultrarelativistic* electron gas ($\varepsilon_F \ll m_e c^2$ or $\gg m_e c^2$, where $m_e \approx 0.5$ MeV/c^2 is the electron mass), we have from the Bernoulli pressure formula either

$$p = \frac{2U}{3V} \quad or \quad p = \frac{U}{3V} \, .$$

For electrons in metals, typically $\varepsilon_F = \mathcal{O}(5)\,\mathrm{eV}$, so that at room temperature we are dealing with the non-relativistic case[1].

An approximation attributable to Sommerfeld, [40], will now be described. Firstly,

$$\langle n(\varepsilon)\rangle_{T,\mu} \quad \text{for} \quad T = 0$$

is given by a step function, i.e.

$$\langle n(\varepsilon)\rangle_{T\to0,\mu} = 0 \quad \text{for} \quad \varepsilon > \varepsilon_F \quad \text{and} \quad = 1 \quad \text{for all} \quad \varepsilon < \varepsilon_F$$

(neglecting exponentially small errors). Furthermore we can write the integrals (53.1) and (53.2), again neglecting exponentially small errors, in the form

$$\int_0^\infty d\varepsilon \frac{dF}{d\varepsilon} \cdot \langle n(\varepsilon)\rangle_{T,\mu} \, ,$$

where $F(\varepsilon)$ are stem functions,

$$F(\varepsilon) = \int_0^\varepsilon d\varepsilon f(\varepsilon) \, ,$$

of the factors

$$f(\varepsilon) := \varepsilon \cdot g(\varepsilon) \quad \text{and} \quad f(\varepsilon) := g(\varepsilon)$$

appearing in the integrands of equations (53.1) and (53.2). Compared to

$$\langle n(\varepsilon)\rangle_{T,\mu} \, ,$$

a function whose negative slope behaves in the vicinity of $\varepsilon = \mu$ as a (slightly smoothed) Dirac δ function:

$$-\frac{d}{d\varepsilon}\langle n(\varepsilon)\rangle_{T,\mu} \equiv \frac{1}{4k_BT \cdot \left(\cosh \frac{\varepsilon - \mu}{2k_BT}\right)^2} (\approx \delta(\varepsilon - \mu)) \, ,$$

the functions $F(\varepsilon)$, including their derivatives, can be regarded at $\varepsilon \approx \mu$ as approximately constant. On partial integration one then obtains[2]:

$$\int_0^\infty d\varepsilon \frac{dF}{d\varepsilon} \cdot \langle n(\varepsilon)\rangle_{T,\mu} = \int_0^\infty d\varepsilon F(\varepsilon) \cdot \left(-\frac{d}{d\varepsilon}\langle n(\varepsilon)\rangle_{T,\mu}\right) \, ,$$

---

[1] The next chapter considers ultrarelativistic applications.
[2] The contributions which have been integrated out disappear, since $F(0) = 0$, whereas $\langle n(\infty)\rangle_{T,\mu} = 0$.

where one inserts for $F(\varepsilon)$ the Taylor expansion

$$F(\varepsilon) = F(\mu) + (\varepsilon - \mu) \cdot F'(\mu) + \frac{(\varepsilon - \mu)^2}{2} \cdot F''(\mu) + \dots \;.$$

On integration, the second, odd term gives zero (again neglecting exponentially small terms $\mathcal{O}(e^{-\beta \varepsilon_F})$). The third gives

$$\frac{\pi^2}{6} (k_B T)^2 \cdot F'' \;,$$

so that, for example, from (53.2) the result

$$N = \int_0^\mu d\varepsilon g(\varepsilon) + \frac{\pi^2 (k_B T)^2}{6} g'(\mu) + \dots$$

follows, where as usual the terms denoted by dots are negligible. The integral gives

$$\int_0^\mu d\varepsilon g(\varepsilon) = N + (\mu(T) - \varepsilon_F) \cdot g(\varepsilon_F) + \dots \;,$$

so that:

$$\mu(T) = \varepsilon_F - \frac{\pi^2 (k_B T)^2}{6} \cdot \frac{g'(\varepsilon_F)}{g(\varepsilon_F)} + \dots \;.$$

From (53.1) we also obtain

$$U(T, V, N) = \int_0^\mu d\varepsilon \varepsilon g(\varepsilon) + \frac{\pi^2 (k_B T)^2}{6} \cdot [\varepsilon g(\varepsilon)]'_{|\varepsilon = \mu} + \dots \;.$$

Inserting the result for $\mu(T)$, after a short calculation we thus obtain (with $U_0 := \int_0^{\varepsilon_F} d\varepsilon \varepsilon \cdot g(\varepsilon)$):

$$U(T, V, N) = U_0 + \frac{\pi^2 (k_B T)^2}{6} g(\varepsilon_F) + \dots \;.$$

By differentiating U with respect to T it follows that
electrons in a metal give a contribution to the heat capacity:

$$C_V = \frac{\partial U}{\partial T} = \gamma k_B T \;,$$

which is linear in T and where the coefficient γ is proportional to the density of states at the Fermi energy ε_F:

$$\gamma \propto g(\varepsilon_F) \;.$$

For free electrons
$$g(\varepsilon) \propto \varepsilon^{\frac{1}{2}} ,$$
so that
$$\frac{g'(\varepsilon_F)}{g(\varepsilon_F)} = \frac{1}{2\varepsilon_F^2} , \quad \text{and} \quad \mu(T) = \varepsilon_F \cdot \left(1 - \frac{\pi^2(k_BT)^2}{12(\varepsilon_F)^2} + \dots \right) ,$$

with negligible terms $+\dots$. This corresponds to quite a small *reduction* in $\mu(T)$ with increasing temperature T. This is quite small because at room temperature the ratio
$$\left(\frac{k_BT}{\varepsilon_F}\right)^2$$
itself is only of the order of $(10^{-2})^2$, as $\varepsilon_F \approx 3\,\text{eV}$ corresponds to a temperature of $3 \cdot 10^4$ K (i.e., hundred times larger than room temperature). In some metals such as Ni and Pd,
$$(k_BT)^2 \frac{g'}{g}$$
is indeed of the same order of magnitude, but in these metals $g'(\varepsilon_F)$ has a negative sign – in contrast to the case of free electrons – so that at room temperature $\mu(T)$ is here slightly *larger* than at $T = 0$. (However, in some compounds γ is larger by several orders of magnitude than usual (even at zero temperature) so that one speaks of "heavy fermions" in these compounds.)

Electrons in conventional metals behave at room temperature (or generally for $(k_BT)^2 \ll \varepsilon_F^2$) as a so-called degenerate Fermi gas.

The essential results of the previous paragraphs, apart from factors of the order of $\mathcal{O}(1)$, can be obtained by adopting the following simplified picture: *Only the small fraction*
$$\frac{k_BT}{\varepsilon_F}$$
of electrons with energies around $\varepsilon = \mu \approx \varepsilon_F$ are at all thermally active.

Thus, multiplying the classical result for the heat capacity
$$C_V = \frac{3}{2}Nk_B \quad \text{by} \quad \frac{k_BT}{\varepsilon_F} ,$$

we obtain the above linear dependence of the heat capacity on temperature for electrons in a metal, apart from factors of the order of $\mathcal{O}(1)$. In particular $g(\varepsilon_F)$ can be approximated by
$$\frac{N}{\varepsilon_F} .$$

In order to calculate the zero-point energy U_0 one must be somewhat more careful. Indeed,
$$U_0 = \int_0^{\varepsilon_F} d\varepsilon \varepsilon \cdot g(\varepsilon) .$$

For free electrons with spin g-factor 2 we have:

$$U_0 = 2V \int_0^{k_F} \frac{d^3k}{(2\pi)^3} \frac{\hbar^2 k^2}{2m} \ .$$

Using

$$d^3k = 4\pi k^2 dk$$

we obtain the non-relativistic *zero-point pressure*

$$p_0 = \frac{2U_0}{3V} \equiv \frac{2}{15\pi^2} \varepsilon_F k_F^3 \ , \quad \text{with} \quad \varepsilon_F = \frac{\hbar^2 k_F^2}{2m} \ .$$

In the result for a classical ideal gas,

$$p = \frac{N}{V} k_B T \ ,$$

one thus has to replace not only the thermal energy $k_B T$ by ε_F but also the number density

$$\frac{N}{V} \quad \text{by} \quad k_F^3$$

(i.e. essentially by the reciprocal of the third power of the separation of two electrons at the Fermi energy). In this way one again obtains the correct result, apart from dimensionless constants.

The *zero-point pressure* of an electron gas (also referred to as *degeneracy pressure*) is the phenomenon preventing the negatively charged electrons in metals from bonding directly with the positively charged atomic nuclei. This *degeneracy pressure* also plays an important part in the following section.

53.2 Some Semiquantitative Considerations on the Development of Stars

The sun is a typical *main sequence star* with a radius $R \approx 10^6$ km and a surface temperature $T \approx 6000$ K; the mass of the sun is denoted below by M_0. In contrast to the main sequence stars, so-called *white dwarfs*, e.g., Sirius B, typically have a mass $M \leq 1.4\ M_0$ of the same order as the sun, but radii about two orders of magnitude smaller, with $R \approx 10^4$ km. So-called *neutron stars* have somewhat larger mass, $M \geq 1.4\ M_0$, but $R \approx 10$ km, and so-called *black holes*, which (roughly put) "suck in" all surrounding matter and radiation below a critical distance, have a mass $M \geq (3\ \text{to}\ 7)\ M_0$ or even $\gg M_0$. The *attractive* force due to *gravitation* is opposed in the interior of the star by a corresponding *repulsive* force or *internal pressure*(see below).

Equilibrium between this internal pressure and the gravitational attraction in a spherical shell between r and $r + dr$ is found in general from the following identity:

$$4\pi r^2 \cdot [p(r + \mathrm{d}r) - p(r)] = -\gamma \frac{M(r)}{r^2} 4\pi r^2 \mathrm{d}r \cdot \varrho_M(r) \ .$$

Here $\varrho_M(r)$ is the mass density, from which we obtain, after omitting the index M, the differential equation:

$$\frac{\mathrm{d}p}{\mathrm{d}r} = -\gamma \frac{M(r)\varrho(r)}{r} \ ,$$

where $M(r)$ is the total mass up to a radius r:

$$M(r) = \int\limits_0^r \mathrm{d}\tilde{r} 4\pi \tilde{r}^2 \varrho(\tilde{r}) \ .$$

Main sequence stars lie on a diagonal line of negative slope in a so-called *Hertzsprung-Russel diagram*, this being a plot of *luminosity* L versus mass M. For these stars the pressure p is determined from the ideal gas equation:

$$p = \frac{N}{V} k_B T \ .$$

The behavior of the function $M(r)$ can be roughly characterized as follows: In a small core region around the centre of a star, which we are not interested in at present, temperatures are extremely high (10^7 K and higher) due to fusion processes (hydrogen is "burnt" to form helium), whereas outside the core in the remainder of the star including its surface region there is a roughly *constant*[3], relatively moderate temperature, e.g., $T \approx 6000$ K. Using this rough approximation we can replace the above differential equation by an average relationship between pressure p, particle density

$$n_V := \frac{N}{V}$$

and temperature T where

$$\bar{p} = \bar{n}_V \cdot k_B T \ .[4]$$

Thus,

$$\frac{\mathrm{d}p}{\mathrm{d}r} \approx \frac{\bar{p}}{R} \ .$$

---

[3] This is a crude approximation, which is nonetheless essentially true for our problem.

[4] In the following the "bar" indicating an average will usually be omitted.

Writing

$$\bar{\varrho} = \bar{n}_V \cdot m_{\text{proton}}$$

we obtain the following sequence of equations:

$$\frac{\bar{p}}{\bar{\varrho} \cdot c^2} = \frac{k_B T}{m_{\text{proton}} \cdot c^2} \overset{!}{=} \gamma \frac{M/c^2}{R} =: \frac{\mathcal{R}(M)}{R} \approx 10^{-6} \ ,$$

i.e. the so-called *Schwarzschild radius* of the sun,

$$\mathcal{R}(M_0) = \gamma M_0/c^2 \ ,$$

has an approximate value of only 1 km.

For main sequence stars the ratio

$$\frac{\bar{p}}{\bar{\varrho}c^2} = \frac{\mathcal{R}(M)}{R}$$

tells us how small *general relativistic space curvature effects* are, i.e. $\mathcal{O}(10^{-6})$. (We shall see later that in the case of white dwarfs such effects are also small: $\mathcal{O}(10^{-4})$; not until we come to neutron stars do the effects reach the order of magnitude of 1.)

After exhausting the original nuclear fuel (i.e., when hydrogen in the core region has been fully converted into helium) an accelerating sequence of processes occurs, commencing with the conversion of helium into heavier elements and ending with iron. During this sequence the temperature T decreases gradually, and as a consequence, as we see from the above series of equations, the radius of the star and its luminosity increase, whereas the total mass remains approximately constant, since the core represents only a small fraction of the total mass of the star. As a consequence, a so-called *red giant* is formed, which is a type of star such as the bright twinkling red star of *Betelgeuze* in the upper left part of the constellation of Orion. Finally a so-called *supernova explosion* occurs, where the gas cloud of the star is almost completely repelled and the remaining rest mass *collapses* into a) a white dwarf for $M \lesssim 1.4 M_0$, b) a neutron star for $M \gtrsim 1.4 M_0$ or c) a black hole for $M \gtrsim (3 \text{ to } 7) M_0$ or $\gg M_0$[5]. (These numbers must be regarded as only very approximate, especially with regard to black holes. The point is, however, that the gravitational attraction can be counteracted by the *degeneracy pressure* of the electron gas in a white dwarf or the neutron gas in a neutron star, but no longer in the case of a black hole[6].)

---

[5] e.g., $10^6 M_0$

[6] or perhaps by quantum fluctuations

a) For *white dwarfs* we may write (using a non-relativistic approach):

$$p = p_0^{(e)} = \frac{2U_0^{(e)}}{3V} \approx n_V \varepsilon_F^{(e)} \ , \quad \varrho_M = m_p \cdot n_V \ ,$$

$$\varepsilon_F^{(e)} \approx \frac{\hbar^2}{2m_e} \left(\frac{1}{d_{1,2}^2} \right)^2 \approx \frac{\hbar^2}{2m_e} \left(n_V^{(e)} \right)^{\frac{2}{3}} \ .$$

The indices $^{(e)}$ and $^{(p)}$ (see below) refer to electrons and protons respectively. Now define the *mixed density*

$$\varrho_0^{(p,e)} := \frac{m_p}{\left[\frac{\hbar}{m_e c} \right]^3} \ .$$

The *proton* mass m_p occurs in the numerator of this expression. However, in the denominator we have the third power of the Compton wavelength of the *electron*, since the electrons determine the pressure and density in a white dwarf, whereas the protons determine its mass. Thus for a white dwarf we have (apart from numerical factors of the order of unity):

$$\frac{p}{\varrho_M c^2} \approx \frac{m_e}{m_p} \left(\frac{\varrho_M}{\varrho_0^{(p,e)}} \right)^{\frac{2}{3}} \overset{!}{=} \frac{\mathcal{R}(M)}{R} \approx 10^{-4} \ , \quad \text{since} \quad R \approx 10^4 \,\text{km} \ .$$

Incidentally the surface temperature of a white dwarf is possibly very much higher than for main sequence stars, e.g., $T \approx 27000$ K in a standard case. However, for the *degenerate electron gas* theory to be applicable, it is only important that the condition $\varepsilon_F \gg k_B T$ nevertheless holds well:

$$\varepsilon_F \approx \frac{m_p c^2}{2000} \approx 10^5 \,\text{eV} \ , \quad \hat{=} 10^9 \,\text{K} \ .$$

Here we have used the fact that the proton mass is approximately 2000 times that of an electron, i.e., $m_p c^2 \approx 931$ MeV, whereas $m_e c^2$ is only 0.511 MeV.

b) *Neutron stars*: If the imploding main-sequence star is heavier than about 1.4 times the mass of the sun, the electrons can no longer withstand the gravitational attraction, not even using a relativistic calculation. But the electrons react with the equally abundant protons in an *inverse β-process* to become neutrons. (Normal β-decay is indeed $n \rightarrow p + e + \bar{\nu}_e$, with an electron-antineutrino, $\bar{\nu}_e$. However, the theory of *equilibrium in a chemical reaction*, which we shall go into later, also allows transitions primarily to occur in the opposite direction when an electron and proton become *squashed* to a separation of the order of 10^{-13} cm, i.e., $p + e \rightarrow n + \nu_e$). In 1987 a supernova in the *Large Magellanic Cloud* occurred accompanied by a neutrino shower that scientists in Japan were able to observe. For such an explosion the remaining mass collapses to an object with a radius of only about 10 km to form a neutron star. Since the moment of inertia

$$J = \frac{2}{5} M R^2$$

has decreased by, say, 10 orders of magnitude, the angular velocity $\boldsymbol{\omega}$ correspondingly increases by as many orders of magnitude, due to the law of conservation of angular momentum. Finally one thus observes the star's remains as a so-called *pulsar* with enormously high values of ω and correspondingly large magnetic field fluctuations, which periodically recur like a *cosmic beacon* as the pulsating star rotates. In any case, since neutron and proton have approximately the same mass, we may write:

$$\frac{p^{(n-\text{star})}}{\varrho_M c^2} = \left(\frac{\varrho_M}{\varrho_0^{(n)}} \right)^{\frac{2}{3}} \overset{!}{=} \frac{\mathcal{R}(M)}{R} \approx 0.1 \quad \text{to} \quad 1 \; .$$

The curvature of space, which Albert Einstein predicted in his *general theory of relativity*, then becomes important. In calculating ϱ_0 one must now insert the Compton wavelength of the neutron, and not that of the electron. This is indicated by the indices $^{(n-\text{star})}$ and $^{(n)}$.

c) *Black holes*: For $M \gg M_0$ even the degeneracy pressure of the neutrons is not sufficient to compensate for gravitational attraction, and a so-called *black hole* forms. In this instance the Schwarzschild radius

$$\mathcal{R}(M) = \gamma \cdot \frac{M}{c^2}$$

has the meaning of an *event horizon*, which we shall not go into here. Instead we refer you to the little RoRoRo-volume by Roman and Hannelore Sexl, "*White Dwarfs – Black Holes*", [42], in which the relationships are excellently shown in a semi-quantitative way at high-school or undergraduate level.[7]

53.3 Bose-Einstein Condensation

After having considered an ideal Fermi gas we shall now deal with an ideal (i.e. interaction-free) Bose gas. We have

$$N(\equiv \langle N \rangle_{T,\mu}) = \sum_{j=0,1,2,\dots} \langle n_j \rangle_{T,\mu} \; ,$$

---

[7] The book by Sexl and Urbantke, [43], is more advanced. *Black holes* are treated particularly thoroughly in the very "fat" book by *Misner, Thorne and Wheeler*, [44].

where the index $j = 0, 1, 2, \ldots$ refers to single particle modes, and

$$\langle n_j \rangle_{T,\mu} = \frac{1}{e^{\frac{\varepsilon_j - \mu}{k_B T}} - 1}$$

is valid. Here, ε_j are the single-particle energies, where the lowest energy is given by $\varepsilon_{j=0} \equiv 0$. Dividing by the very large, but finite volume of the system V, we obtain

$$n_V(T, V, \mu) := \frac{N}{V} = \frac{V^{-1}}{e^{-\frac{\mu}{k_B T}} - 1} + \int_{0+}^{\infty} \frac{d\varepsilon \, \tilde{g}(\varepsilon)}{e^{\frac{\varepsilon - \mu}{k_B T}} - 1} \, . \tag{53.3}$$

0^+ is an arbitrarily small positive number. The integral on the right-hand-side of the equation replaces the sum

$$\sum_{j=1,2,\ldots} \langle n_j \rangle_{T,\mu}$$

in the so-called *thermodynamic limit* $V \to \infty$, whereas the first term on the right, which belongs to $j = 0$, gives zero in this limit, as long as the chemical potential μ is still negative. The quantity

$$\tilde{g}(\varepsilon) := \frac{g(\varepsilon)}{V}$$

remains finite in the *thermodynamic limit*; and for Bose particles (with integral spin s) the following is valid:

$$\tilde{g}(\varepsilon)d\varepsilon = (2s + 1) \cdot \frac{d^3k}{(2\pi)^3} \, .$$

With

$$d^3k = 4\pi k^2 dk \quad \text{and} \quad \varepsilon(k) := \frac{\hbar^2 k^2}{2M_B} \, ,$$

where M_B is the mass of the Bose particle, we obtain a result in the form

$$\tilde{g}(\varepsilon) = (2s + 1) \cdot c_M \cdot \varepsilon^{\frac{1}{2}} \, ,$$

where c_M is a constant with dimensions,

$$c_M = \frac{M^{\frac{3}{2}}}{2^{\frac{1}{2}} \pi^2 \hbar^3} \, .$$

Thus,

$$n_V = \frac{V^{-1}}{e^{-\frac{\mu}{k_B T}} - 1} + (2s + 1) \cdot c_M \cdot \int_{0+}^{\infty} \frac{\varepsilon^x d\varepsilon}{e^{\frac{\varepsilon - \mu}{k_B T}} - 1} \, , \quad \text{with} \quad x := \frac{1}{2} \, .$$

We shall now consider the limit $\mu \to 0$ for negative μ. The following is strictly valid:

$$\int\limits_0^\infty \frac{\varepsilon^x \mathrm{d}\varepsilon}{e^{\frac{\varepsilon-\mu}{k_BT}} - 1} \le \int\limits_0^\infty \frac{\varepsilon^x \mathrm{d}\varepsilon}{e^{\frac{-\mu}{k_BT}} - 1} \equiv \Gamma(x+1) \cdot \zeta(x+1) \cdot (k_BT)^{x+1} . \qquad (53.4)$$

If $x > (-1)$, we have the *gamma function*

$$\Gamma(x+1) = \int\limits_0^\infty t^x e^{-t} \mathrm{d}t , \quad \text{and} \quad \zeta(s) , \quad \text{for} \quad s > 1 ,$$

the so-called *Riemann zeta function*

$$\zeta(s) = \sum_{n=1}^\infty \frac{1}{n^s} ;$$

($\zeta(\frac{3}{2})$ is $2.612\ldots$).

Thus, as long as the density remains below the critical limit $n_c(T)$, which results at a given temperature from the above inequality for $\mu \to 0$, everything is "normal", i.e. the first term on the right-hand-side of (53.3) can be neglected in the thermodynamic limit, and μ (< 0) is determined from the equation

$$n_V(T,\mu) = (2s+1) \cdot c_M \cdot \int\limits_0^\infty \frac{\varepsilon^{\frac{1}{2}} \mathrm{d}\varepsilon}{e^{\frac{\varepsilon-\mu}{k_BT}} - 1} .$$

The critical density given above is $\propto T^{x+1}$. On the other hand, at a given density one thus has a critical temperature $T_c(n_V)$.

However, if at a given temperature the critical density $n_c(T)$ is *exceeded* or at a given density n_V the temperature is below T_c, i.e.,

$$n_V \equiv n_c(T) + \Delta n_V , \quad \text{with} \quad \Delta n_V > 0 ,$$

then *the chemical potential remains "held" constant at zero*,

$$\mu(T, n_V) \equiv 0 , \quad \forall T \le T_c .$$

Also,

$$\Delta n_V = \frac{V^{-1}}{e^{-\frac{\mu}{k_BT}} - 1} , \quad \text{i.e.} \quad \approx \frac{k_BT}{-V \cdot \mu} , \quad \text{and}$$

$$-\mu = \frac{k_BT}{V \Delta n_V} \to 0 \quad \text{for} \quad V \to \infty .$$

Thus in the thermodynamic limit the behavior is not smooth at T_c, but shows a discontinuity in the derivative

$$\frac{\mathrm{d}\mu(T)}{\mathrm{d}T} \ .$$

One can easily ascertain the order of magnitude of the critical temperature: Whereas for metals (Fermi gases) the following relation holds between Fermi energy and density (apart from a factor of the order of unity)

$$\varepsilon_F \approx \frac{\hbar^2}{2m_e} n_V^{\frac{2}{3}} \ ,$$

for Bose particles we have

$$k_B T_c \approx \frac{\hbar^2}{2M_B} n_V^{\frac{2}{3}}$$

(more exactly: $k_B T_c = \frac{\hbar^2}{2M_B} \frac{n_V^{\frac{2}{3}}}{2.612...}$). Both expressions correspond to each other in the substitution $\varepsilon_F \rightarrow k_B T_c$ with simultaneous replacement of the particle mass $m_e \rightarrow M_B$. The factor

$$n_V^{\frac{2}{3}}$$

is therefore common to both, because, e.g.,

$$\frac{\hbar^2}{2m_e} n_V^{\frac{2}{3}}$$

gives the characteristic value of the kinetic energy of the electrons in the region of the Fermi energy. These are the same semi-quantitative considerations as in the previous section on star development.

The first term on the right-hand-side of equation (53.3) can be assigned to the *superfluid component*. It relates to condensed particles in their ground state. The fraction of this so-called *condensate* is 100% at $T = 0\,\mathrm{K}$, decreasing continuously to 0% as $T \rightarrow T_c$. The second term is the *normal fluid*. It distributes itself over the single-particle excited states corresponding to grand canonical Boltzmann-Gibbs statistics for bosons.

At normal pressure He^4 becomes liquid at $4.2\,\mathrm{K}$ and *superfluid* at $2.17\,\mathrm{K}$. The fact that the superfluid component possesses no internal friction can be experimentally demonstrated by the well-known "fountain effect" and other similar effects. However, if one calculates the critical temperature T_c from the above *exact formula*, one obtains 3.5 K instead of $2.17\,\mathrm{K}$, and at very low temperatures only 8% of the liquid is condensed, not 100%: The reason for these quantitative discrepancies lies in interaction effects which are neglected in the theory of an *ideal Bose gas*. Pure Bose-Einstein condensation, i.e., with negligible particle interaction, has only recently[8] been found at ultralow temperatures ($T_c \lesssim 10^{-7}$ K) in alkali gases.

The other noble gases (*Ar, Kr, Xe*) do not show superfluid behavior, because they first become solid. *He* on the other hand remains liquid at normal pressure even at very low temperatures, because the kinetic energy of the atoms is too large for solidification to occur. In contrast to He^4 the He^3 isotope is a fermion, not a boson. Therefore, it was a great surprise when in 1972 Osheroff *et al.*, [26], found superfluidityy in He^3. All three authors of that paper were awarded the physics Nobel prize of 1996 honoring their detection that also He^3 becomes superfluid, however, at temperatures about three orders of magnitude smaller than He^4: $T_c = 2.6$ mK[9]for He^3. As we shall see later, this occurs by the formation of so-called *Cooper pairs* each consisting of two fermions, which themselves form a pair condensate[10]. Legget was able to interpret the experiments of Osheroff *et al.* theoretically and was awarded the Nobel prize in 2003 for this achievement, together with Ginzburg and Abrikosov, who were rewarded for their work in the field of superconductivity (see next section).

53.4 Ginzburg-Landau Theory of Superconductivity

The phenomena of *superconductivity* and *superfluidity* are in fact very closely connected: a superconducting system can be thought of as a *charged superfluid* (see below), though the charge carriers are not of the elementary value $q_e = e$, as one had believed up to the introduction of the BCS theory in 1957[11]. Instead, they correspond to $q_e = 2e$, i.e. to Cooper pairs, which were proposed just before the BCS theory, [25]. However, aside from that, a phenomenological theory of superconductivity had already been proposed in 1950 by Ginzburg and Landau, [46], which proved to be very fruitful and correct in all details, and which lead amongst other things to the flux line lattice theory of Abrikosov, [47], being established, for which – as already mentioned – the Nobel prize in 2003 was awarded (Abrikosov, Ginzburg, Legget).

In the following section we shall describe Ginzburg and Landau's theory of superconductivity: In this theory the superconducting condensate is described by a complex so-called order parameter function $\Psi(\boldsymbol{r}, t)$. The name of this function reminds one of quantum mechanics; however, the capital letter

---

[8] Cornell, Ketterle and Wiemann were awarded the Nobel prize in 2001 for work they had performed on the Bose-Einstein condensation of ultracold gases of alkali atoms in 1995.

[9] These are *low* temperatures, but not yet the *ultralow* ones mentioned above.

[10] The complexity of the order parameter in He^3 is described in a comprehensive book by Vollhardt and Wölfle, [45].

[11] Named after Bardeen, Cooper und Schrieffer. The BCS theory, see [25], was proposed in 1957, almost half a century after the experimental discovery of the phenomenon (1911) by Kammerlingh-Onnes in Leiden.

suggests that Ψ is not to be thought of as probability amplitude, but rather as a classical quantity. For stationary states the time dependence will not be explicitly mentioned. $n_s(\boldsymbol{r}) = |\Psi(\boldsymbol{r})|^2$ is the *density of the superconducting condensate* ("pair density") and

$$\boldsymbol{j}_s(\boldsymbol{r}) = \frac{q_e}{m^{\text{eff}}} \cdot \mathcal{R}e\left\{ \Psi^*(-\mathrm{i}\hbar\nabla - q_e\boldsymbol{A})\Psi \right\}$$

$$= \frac{q_e}{m^{\text{eff}}} \cdot \left\{ \frac{\hbar}{2\mathrm{i}}(\Psi^*\nabla\Psi - \Psi\nabla\Psi^*) - q_e n_s \boldsymbol{A} \right\}$$

is the *supercurrent density*. Separating $\Psi(\boldsymbol{r})$ into modulus and phase,

$$\Psi = |\Psi(\boldsymbol{r})| \cdot \mathrm{e}^{\mathrm{i}\phi(\boldsymbol{r})} \;,$$

we have

$$\boldsymbol{j}_s = q_e \cdot n_s(\boldsymbol{r}) \cdot \frac{\hbar\nabla - q_e\boldsymbol{A}}{m^{\text{eff}}}\phi(\boldsymbol{r}) \;,$$

an expression, whose gauge-invariance[12] can be explicitly seen; m^{eff} is the effective mass of the carriers of the superconductivity.

The free energy $F(T,V)$ is written as a power series in $|\Psi|^2$. Neglecting terms which do not influence the onset of superconductivity we have:

$$F(T,V) = \min_{\Psi,\Psi^*,\boldsymbol{A}}$$

$$\left\{ \int_V \frac{\mathrm{d}^3r}{2}\left[\frac{1}{2m^{\text{eff}}}|\left(-\mathrm{i}\hbar\nabla - q_e\boldsymbol{A}\right)\Psi|^2 + \alpha \cdot (T - T_0) \cdot |\Psi|^2 \frac{\beta}{2}|\Psi|^4 + \ldots \right] \right.$$

$$\left. + \int_{\mathcal{R}^3} \mathrm{d}^3r \frac{(\text{curl}\boldsymbol{A})^2}{2\mu_0} \right\} \;. \tag{53.5}$$

Here, μ_0 is the permeability of free space; α and β are positive constants, and differentiation should be carried out independently with respect to Ψ and Ψ^* (i.e. with respect to the real part and imaginary parts of $\Psi(\boldsymbol{r})$), as well as with respect to rotation of $\boldsymbol{A}$, i.e. with respect to the magnetic induction

$$\boldsymbol{B} = \text{curl}\boldsymbol{A} \;.$$

In (53.5), the last integral over $\mathcal{R}^3$ is the magnetic field energy, whereas the first integral (delimited by square brackets) represents the free energy of the condensate. The important term,

$$\alpha \cdot (T - T_0)|\Psi|^2 \;,$$

which shows a change of sign at the *critical temperature* T_0, has been introduced by the authors in an *ad hoc* way, and is justified by the results which follow (see below).

---

[12] Invariance w.r.t. *gauge transformations*, $\boldsymbol{A} \to \boldsymbol{A}+\nabla f(\boldsymbol{r}); \Psi \to \Psi\cdot\exp(\mathrm{i}q_e f(\boldsymbol{r})/\hbar)$, simultaneously and for any $f(\boldsymbol{r})$.

Similar to Lagrangian formalism in classical mechanics, minimizing the free energy with respect to Ψ^* provides the following Euler-Langrange equation for the variation problem (53.5):

$$\frac{1}{2m^{\text{eff}}}(-i\hbar\nabla - q_e\boldsymbol{A})^2\Psi(\boldsymbol{r}) + \alpha\cdot(T-T_0)\Psi + \beta\cdot|\Psi|^2\Psi + \ldots = 0\ . \quad (53.6)$$

(Minimizing with respect to Ψ does not give anything new, only the complex conjugate result.) Minimizing F with respect to $\boldsymbol{A}$ on the other hand leads to the *Maxwell equation*

$$\text{curl curl}\boldsymbol{A} = \mu_0\boldsymbol{j}_s\ , \quad \text{since} \quad \boldsymbol{B} = \mu_0\boldsymbol{H} = \text{curl}\boldsymbol{A}\ , \quad \text{and thus:} \quad \text{curl}\boldsymbol{H} \equiv \boldsymbol{j}_s\ .$$

Solving equation (53.6) for $\boldsymbol{A} \equiv 0$ assuming spatially homogeneous states and neglecting higher terms, one obtains for $T \geq T_0$ the trivial result $\Psi \equiv 0$, while for $T < 0$ the non-trivial expression

$$|\Psi| = \sqrt{\frac{\alpha^2}{2\beta}(T_0 - T)}$$

results. In the first case the free energy is zero, while in the second case it is given by

$$F(T,V) = -\frac{V\cdot\alpha^2}{4\beta}\cdot(T_0-T)^2\ .$$

On passing through T_0 the heat capacity

$$C := -\frac{\partial^2 F}{\partial T^2}$$

therefore changes discontinuously by an amount

$$\Delta C = V\frac{\alpha^2}{2\beta}\ .$$

At $T = T_0$ a continuous phase change[13] thus takes place, as for the case of Bose-Einstein condensation, in which the order parameter Ψ increases smoothly from zero (for $T \geq T_0$) to finite values (for $T < T_0$), whereas the heat capacity increases discontinuously, as mentioned.

Two characteristic lengths result from the Ginzburg-Landau theory of superconductivity. These are:

a) the so-called *coherence length* $\xi(T)$ of the order parameter, and
b) the so-called *penetration depth* $\lambda(T)$ of the magnetic induction.

---

[13] A discontinuous change of the specific heat is allowed by a continuous phase transition. It is only necessary that the order parameter changes continuously.

One obtains the *coherence length* $\xi(T)$ for the density of Cooper pairs by assuming that $\Psi = \Psi_0 + \delta\Psi(\boldsymbol{r})$ (where, as above, $\boldsymbol{A} \equiv 0$ and $\Psi_0 = \sqrt{\frac{\alpha}{\beta} \cdot (T_0 - T)}$).

Using (53.6) we obtain:

$$-\frac{\hbar^2}{2m^{\text{eff}}} \frac{\mathrm{d}^2(\delta\Psi)}{\mathrm{d}x^2} + \left[\alpha \cdot (T - T_0) + 3\beta|\Psi_0|^2\right] \cdot \delta\Psi = 0 ,$$

which by assuming

$$\delta\Psi(\boldsymbol{r}) \propto \mathrm{e}^{-\frac{x}{\xi}}$$

leads to

$$\xi(T) = \sqrt{\frac{4m^{\text{eff}}}{\hbar^2\alpha \cdot (T_0 - T)}} .$$

On the other hand one obtains the *penetration depth* $\lambda(T)$ of the magnetic field assuming $\boldsymbol{A} \neq 0$ and $\Psi \equiv \Psi_0$. Thus

$$\operatorname{curl} \operatorname{curl}\boldsymbol{A}(= \operatorname{grad} \operatorname{div}\boldsymbol{A} - \nabla^2\boldsymbol{A}) = \mu_0\boldsymbol{j}_s = \ldots - \mu_0 \cdot q_e^2 \cdot |\Psi_0|^2\boldsymbol{A} ,$$

so that with the assumption

$$\boldsymbol{A} \propto \boldsymbol{e}_y \cdot \mathrm{e}^{-\frac{x}{\lambda}}$$

the relation

$$\lambda(T) = \sqrt{\frac{\beta}{\alpha\mu_0 q_e^2 \cdot (T_0 - T)}}$$

results. Thus the magnetic induction inside a superconductor is compensated completely to zero by surface currents, which are only non-zero in a thin layer of width $\lambda(T)$ (*Meissner-Ochsenfeld effect, 1933*). This is valid however only for sufficiently weak magnetic fields.

In order to handle stronger fields, according to Abrikosov we must distinguish between *type I* and *type II* superconductors, depending on whether $\xi > \lambda\sqrt{2}$ is valid or not. The difference therefore does not depend on the temperature. For type II superconductors between two critical magnetic fields H_{c_1} and H_{c_2} it is energetically favorable for the magnetic induction to penetrate inside the superconductor in the form of so-called *flux tubes*, whose diameter is given by 2λ, whereas the region in the center of these flux tubes where the superconductivity vanishes has a diameter of *only* 2ξ. The superconductivity does not disappear until the field H_{c_2} is exceeded.

The function $\Psi(\boldsymbol{r}, t)$ in the Ginzburg-Landau functional (53.5) corresponds to the *Higgs boson* in the field theory of the electro-weak interaction, whereas the vector potential $\boldsymbol{A}(\boldsymbol{r}, t)$ corresponds to the standard fields $W^{\pm}$ and Z occurring in this field theory. These massless particles "receive" a mass $M_{W^{\pm}, Z} \approx 90\,\text{GeV}/\text{c}^2$ via the so-called *Higgs-Kibble mechanism*, which corresponds to the *Meissner-Ochsenfeld effect* in superconductivity. This correspondence rests on the possibility of translating the magnetic field penetration depth λ into a mass M_λ, in which one interprets λ as the Compton wavelength of the mass,

$$\lambda =: \frac{\hbar}{M_\lambda c} \ .$$

It is certainly worth taking note of such relationships between low temperature and high energy physics.

53.5 Debye Theory of the Heat Capacity of Solids

In the following consider the contributions of phonons, magnons and similar bosonic *quasiparticles* to the heat capacity of a solid. Being bosonic *quasiparticles* they have the particle-number expectation value

$$\langle n(\varepsilon) \rangle_{T,\mu} = \frac{1}{e^{\beta \cdot (\varepsilon - \mu)} - 1} \ .$$

But since for all these quasiparticles the rest mass vanishes such that they can be generated in arbitrary number without requiring work μdN, we also have $\mu \equiv 0$.

Phonons are the quanta of the *sound-wave field*,

$$\boldsymbol{u}(\boldsymbol{r}, t) \propto e^{i(\boldsymbol{k} \cdot \boldsymbol{r} - \omega_k \cdot t)} \ ,$$

magnons are the quanta of the *spin-wave field* $(\delta \boldsymbol{m} \propto e^{i(\boldsymbol{k} \cdot \boldsymbol{r} - \omega_k \cdot t)})$, where the so-called *dispersion relations* $\omega_k (\equiv \omega(k))$ for the respective wave fields are different, i.e., as follows.

For wavelengths

$$\lambda := \frac{2\pi}{k} \ ,$$

which are much larger than the distance between nearest neighbors in the system considered, we have for phonons:

$$\omega_k = c_s \cdot k + \dots \ ,$$

where c_s is the longitudinal or transverse sound velocity and terms of higher order in k are neglected. Magnons in *antiferromagnetic* crystals also have a linear dispersion relation $\omega \propto k$, whereas magnons in *ferromagnetic* systems have a quadratic dispersion,

$$\omega_k = D \cdot k^2 + \dots \ ,$$

with so-called *spin-wave stiffness* D. For the excitation energy ε_k and the excitation frequency ω_k one always has of course the relation

$$\varepsilon_k \equiv \hbar \omega_k .$$

For the internal energy U of the system one thus obtains (apart from an arbitrary additive constant):

$$U(T, V, N) = \int_0^\infty \hbar\omega \frac{1}{e^{\frac{\hbar\omega}{k_B T}} - 1} \cdot g(\omega)d\omega , \quad \text{where}$$

$$g(\omega)d\omega = \frac{V \cdot d^3k}{(2\pi)^3} = \frac{V \cdot k^2}{2\pi^2} \left(\frac{dk}{d\omega}\right) d\omega , \quad \text{with}$$

$$\frac{dk}{d\omega} = \left(\frac{d\omega}{dk}\right)^{-1} .$$

From the *dispersion relation* $\omega(\boldsymbol{k})$ it follows that for $\omega \to 0$:

$$g(\omega)d\omega = \begin{cases} \dfrac{V \cdot (\omega^2 + \dots)d\omega}{2\pi^2 c_s^3} & \text{for phonons with sound velocity} c_s , \\[3mm] \dfrac{V \cdot \left(\omega^{\frac{1}{2}} + \dots\right)d\omega}{4\pi^2 D^{\frac{3}{2}}} & \text{for magnons in } ferromagnets , \end{cases}$$

where the terms $+ \dots$ indicate that the above expressions refer to the asymptotes for $\omega \to 0$.

For magnons in *antiferromagnetic* systems a similar formula to that for *phonons* applies; the difference is only that the magnon contribution can be quenched by a strong magnetic field. In any case, since for fixed $\boldsymbol{k}$ there are two linearly independent transverse sound waves with the same sound velocity $c_s^{(\perp)}$ plus a longitudinal sound wave with higher velocity $c_s^{(\|)}$, one uses the *effective* sonic velocity given by

$$\frac{1}{\left(c_s^{(\text{eff})}\right)^3} := \frac{2}{\left(c_s^{(\perp)}\right)^3} + \frac{1}{\left(c_s^{(\|)}\right)^3} .$$

However, for accurate calculation of the contribution of phonons and magnons to the internal energy $U(T, V, N)$ one needs the complete behavior of the density of excitations $g(\omega)$, of which, however, e.g., in the case of phonons, only (i) the behavior at low frequencies, i.e. $g(\omega) \propto \omega^2$, and (ii) the so-called *sum rule*, e.g.,

$$\int_0^{\omega_{\text{max}}} d\omega g(\omega) = 3N ,$$

are exactly known [14], where ω_{max} is the maximum eigenfrequency.

---

[14] The *sum rule* states that the total number of eigenmodes of a system of N coupled harmonic oscillators is $3N$.

In the second and third decade of the twentieth century the Dutch physicist Peter Debye had the brilliant idea of replacing the exact, but matter-dependent function $g(\omega)$ by a matter-independent approximation, the so-called *Debye approximation* (see below), which interpolates the essential properties, (i) and (ii), in a simple way, such that

a) not only the low-temperature behavior of the relevant thermodynamic quantities, e.g., of the phonon contribution to $U(T, V, N)$,
b) but also the high-temperature behavior can be calculated exactly and analytically,
c) and in-between a reasonable interpolation is given.

The Debye approximation extrapolates the ω^2-behavior from low frequencies to the whole frequency range and simultaneously introduces a *cut-off frequency* ω_{Debye}, i.e., in such a way that the above-mentioned *sum rule* is satisfied.

Thus we have for phonons:

$$g(\omega) \xrightarrow{\approx} g_{\mathrm{Debye}}(\omega) = \frac{V \cdot \omega^2}{2\pi^2 \left(c_s^{\mathrm{eff}}\right)^3} , \tag{53.7}$$

i.e., for all frequencies

$$0 \leq \omega \leq \omega_{\mathrm{Debye}} ,$$

where the *cut-off frequency* ω_{Debye} is chosen in such a way that the *sum rule* is satisfied, i.e.

$$\frac{V \omega_{\mathrm{Debye}}^3}{6\pi^2 \left(c_s^{\mathrm{eff}}\right)^3} \stackrel{!}{=} 3N .$$

Furthermore, the integral

$$U(T, V, N) = \int\limits_0^{\omega_{\mathrm{Debye}}} \mathrm{d}\omega g_{\mathrm{Debye}}(\omega) \cdot \frac{\hbar\omega}{e^{\frac{\hbar\omega}{k_B T}} - 1}$$

can be evaluated for both low and high temperatures, i.e., for

$$k_B T \ll \hbar\omega_{\mathrm{Debye}} \quad \text{and} \quad \gg \hbar\omega_{\mathrm{Debye}} ,$$

viz in the first case after neglecting exponentially small terms if the upper limit of the integration interval, $\omega = \omega_{\mathrm{Debye}}$, is replaced by ∞. In this way one finds the low-temperature behavior

$$U(T, V, N) = \frac{9N\pi^4}{15} \cdot \hbar\omega_{\mathrm{Debye}} \cdot \left(\frac{k_B T}{\hbar\omega_{\mathrm{Debye}}}\right)^4 .$$

The low-temperature contribution of phonons to the heat capacity

$$C_V = \frac{\partial U}{\partial T}$$

is thus $\propto T^3$.

Similar behavior, $U \propto V \cdot T^4$ (the *Stefan-Boltzmann law*) is observed for a *photon gas*, i.e., in the context of black-body radiation; however this is valid at all temperatures, essentially since for photons (in contrast to phonons) the value of N is not defined. Generally we can state:

a) The low-temperature contribution of *phonons*, i.e., of sound-wave quanta, thus corresponds essentially to that of light-wave quanta, *photons*; the velocity of light is replaced by an effective sound-wave velocity, considering the fact that light-waves are always transverse, whereas in addition to the two transverse sound-wave modes there is also a longitudinal sound-wave mode.

b) In contrast, the high-temperature phonon contribution yields Dulong and Petits's law; i.e., for

$$k_B T \gg \hbar \omega_{\text{Debye}}$$

one obtains the exact result:

$$U(T, V, N) = 3 N k_B T .$$

This result is independent of the material properties of the system considered: once more essentially universal behavior, as is common in thermodynamics.

In the same way one can show that magnons in *ferromagnets* yield a low-temperature contribution to the internal energy

$$\propto V \cdot T^{\frac{5}{2}}$$

which corresponds to a low-temperature contribution to the heat capacity

$$\propto T^{\frac{3}{2}} .$$

This results from the quadratic dispersion relation, $\omega(k) \propto k^2$, for magnons in *ferromagnets*. In contrast, as already mentioned, magnons in *antiferromagnets* have a linear dispersion relation, $\omega(k) \propto k$, similar to phonons. Thus in antiferromagnets the low-temperature magnon contribution to the specific heat is $\propto T^3$ as for phonons. But by application of a strong magnetic field the magnon contribution can be suppressed.

- In an earlier section, 53.1, we saw that electrons in a metal produce a contribution to the heat capacity C which is proportional to the temperature T. For sufficiently low T this contribution always dominates over all other contributions. However, a linear contribution, $C \propto T$, is not characteristic for metals but it also occurs in *glasses* below ~ 1 K. However in glasses this linear term is not due to the electrons but to so-called two-level "tunneling states" of local atomic aggregates. More details cannot be given here.

53.6 Landau's Theory of 2nd-order Phase Transitions

The Ginzburg-Landau theory of superconductivity, which was described in an earlier subsection, is closely related to Landau's theory of second-order phase transitions [48]. The *Landau theory* is described in the following. One begins with a *real or complex scalar (or vectorial or tensorial) order parameter $\eta(\boldsymbol{r})$*, which marks the onset of order at the critical temperature, e.g., the onset of superconductivity. In addition, the fluctuations of the vector potential $\boldsymbol{A}(\boldsymbol{r})$ of the magnetic induction $\boldsymbol{B}(\boldsymbol{r})$ are important.

In contrast, most phase transitions are *first-order*, e.g., the liquid-vapor type or magnetic phase transition below the critical point, and of course the transition from the liquid into the solid state, since at the phase transition a discontinuous change in (i) density $\Delta\varrho$, (ii) magnetization ΔM, and/or (iii) entropy ΔS occurs, which is related to a *heat of transition*

$$\Delta l = T \cdot \Delta S .$$

These discontinuities always appear in first-order derivatives of the relevant thermodynamic potential. For example, we have

$$S = -\frac{\partial G(T, p, N)}{\partial T} \quad \text{or} \quad M = -\frac{\partial F_g(T, H)}{\partial H} ,$$

and so it is natural to define a first-order phase transition as a transition for which at least one of the derivatives of the relevant thermodynamic potentials is discontinuous.

In contrast, for a *second-order phase transition*, i.e., at the critical point of a liquid-vapor system, or at the Curie temperature of a ferromagnet, the Néel temperature of an antiferromagnet, or at the onset of superconductivity, all first-order derivatives of the thermodynamic potential are *continuous*.

At these critical points considered by Landau's theory, which we are going to describe, there is thus neither a heat of transformation nor a discontinuity in density, magnetization or similar quantity. In contrast discontinuities and/or divergencies only occur for second-order (or higher) derivatives of the thermodynamic potential, e.g., for the heat capacity

$$-\frac{T \cdot \partial^2 F_g(T, V, H, N, \dots)}{\partial T^2}$$

and/or the magnetic susceptibility

$$\chi = -\frac{\partial^2 F_g(T, V, H, N, \dots)}{\partial H^2} .$$

Thus we have the following definition due to Ehrenfest:
For n-th order phase transitions at least one n-th order derivative,

$$\frac{\partial^n F_g}{\partial X_{i_1} \dots \partial X_{i_n}} ,$$

of the relevant thermodynamic potential, e.g., of $F_g(T, V, N, H, \ldots)$, where the X_{i_k} are one of the variables of this potential, is discontinuous and/or divergent, whereas all derivatives of lower order are continuous.

Ehrenfest's definition is mainly mathematical. Landau recognized that the above examples, for which the ordered state is reached by falling below a critical temperature T_c, are second-order phase transitions in this sense and that here the symmetry in the ordered state always forms a subgroup of the symmetry group of the high-temperature *disordered* state (e.g., in the ferromagnetic state one only has a rotational symmetry restricted to rotations around an axis parallel to the magnetization, whereas in the disordered phase there is full rotational symmetry.

Further central notions introduced by Landau into the theory are

a) as mentioned, the *order parameter* η, i.e., a real or complex scalar or vectorial (or tensorial) quantity[15] which vanishes everywhere in the disordered state, i.e., at $T > T_c$, increasing continuously nonetheless at $T < T_c$ to finite values; and

b) the *conjugate field*, h, associated with the order parameter, e.g., a *magnetic field* in the case of a ferromagnetic system, i.e., for $\eta \equiv M$.

On the basis of phenomenological arguments Landau then assumed that for the Helmholtz free energy $F(T, V, h)$ of the considered systems with positive coefficients A, α and b (see below) in the vicinity of T_c, in the sense of a Taylor expansion, the following expression should apply (where the second term on the r.h.s., with the change of sign at T_c, is an important point of Landau's *ansatz*, formulated for a real order parameter):

$$F(T, V, h) =$$
$$\min_{\eta} \int_V d^3r \left[\frac{1}{2} \left\{ A \cdot (\nabla \eta)^2 + \alpha \cdot (T - T_c) \cdot \eta^2 + \frac{b}{2} \eta^4 \right\} - h \cdot \eta \right] . \qquad (53.8)$$

By minimization w.r.t. η, for $\nabla \eta \overset{!}{=} 0$ plus $h \overset{!}{=} 0$, one then obtains similar results as for the above Ginzburg-Landau theory of superconductivity, e.g.,

$$\eta(T) \equiv \eta_0(T) := 0 \quad \text{at} \quad T > T_c , \quad \text{but}$$

$$|\eta(T)| \equiv \eta_0(T) := \sqrt{\frac{\alpha \cdot (T_c - T)}{b}} \quad \text{at} \quad T < T_c ,$$

and for the susceptibility

$$\chi := \frac{\partial \eta}{\partial h}\bigg|_{h \to 0} = \frac{1}{\alpha \cdot (T - T_c)} \quad \text{at} \quad T > T_c \quad \text{and}$$

$$\chi = \frac{1}{2\alpha \cdot (T_c - T)} \quad \text{at} \quad T < T_c .$$

---

[15] For *tensor* order parameters there are considerable complications.

A k-dependent susceptibility can also be defined. With

$$h(\boldsymbol{r}) = h_k \cdot \mathrm{e}^{\mathrm{i}\boldsymbol{k}\cdot\boldsymbol{r}}$$

and the ansatz

$$\eta(\boldsymbol{r}) := \eta_0(T) + \eta_k \cdot \mathrm{e}^{\mathrm{i}\boldsymbol{k}\cdot\boldsymbol{r}}$$

we obtain for $T > T_c$:

$$\chi_k(T) := \frac{\partial \eta_k}{\partial h_k} = \frac{1}{2Ak^2 + \alpha \cdot (T - T_c)}$$

and for $T < T_c$:

$$\chi_k(T) = \frac{1}{2Ak^2 + 2\alpha \cdot (T_c - T)} \, ,$$

respectively.

Thus one can write

$$\chi_k(T) \propto \frac{1}{k^2 + \xi^{-2}} \, ,$$

with the so-called *thermal coherence length*[16]

$$\xi(T) = \sqrt{\frac{2A}{\alpha \cdot (T - T_c)}} \quad \text{and}$$

$$= \sqrt{\frac{A}{\alpha \cdot |T - T_c|}} \quad \text{for} \quad T > T_c \quad \text{and} \quad T < T_c \, ,$$

respectively. But it does not make sense to pursue this ingeniously simple theory further, since it is as good or bad as all *molecular field theories*, as described next.

53.7 Molecular Field Theories; Mean Field Approaches

In these theories complicated *bilinear* Hamilton operators describing interacting systems, for example, in the so-called Heisenberg model,

$$\mathcal{H} = -\sum_{l,m} J_{l,m} \hat{\boldsymbol{S}}_l \cdot \hat{\boldsymbol{S}}_m \, ,$$

---

[16] The concrete meaning of the *thermal coherence length* $\xi(T)$ is based on the fact that in a snapshot of the momentary spin configuration spins at two different places, if their separation $|\boldsymbol{r} - \boldsymbol{r}'|$ is much smaller than $\xi(T)$, are almost always parallel, whereas they are uncorrelated if their separation is $\gg \xi$.

are approximated by *linear* molecular field operators[17] e.g.,

$$\mathcal{H} \xrightarrow{\approx} \mathcal{H}^{\mathrm{MF}} := -\sum_{l} \left\{ 2 \sum_{m} J_{l,m} \left\langle \hat{\boldsymbol{S}}_m \right\rangle_T \right\} \cdot \hat{\boldsymbol{S}}_l + \sum_{l,m} J_{l,m} \left\langle \hat{\boldsymbol{S}}_m \right\rangle_T \cdot \left\langle \hat{\boldsymbol{S}}_l \right\rangle_T .$$

(53.9)

The last term in $\mathcal{H}^{\mathrm{MF}}$, where the factor 2 is missing, in contrast to the first term in the summation, is actually a temperature-dependent constant, and only needed if energies or entropies of the system are calculated. Moreover, the sign of this term is opposite to that of the first term; in fact, the last term is some kind of double-counting correction to the first term.

The approximations leading from Heisenberg's model to the molecular field theory are detailed below.

Furthermore, from molecular field theories one easily arrives at the Landau theories by a Taylor series leading from discrete sets to continua, e.g.,

$$S(\boldsymbol{r}_l) \to S(\boldsymbol{r}) := S(\boldsymbol{r}_l) + (\boldsymbol{r} - \boldsymbol{r}_l) \cdot \nabla S(\boldsymbol{r})_{|\boldsymbol{r}=\boldsymbol{r}_l}$$

$$+ \frac{1}{2!} \sum_{i,k=1}^{3} (x_i - (\boldsymbol{r}_l)_i) \cdot (x_k - (\boldsymbol{r}_l)_k) \cdot \frac{\partial^2 S(\boldsymbol{r})}{\partial x_i \partial x_k}_{|\boldsymbol{r}=\boldsymbol{r}_l} + \dots .$$

The name *molecular field theory* is actually quite appropriate. One can in fact write

$$\mathcal{H}^{\mathrm{MF}} \equiv -g\mu_B \sum_{l} H_l^{\mathrm{MF}}(T) \hat{\boldsymbol{S}}_l + \dots .$$

In this equation $H_l^{\mathrm{MF}}(T)$ is the effective magnetic field given by

$$2 \sum_{m} J_{l,m} \left\langle \hat{\boldsymbol{S}}_m \right\rangle_T / (g\mu_B) ;$$

g is the Landé factor,

$$\mu_B = \frac{\mu_0 e \hbar}{2 m_e}$$

the *Bohr magneton*, and e and m_e are the charge and mass of an electron respectively; μ_0 and

$$\hbar = \frac{h}{2\pi}$$

are, as usual, the vacuum permeability and the reduced Planck constant.

Van der Waals' theory is also a kind of molecular field theory, which in the vicinity of the critical point is only *qualitatively* correct, but *quantitatively* wrong. The approximation[18] neglects fluctuations (i.e., the last term in the

---

[17] Here operators are marked by the hat-symbol, whereas the thermal expectation value $\langle \hat{\boldsymbol{S}}_l \rangle_T$ is a real vector.

[18] One may also say *incorrectness*, since neglecting all fluctuations can be a very *severe* approximation.

following identity):

$$\hat{S}_l \cdot \hat{S}_m \equiv \left\langle \hat{S}_l \right\rangle_T \cdot \hat{S}_m + \hat{S}_l \cdot \left\langle \hat{S}_m \right\rangle_T + \left(\hat{S}_l - \left\langle \hat{S}_l \right\rangle_T \right) \cdot \left(\hat{S}_m - \left\langle \hat{S}_m \right\rangle_T \right) .$$

Even the factor 2 in front of $\sum_m \ldots$ is derived in this way, i.e., from

$$J_{l,m} = J_{m,l} .$$

However, neglecting fluctuations is grossly incorrect in the critical region, which is often not small (e.g., for typical ferromagnets it amounts to the upper $\sim 20\%$ of the region below T_c; in contrast, for conventional 3d-superconductivity the critical region is negligible).

Generally the neglected terms have a drastic influence on the critical exponents. In the critical region the order parameter does not converge to zero

$$\propto (T_c - T)^{\frac{1}{2}} ,$$

as the Landau theory predicts, but $\propto (T_c - T)^\beta$, where for a $d = 3$-dimensional system the value of β is $\approx \frac{1}{3}$ instead of

$$\beta_{\text{Landau}} \equiv \frac{1}{2} .$$

We also have, both above and below T_c, with different coefficients:

$$\chi \propto |T_c - T|^{-\gamma} , \quad \text{with} \quad \gamma = \gamma_{3d} \approx \frac{4}{3}$$

instead of the value from the Landau theory,

$$\gamma_{\text{Landau}} \equiv 1 .$$

Similarly we have

$$\xi \propto |T_c - T|^{-\nu} , \quad \text{with} \quad \nu = \nu_{3d} \approx \frac{2}{3}$$

instead of the Landau value

$$\nu_{\text{Landau}} \equiv \frac{1}{2} .$$

For the twodimensional *Ising model*, a model which only differs from the Heisenberg model by the property that the vectorial spin operators $\hat{S}_l$ and $\hat{S}_m$ are replaced by the z-components $(\hat{S}_z)_l$ and $(\hat{S}_z)_m$, the difference between the critical exponents and the Landau values is even more drastic than in $d=3$, as detailed in the following.

For $d = 2$ the exact values of the critical exponents are actually known:

$$\beta(2d)_{\text{Ising}} \equiv \frac{1}{8} , \quad \gamma(2d)_{\text{Ising}} \equiv \frac{7}{4} \quad \text{and} \quad \nu(2d)_{\text{Ising}} \equiv 1 ,$$

and the critical temperature is reduced by the fluctuations by almost 50% compared to the molecular field approximation. (For the $d = 2$-dimensional *Heisenberg* model the phase transition is even completely *suppressed* by fluctuations).

It is no coincidence[19] that the quantitative values of the critical exponents and other characteristic quantities of the critical behavior are *universal* in the sense that they are (almost) independent of the details of the interactions. For interactions which are sufficiently short-ranged, these quantities depend only on (i) the *dimensionality d* of the system, and (ii) the *symmetry* of the order parameter, e.g., whether one is dealing with uniaxial symmetry, as for the Ising model, or isotropic symmetry, as for the Heisenberg model. In this context one defines *universality classes* of systems with the same critical behavior.

For different models one can also define a kind of *molecular field approximation*, usually called mean field approximation, by replacing a sum of *bilinear* operators by a self-consistent temperature-dependent *linear* approximation in which fluctuations are neglected. In this context one should mention the Hartree-Fock approximation[20], see Part III, and the Hartree-Fock-Bogoliubov approximation for the normal and superconducting states of a system with many electrons. Without going into details we simply mention that the results of all these approximations are similar to the Landau theory. However, in the following respect the mean field theories are somewhat more general than the Landau theory: they lead to quantitative predictions for the order parameter, for T_c, and for Landau's phenomenological coefficients A, α and b. In this context one should also mention the BCS theory of superconductivity (which we do not present, see [25]), since not only the Ginzburg-Landau theory of superconductivity can be derived from it but also one sees in particular that the carriers of superconductivity have the charge $2e$, not e.

53.8 Fluctuations

In the preceding subsections we have seen that thermal fluctuations are important in the neighborhood of the critical temperature of a second-order phase transition, i.e. near the critical temperature of a liquid-gas transition or for a ferromagnet. For example, it is plausible that density fluctuations become very large if the isothermal compressibility diverges (which is the case at T_c).

---

[19] The reason is again the universality of the phenomena within regions of diameter ξ, where ξ is the thermal correlation length.

[20] Here the above *bilinear* operators are "bilinear" expressions (i.e., terms constructed from four operators) of the "linear" entities $\hat{c}_i^+ \hat{c}_j$ (i.e., a basis constructed from the products of two operators), where $\hat{c}_i^+$ and $\hat{c}_j$ are Fermi creation and destruction operators, respectively.

We also expect that the fluctuations in magnetization for a ferromagnet are especially large when the isothermal susceptibility diverges, and that especially strong energy fluctuations arise when the isothermal specific heat diverges.

These qualitative insights can be formulated quantitatively as follows (here, for simplicity, only the second and third cases are considered):

$$\langle |\eta_k|^2 \rangle_T - |\langle \eta_k \rangle_T|^2 \equiv k_B T \cdot \chi_k(T) \quad \text{and} \tag{53.10}$$

$$\langle \mathcal{H}^2 \rangle_T - (\langle \mathcal{H} \rangle_T)^2 \equiv k_B T^2 \cdot C_{(V,N,H,...)} , \tag{53.11}$$

where

$$C_{(V,N,H,...)} = \frac{\partial U(T,V,N,H,...)}{\partial T} , \quad \text{with} \quad U(T,V,N,H,...) = \langle \mathcal{H} \rangle_T ,$$

is the isothermal heat capacity and $\chi_k(T)$ the k-dependent magnetic susceptibility (see below).

For simplicity our proof is only performed for $k = 0$ and only for the Ising model. The quantity η is (apart from a proportionality factor) the value of the saturation magnetization for $H \to 0^+$, i.e.,

$$\eta(T) := M_s(T) .$$

We thus assume that

$$\mathcal{H} = -\sum_{l,m} J_{l,m} s_l s_m - h \sum_l s_l , \quad \text{with} \quad s_l = \pm 1 , \quad \text{and define} \quad M := \sum_l s_l ,$$

i.e., $\mathcal{H} = \mathcal{H}_0 - hM$.

In the following, "tr" means "trace". We then have

$$\langle M \rangle_T = \frac{\text{tr}\left\{ M \cdot e^{-\beta \cdot (\mathcal{H}_0 - hM)} \right\}}{\text{tr}\left\{ e^{-\beta \cdot (\mathcal{H}_0 - hM)} \right\}} ,$$

and hence

$$\chi = \frac{\partial \langle M \rangle_T}{\partial h} = \beta \cdot \left(\frac{\text{tr}\left\{ M^2 e^{-\beta(\mathcal{H}_0 - hM)} \right\}}{\text{tr}\left\{ e^{-\beta(\mathcal{H}_0 - hM)} \right\}} - \left[\frac{\text{tr}\left\{ M e^{-\beta(\mathcal{H}_0 - hM)} \right\}}{\text{tr}\left\{ e^{-\beta(\mathcal{H}_0 - hM)} \right\}} \right]^2 \right) ,$$

which is (53.10). Equation (53.11) can be shown similarly, by differentiation w.r.t.

$$\beta = \frac{1}{k_B T}$$

of the relation

$$U = \langle \mathcal{H} \rangle_T = \frac{\text{tr}\left\{ \mathcal{H} e^{-\beta \mathcal{H}} \right\}}{\text{tr}\left\{ e^{-\beta \mathcal{H}} \right\}} .$$

A similar relation between fluctuations and *response* is also valid in dynamics – which we shall not prove here since even the formulation requires considerable effort ($\rightarrow$ *fluctuation-dissipation theorem*, [49], as given below):

Let

$$\Phi_{\hat{A},\hat{B}}(t) := \frac{1}{2} \cdot \left\langle \hat{A}_H(t)\hat{B}_H(0) + \hat{B}_H(0)\hat{A}_H(t) \right\rangle_T$$

be the so-called fluctuation function of two observables $\hat{A}$ and $\hat{B}$, represented by two Hermitian operators in the Heisenberg representation, e.g.,

$$A_H(t) := e^{i\frac{\mathcal{H}t}{\hbar}} \hat{A} e^{-i\frac{\mathcal{H}t}{\hbar}} .$$

The Fourier transform $\varphi_{\hat{A},\hat{B}}(\omega)$ of this fluctuation function is defined through the relation

$$\Phi_{\hat{A},\hat{B}}(t) =: \int e^{i\omega t} \frac{d\omega}{2\pi} \varphi_{\hat{A},\hat{B}}(\omega) .$$

Similarly for $t > 0$ the *generalized dynamic susceptibility* $\chi_{\hat{A},\hat{B}}(\omega)$ is defined as the Fourier transform of the *dynamic response function*

$$X_{\hat{A},\hat{B}}(t - t') := \frac{\delta\langle\hat{A}\rangle_T(t)}{\delta h_{\hat{B}}(t')} ,\ ^{21}$$

where

$$\mathcal{H} = \mathcal{H}_0 - h_B(t')\hat{B} ,$$

i.e., the Hamilton operator $\mathcal{H}_0$ of the system is *perturbed*, e.g., by an alternating magnetic field $h_B(t')$ with the associated operator $\hat{B} = \hat{S}_z$, and the *response* of the quantity $\hat{A}$ on this perturbation is observed.

Now, the dynamic susceptibility $\chi_{\hat{A},\hat{B}}(\omega)$ has two components: a *reactive* part

$$\chi'_{\hat{A},\hat{B}}$$

and a *dissipative* part

$$\chi''_{\hat{A},\hat{B}} ,$$

i.e.,

$$\chi_{\hat{A},\hat{B}}(\omega) = \chi'_{\hat{A},\hat{B}}(\omega) + i\chi''_{\hat{A},\hat{B}}(\omega) ,$$

where the reactive part is an *odd* function and the dissipative part an *even* function of ω. The dissipative part represents the *losses* of the response process.

Furthermore, it is generally observed that the larger the dissipative part, the larger the fluctuations. Again this can be formulated quantitatively using the *fluctuation-dissipation theorem* [49]. All expectation values and the

---

[21] This quantity is only different from zero for $t \geq t'$.

quantities $\hat{A}_H(t)$ etc. are taken with the unperturbed Hamiltonian $\mathcal{H}_0$:

$$\varphi_{\hat{A},\hat{B}}(\omega) \equiv \hbar \cdot \coth \frac{\hbar\omega}{2k_BT} \cdot \chi''_{\hat{A},\hat{B}}(\omega) \ . \tag{53.12}$$

In the "classical limit", $\hbar \to 0$, the product of the first two factors on the r.h.s. of this theorem converges to

$$\frac{2k_BT}{\omega} \ ,$$

i.e., $\propto T$ as for the static behavior, in agreement with (53.10). In this limit the theorem is also known as the Nyquist theorem; but for the true value of $\hbar$ it also covers quantum fluctuations. Generally, the *fluctuation-dissipation theorem* only applies to *ergodic systems*, i.e., if, after the onset of the perturbation, the system under consideration comes to thermal equilibrium within the time of measurement.

Again this means that without generalization the theorem does not apply to "glassy" systems.

53.9 Monte Carlo Simulations

Many of the important relationships described in the previous sections can be visualized directly and evaluated numerically by means of computer simulations. This has been possible for several decades. In fact, so-called *Monte Carlo simulations* are very well known, and as there exists a vast amount of literature, for example [51], we shall not go into any details, but only describe the principles of the *Metropolis algorithm*, [50].

One starts at time t_ν from a configuration $X(x_1, x_2, \ldots)$ of the system, e.g., from a spin configuration (or a fluid configuration) of all spins (or all positions plus momenta) of all N particles of the system. These configurations have the energy $E(X)$. Then a new state X' is proposed (but not yet accepted) by some systematic procedure involving random numbers. If the proposed new state has a lower energy,

$$E(X') < E(X) \ ,$$

then it is always accepted, i.e.,

$$X(t_{\nu+1}) = X' \ .$$

In contrast, if the energy of the proposed state is enhanced or at least as high as before, i.e., if

$$E(X') = E(X) + \Delta E \ , \quad \text{with} \quad \Delta E \geq 0 \ ,$$

then the suggested state is only accepted if it is not too unfavorable. This means precisely: a random number $r \in [0, 1]$, independently and identically distributed in this interval, is drawn, and the proposed state is accepted iff

$$e^{-\Delta E/(k_B T)} \geq r \ .$$

In the case of acceptance (or rejection, respectively), the next state is the proposed one (or the old one),

$$X(t_{\nu+1}) \equiv X' \quad (\text{or} \quad X(t_{\nu+1}) \equiv X) \ .$$

Through this algorithm one obtains a sequence $X(t_\nu) \to X(t_{\nu+1}) \to \dots$ of random configurations of the system, a so-called Markoff chain, which is equivalent to classical thermodynamics i.e., after n equilibration steps, thermal averages at the considered temperature T are identical to chain-averages,

$$\langle f(X) \rangle_T = M^{-1} \sum_{\nu \equiv n+1}^{n+M} f(X(t_\nu)) \ ,$$

and ν is actually proportional to the time.

In fact, it can be proved that the Metropolis algorithm leads to thermal equilibrium, again provided that the system is *ergodic*, i.e., that the dynamics do not show glassy behavior.

Monte Carlo calculations are now a well-established method, and flexible enough for dealing with a classical problem, whereas the inclusion of quantum mechanics, i.e., at low temperatures, still poses difficulties.

54 Applications II: Phase Equilibria in Chemical Physics

Finally, a number of sections on chemical thermodynamics will now follow.

54.1 Additivity of the Entropy; Partial Pressure; Entropy of Mixing

For simplicity we start with a closed fluid system containing two phases. In thermal equilibrium the entropy is maximized:

$$S(U_1, U_2, V_1, V_2, N_1, N_2) \overset{!}{=} \max. \tag{54.1}$$

Since (i) $V_1 + V_2 = $ constant, (ii) $N_1 + N_2 = $ constant and (iii) $U_1 + U_2 = $ constant we may write:

$$dS = \left(\frac{\partial S}{\partial U_1} - \frac{\partial S}{\partial U_2} \right) \cdot dU_1 + \left(\frac{\partial S}{\partial V_1} - \frac{\partial S}{\partial V_2} \right) \cdot dV_1 + \left(\frac{\partial S}{\partial N_1} - \frac{\partial S}{\partial N_2} \right) \cdot dN_1 .$$

Then with

$$\frac{\partial S}{\partial U} = \frac{1}{T} , \quad \frac{\partial S}{\partial V} = \frac{p}{T} \quad \text{and} \quad \frac{\partial S}{\partial N} = -\frac{\mu}{T}$$

it follows that in thermodynamic equilibrium because $dS \overset{!}{=} 0$:

$$T_1 = T_2 , \quad p_1 = p_2 \quad \text{and} \quad \mu_1 = \mu_2 .$$

Let the partial systems "1" and "2" be *independent,* i.e., the probabilities ϱ for the states of the system *factorize*:

$$\varrho(x\text{"}1 + 2\text{"}) = \varrho_1(x_1) \cdot \varrho_2(x_2) .$$

Since

$$\ln(a \cdot b) = \ln a + \ln b .$$

we obtain

$$\frac{S}{k_B} = - \sum_{\forall \text{states of "}1 + 2\text{"}} \varrho(\text{"}1 + 2\text{"}) \cdot \ln \varrho(\text{"}1 + 2\text{"})$$

$$= -\sum_{x_1} \varrho_1(x_1) \cdot \sum_{x_2} \varrho_2(x_2) \cdot \ln \varrho_1(x_1)$$

$$- \sum_{x_2} \varrho_2(x_2) \cdot \sum_{x_1} \varrho_1(x_1) \cdot \ln \varrho_2(x_2) \, ,$$

and therefore with

$$\sum_{x_i} \varrho_i(x_i) \equiv 1 \text{ we obtain}:$$

$$\frac{S_{\text{"}1+2\text{"}}}{k_B} = \sum_{i=1}^{2} \frac{S_i}{k_B} \, , \tag{54.2}$$

implying the *additivity of the entropies of independent partial systems*. This fundamental result was already known to Boltzmann.

To define the notions of *partial pressure* and *entropy of mixing*, let us perform a *thought experiment* with two *complementary semipermeable membranes*, as follows.

Assume that the systems 1 and 2 consist of two different well-mixed fluids (particles with attached fluid element, e.g., ideal gases with vacuum) which are initially contained in a common rectangular 3d volume V.

For the two *semipermeable membranes*, which form rectangular 3d cages, and which initially both coincide with the common boundary of V, let the first membrane, SeM_1, be *permeable* for particles of kind 1, but *nonpermeable* for particles of kind 2; the *permeability properties* of the second semipermeable membrane, SeM_2, are just the opposite: the second membrane is *nonpermeable* (*permeable*) for particles of type 1 (type 2).[1]

On adiabatic (loss-less) separation of the two *complementary semipermeable* cages, the two kinds of particles become separated ("de-mixed") and then occupy equally-sized volumes, V, such that the respective pressures p_i are well-defined. In fact, these (measurable!) pressures in the respective cages (after separation) define the *partial pressures* p_i.

The above statements are supported by Fig. 54.1 (see also Fig. 54.2):

Compared with the original *total pressure* p, the *partial pressures* are *reduced*. Whereas for ideal gases we have $p_1 + p_2 \equiv p$ (since $p_i = \frac{N_i p}{N_1 + N_2}$), this is generally *not* true for interacting systems, where typically $p_1 + p_2 > p$, since through the separation an important part of the (negative) internal pressure, i.e., the part corresponding to the attraction by particles of a different kind, ceases to exist.

---

[1] In Fig. 54.1 the permeabilty properties on the l.h.s. (component 1) are somewhat different from those of the text.

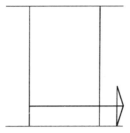

Fig. 54.1. Partial pressures and adiabatic de-mixing of the components of a fluid. The volume in the middle of the diagram initially contains a fluid mixture with two different components "1" and "2". The wall on the l.h.s. is non-permeable for both components, whereas that on the r.h.s. is *semipermeable*, e.g. non-permeable only to the first component (see e.g. Fig. 54.2). The compressional work is therefore reduced from $\delta A = -p \cdot dV$ to $\delta A = -p_1 \cdot dV$, where p_1 is the partial pressure. Moreover, by moving the l.h.s. wall to the right, the two fluid-components can be *demixed*, and if afterwards the size of the volumes for the separate components is the same as before, then $S(T,p)_{|\text{fluid}} \equiv \sum_{i=1,2} S_i(T,p_i)$

Since the separation is done reversibly, in general we have

$$S_{\text{"1 + 2"}}(p) \equiv S_1(p_1) + S_2(p_2)$$

(note the different pressures!), which is not in contradiction with (54.2), but should be seen as an additional specification of the additivity of partial entropies, which can apply even if the above partial pressures do not add-up to p.

Now assume that in the original container not two, but k components (i.e., k different ideal gases) exist. The entropy of an ideal gas as a function of temperature, pressure and particle number has already been treated earlier. Using these results, with

$$S(T,p,N_1,\dots N_k) = \sum_{i=1}^{k} S_i(T,p_i,N_i) \quad \text{and}$$

$$p_i = \frac{N_i}{N} \cdot p\,, \quad \text{i.e.,} \quad p_i = c_i \cdot p\,,$$

with the concentrations $c_i = \frac{N_i}{N}$, we have

$$\frac{S(T,p,N_1,N_2,\dots,N_k)}{k_B} = \sum_{i=1}^{k} N_i \cdot \left\{ \ln \frac{5k_B T}{2p} - \ln c_i + \frac{s_i^{(0)}}{k_B} \right\}. \qquad (54.3)$$

Here $s_i^{(0)}$ is a non-essential *entropy constant*, which (apart from constants of nature) only depends on the logarithm of the mass of the molecule considered (see above).

In (54.3) the term in $-\ln c_i$ is more important. This is the *entropy of mixing*. The presence of this important quantity does not contradict the additivity of entropies; rather it is a consequence of this fundamental property, since the partial entropies must be originally calculated with the partial pressures $p_i = c_i \cdot p$ and not the total pressure p, although this is finally introduced via the above relation.

It was also noted earlier that in fluids the *free enthalpy per particle*,

$$g_i(T, p_i) = \frac{1}{N} \cdot G_i(T, p_i, N_i) \,,$$

is identical with the chemical potential $\mu_i(T, p_i)$, and that the entropy *per particle*, $s_i(T, p_i)$, can be obtained by derivation w.r.t. the temperature T from $g_i(T, p_i)$, i.e.,

$$s_i(T, p_i) = -\frac{\partial g(T, p_i)}{\partial T} \,.$$

From (54.3) we thus obtain for the *free enthalpy per particle* of type i in a mixture of fluids:

$$g_i(T, p) = g_i^{(0)}(T, p) + k_B T \cdot \ln c_i, \tag{54.4}$$

where the function $g_i^{(0)}(T, p)$ depends on temperature and pressure, which corresponds to $c_i \equiv 1$, i.e., to the pure compound. This result is used in the following. It is the basis of the law of mass action, which is treated below.

54.2 Chemical Reactions; the Law of Mass Action

In the following we consider so-called *uninhibited* [2] *chemical reaction equilibria* of the form

$$\nu_1 A_1 + \nu_2 A_2 \leftrightarrow -\nu_3 A_3 \,.$$

Here the ν_i are suitable positive or negative integers (the sign depends on i, and the negative integers are always on the r.h.s., such that $-\nu_3 \equiv |\nu_3|$); e.g., for the so-called *detonating gas* reaction, the following formula applies: $2H + O \leftrightarrow H_2O$, whereas the corresponding *inhibited reaction* is $2H_2 + O_2 \leftrightarrow 2H_2O$ (slight differences, but remarkable effects! In the "uninhibited" case one is dealing with O atoms, in the "inhibited" case, however, with the usual O_2 molecule.)

The standard form of these reactions that occur after *mixing* in fluids is:

$$\sum_{i=1}^{k} \nu_i A_i = 0 \,,$$

---

[2] The term *uninhibited reaction equilibrium* means that the reactions considered are in thermodynamic equilibrium, possibly after the addition of suitable catalyzing agencies (e.g., *Pt* particles) which decrease the inhibiting barriers.

where without lack of generality it is assumed that only ν_k is negative. (Of course we choose the smallest possible integral values for the $|\nu_i|$).

The concentration ratios c_i then result from (54.3) and (54.4) according to the principle that the entropy should not change in course of the equilibrium reaction. In this way from the related free enthalpy condition one obtains the *Law of Mass Action*:

$$\frac{c_1^{\nu_1} \cdot c_2^{\nu_2} \cdot \ldots \cdot c_{k-1}^{\nu_{k-1}}}{c_k^{-\nu_k}} \equiv f(T,p) , \qquad (54.5)$$

where from the ideal gas equation (see the preceding subsection) for the *yield function* $f(T,p)$ with a pressure unit p_0 the following general expression results:

$$f(T,p) = \prod_{i=1}^{k} c_i^{\nu_i} \equiv \prod_{i=1}^{k} \left(\frac{p}{p_0}\right)^{-\nu_i} \cdot e^{-\nu_i \frac{\tilde{g}_i^{(0)}(T)}{k_B T}} . \qquad (54.6)$$

For further simplification we have written

$$g_i^{(0)}(T,p) = \tilde{g}^{(0)}(T) + k_B T \cdot \ln \frac{p}{p_0} ,$$

which agrees with

$$s_i = k_B \ln \frac{5 k_B T}{2p} + \ldots .$$

By varying pressure and/or temperature one can thus systematically shift the reaction yield in accordance with (54.6) (e.g., increasing the pressure leads to an increased fraction of components with negative ν_i).

As an application and generalization we shall presently treat the osmotic pressure and the decrease in boiling and solidification temperatures of a liquid by addition of sugar and de-icing salt. Firstly, however, an unusual topic.

54.3 Electron Equilibrium in Neutron Stars

Consider the quantitative answer to the following question: *How many electrons (or protons!) are there in a neutron star?* Firstly we need to know the fraction:

$$N_{\text{electrons}} = N_{\text{protons}} \cong 10^{-5} \cdot N_{\text{neutrons}} ;$$

i.e., the ratio is extremely small, although different from zero. Of course, all absolute numbers would be extremely large[3].

---

[3] How large is N_e? One can estimate this number by inserting typical values for the radius of a neutron star (≈ 10 km) and of a neutron ($\approx 10^{-13}$ cm).

The above result is then obtained. Consider the relevant astrophysical equilibrium reaction, which is the equilibrium *inverse β-decay* process:

$$pp + ee \leftrightarrow nn + \nu\nu_e$$

Here *pp* stands for the proton[4], *nn* for the neutron, *ee* for the electron and $\nu\nu_e$ for the electron-neutrino. (In a β-decay process a neutron emitting an electron plus an electron-antineutrino $\overline{\nu\nu}_e$ decays as follows: $nn \rightarrow pp + ee + \overline{\nu\nu}_e$; the inverse β-decay is the fusion of a proton and an electron under very high pressure, as high as is typical for a neutron star, or even higher, into a neutron plus an outgoing electron-neutrino, i.e., essentially according to the ← part of the previous reaction equation. The electron-neutrino and electron-antineutrino are particles and antiparticles with (almost) vanishing rest mass, i.e., negligible for the present thermodynamics (see below).) In any case, the first reaction can actually be "equilibrated" in both directions; in neutron stars there is thermodynamic equilibrium as above, and additionally the temperature can be neglected, since one is dealing with a degenerate Fermi gas.

The equilibrium condition is

$$\sum_i \nu_i \cdot \mu_i(T, p, N_1, \dots, N_k) \overset{!}{=} 0 ,$$

where the $\mu_i(T, p, N_1, \dots)$ are the respective *chemical potentials*, i.e., the *free enthalpies per particle* for the particle considered, and the ν_i the reaction numbers (not to be confused with the neutrinos).

The chemical potentials of neutrino and anti-neutrino, as already mentioned, can be neglected in the present context. However, for the other particles under consideration, the chemical potential at *zero temperature* (and in this approximation, our *neutron stars* – and also "white dwarfs", see above, can always be treated at the temperatures considered, *viz* as a *degenerate Fermi gas*) is identical with the nonrelativistic kinetic energy per particle:

$$\mu_i \approx \varepsilon_i^{\text{kin.}} = \frac{\hbar^2}{2M} n_V^{\frac{2}{3}} ,$$

as was shown for the electron gas in metals (n_V is the number density and M the mass of the considered particles). Thus

$$-\frac{(n_V)_{\text{neutron}}^{2/3}}{M_{\text{neutron}}} + \left\{ \frac{(n_V)_{\text{proton}}^{2/3}}{M_{\text{proton}}} + \frac{(n_V)_{\text{electron}}^{2/3}}{m_{\text{electron}}} \right\} \overset{!}{=} 0 . \qquad (54.7)$$

---

[4] We write *pp* (instead of *p*) for the proton, *nn* (instead of *n*) for the neutron, *ee* for the electron, and $\nu\nu$ for the neutrino, to avoid confusion with the pressure *p*, the particle density *n* or the particle number *N* as well as with the elementary charge *e* and the reaction numbers ν_i mentioned above.

However, the electron mass, m_e, is 2000 times smaller than the proton mass $M_P (\approx$ neutron mass $M_N)$, whereas the number densities of electrons and protons are equal. Thus the second term of the previous equation can be neglected. As a consequence

$$\frac{(n_V)_{\text{electron}}}{(n_V)_{\text{neutron}}} \approx \left(\frac{m_e}{M_N}\right)^{\frac{3}{2}} \approx 10^{-5} ,$$

as stated.

54.4 Gibbs' Phase Rule

In an earlier section we considered the case of a *single component* ($K = 1$, e.g., H_2O) which could exist in three different phases ($P = 3$, solid, liquid or vapor.)

In contrast, in the second-from-last subsection we treated the case of two or more components ($K \geq 2$) reacting with each other, but in a single phase ($P = 1$), according to the reaction equation $\sum_{i=1}^{k} \nu_i A_i = 0$.

We shall now consider the general case, and ask how many degrees of freedom f, i.e., arbitrary real variables, can be chosen, if K different *components* are in thermodynamic equilibrium for P different phases. The answer is found in *Gibbs' phase rule*:

$$f \equiv K - P + 2 . \tag{54.8}$$

As a first application we again consider the (p, T) *phase diagram* of H_2O with the three phases: solid, liquid and vapor. Within a single phase one has $f \equiv 2 (= 1 - 1 + 2)$ degrees of freedom, e.g., $G = G(T, p)$; on the boundary lines between two phases there remains only 1 degree of freedom $(= 1 - 2 + 2)$ (e.g., the saturation pressure is a unique function of the temperature only, and cannot be varied by choosing additional variables), and finally at the triple point we have $f = 0$, i.e., at this point all variables, temperature and pressure are completely fixed ($f = 1 - 3 + 2 \equiv 0$).

As a second application consider a system with $K = 2$, e.g., two salts. In one solvent (i.e., for $P = 1$) in thermal equilibrium one would have $f = 2 - 1 + 2 = 3$ degrees of freedom. For example, one could vary T, p and c_1, whereas $c_2 = 1 - c_1$ would then be fixed. If there is thermodynamic equilibrium with $P = 2$ phases, e.g., a solid phase plus a fluid phase, the number of degrees of freedom is reduced to $f = 2$; e.g., only T and p can be varied independently, in contrast to c_i. If there is thermodynamic equilibrium of the two components in three phases, then one has only one free variable ($f = 1$).

We shall now proceed to a proof of Gibbs' phase rule:

a) Firstly, consider a system with $P = 1$. Then $K + 1$ variables can be freely chosen, e.g., T, p, c_1, c_2, ..., c_{k-1}, whereas $c_k = 1 - c_1 - c_2 - \ldots - c_{k-1}$ is dependent; thus the *Gibbs' phase rule* is explicitly satisfied: $f = K - 1 + 2$.

b) Now let $P \geq 2$! Then at first one should consider that for every component, $k = 1, \ldots, K$, there are $P - 1$ additional degrees of freedom, i.e. the ratios

$$c_k^{(2)}/c_k^{(1)}\ ,\quad c_k^{(3)}/c_k^{(1)}\ ,\ldots,\quad c_k^{(P)}/c_k^{(1)}\ .$$

Thus one has $(P-1) \cdot K$ additional variables. But there are also $(P-1) \cdot K$ additional constraints, i.e.

$$\mu_k^{(2)} \overset{!}{=} \mu_k^{(1)}\ ,\ldots,\quad \mu_k^{(P)} \overset{!}{=} \mu_k^{(1)}\ .$$

c) A last sequence of constraints must be considered:

$$p_k^{(2)} \overset{!}{=} p_k^{(1)}\ ,\ldots,\quad p_k^{(P)} \overset{!}{=} p_k^{(1)}\ .$$

This gives $P - 1$ constraints on the pressure.

Hence

$$f = K - 1 + 2 + (P - 1) \cdot K - (P - 1) \cdot K - (P - 1) \equiv K - P + 2\ ,\quad \text{q.e.d.}$$

54.5 Osmotic Pressure

Again assume, as in section 54.2, that we are dealing with a *semipermeable membrane* (as commonly occurs in biological cells). Let the semipermeable membrane be nonpermeable for the *solute*, 2, ($\hat{=}$ salt or sugar), but permeable for the solvent, 1.

Now consider a U-tube, which is separated into two parts at the center by a semipermeable membrane and filled with a liquid (water) to different heights in the respective parts.

In the left-hand part of the U-tube the water level is h; in the enriched right-hand part (enriched by salt or sugar, $\Delta c_2 > 0$) the level is enhanced, $h + \Delta h$, with $\Delta h > 0$.

This corresponds to a pressure difference[5] Δp, the so-called *osmotic pressure* $p_{\text{osmotic p.}}$, which is given by

$$\Delta p \equiv p_{\text{osmotic p.}} = \Delta c_2 \cdot \frac{N k_B T}{V} = \frac{\Delta N_2}{V} k_B T\ . \tag{54.9}$$

---

[5] The solvent concentration is only slightly diminished on the right-hand side, in favor of the enhanced solute concentration on this side, see the text.

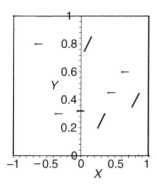

Fig. 54.2. Osmotic pressure (schematically). The volume V is divided into two parts by a *semipermeable* membrane (the Y-axis between 0 and 1) . The membrane is *permeable* for the *solvent* (− symbols), but *non-permeable* for the *solute molecules* (↗ symbols). As a result, the pressure on the r.h.s. is enhanced by an amount called the *osmotic pressure* Δp.

(This expression is analogous to an effective *ideal gas* of ΔN_2 solute molecules suspended in a solvent. For a true ideal gas the molecules would be suspended in a vacuum; but there the pressure does not depend on the *mass* of the molecule. This is also true for the present situation.)

Figure 54.2 above schematically shows a semi-permeable membrane through which solute molecules cannot pass.

In the above derivation, it does not matter that the vacuum mass of the molecule is replaced by an effective mass, for which, however, the value is unimportant if the solute concentrations c_2 (on the l.h.s.) and $c_2 + \Delta c_2$ (on the r.h.s.) are small enough, such that only the interactions with the solvent come into play. (These arguments make a lot of 'microscopic' calculations unnecessary.)

A precise proof again uses the *entropy of mixing* and the *partial pressure* p_i :

$$s_i(T, p_i) = s_i(T, p) - k_B \cdot \ln c_i .$$

Thus we have the *molecular free enthalpy*

$$g_i(T, p_i) = g_i^{(0)}(T, p) + k_B T \cdot \ln c_i ,$$

and because of the equality of chemical potential and the molecular free enthalpy:

$$\mu_i(T, p, c_i) = \mu_i(T, p) + k_B T \cdot \ln c_i .$$

The two equilibrium conditions (*not* three!) for each side of the semipermeable membrane are:

(i) $\mu_1(T, p, c_1(= 1 - c_2)) \equiv \mu_1(T, p + \Delta p, c_1 - \Delta c_2)$,   and

(ii) $T_1 \equiv T_2(= T)$.

Thus

$$g_1^{(0)}(T, p) + k_B T \cdot \ln(1 - c_2) \equiv g_1^{(0)}(T, p) + \Delta p \cdot \left(\frac{\partial g_1^{(0)}}{\partial p} \right) + k_B T \cdot \ln(1 - c_2 - \Delta c_2) .$$

With

$$\frac{\partial g_1^{(0)}}{\partial p} = v_1$$

and the linearizations

$$\ln(1 - c_2) \approx -c_2 \quad \text{and} \quad \ln(1 - c_2 - \Delta c_2) \approx -c_2 - \Delta c_2$$

we obtain for the osmotic pressure:

$$\Delta p = \frac{k_B T}{v_1} \Delta c_2 \;,$$

and finally with

$$v_1 := \frac{V}{N_1} \approx \frac{V}{N_1 + N_2} \;, \quad \text{for} \quad N_2 \ll N_1 : \; \Delta p \approx \frac{\Delta N_2}{V} k_B T \;.$$

This derivation is, in principle, astonishingly simple, and becomes even simpler by using the "effective mass" argument above; a microscopic statistical-mechanical calculation would, in contrast, be unnecessarily complicated.

54.6 Decrease of the Melting Temperature Due to "De-icing" Salt

In the preceding section the addition of sugar on the nonpermeable side (nonpermeable for the sugar but not for the solvent) of a *semipermeable membrane* led to a *pressure difference*, i.e., a higher pressure, higher by the *osmotic pressure*, on the sugar-enriched side. However the *temperatures* were identical on both sides of the interface.

In contrast, we now consider the changes in the *melting temperature of ice* (and the *boiling temperature of a liquid*) by addition of soluble substances, e.g., again some kind of salt or sugar, to the liquid phase[6], i.e., the interface is here the surface of the liquid.

We thus consider, e.g., the phase equilibria A) solid-liquid and B) liquid-vapor, i.e., a) *without* addition and b) *with* addition of the substance considered. For simplicity we only treat case A).

In case Aa) we have:

$$\mu_1^{\text{solid}}(T, p, c_1 = 1) \overset{!}{=} \mu_1^{\text{liquid}}(T, p, c_1 = 1) \;,$$

whereas for Ab):

$$\mu_1^{\text{solid}}(T - \Lambda T, p, c_1 = 1) \overset{!}{=} \mu_1^{\text{liquid}}(T - \Delta T, p, c_1 = 1 - \Delta c_2) \;.$$

---

[6] We assume that the added substance is only soluble in the liquid phase, but this can be changed, if necessary.

Forming a suitable difference we obtain

$$-\frac{\partial\mu_1^{\text{solid}}}{\partial T}\Delta T = -\frac{\partial\mu_1^{\text{liquid}}}{\partial T}\Delta T + k_B T \cdot \ln(1-\Delta c_2)\,,$$

and with $\ln(1-x) \approx -x$ we have:

$$\Delta T = -\frac{k_B T \cdot \Delta c_2}{\frac{\partial\mu_1^{\text{l.}}}{\partial T}-\frac{\partial\mu_1^{\text{s.}}}{\partial T}}\,.$$

Hence one obtains, with

$$\mu = \frac{G}{N} \quad \text{and} \quad \frac{\partial G}{\partial T} = -S\,, \quad \text{as well as} \quad \frac{S}{N} \equiv s\,:$$

$$\Delta T = -\frac{k_B T \cdot \Delta c_2}{s_1^{\text{s.}} - s_1^{\text{l.}}} \equiv +\frac{k_B T^2 \cdot \Delta c_2}{l_{s.\to l.}}\,, \tag{54.10}$$

where additionally the *molecular heat of melting*

$$l_{s.\to l.} := T \cdot \left(s_1^{\text{l.}} - s_1^{\text{s.}}\right)$$

has been introduced.

As a result we can state that the melting temperature (and analogously the boiling temperature) of the ice (and of the heated liquid) is *decreased* by the addition of salt (or sugar) to the liquid phases. In the boiling case the relevant equation is

$$\mu_1^{\text{l.}}(T-\Delta T, p, c_1 - \Delta c_2) \overset{!}{=} \mu_1^{\text{vapor}}(T-\Delta T, p, c_1))\,.$$

In this case too it is essentially the enhancement of the entropy by mixing which is responsible for the effect.

54.7 The Vapor Pressure of Spherical Droplets

Now consider an ensemble of spherical droplets of radius R with a sufficiently large value of R such that the number N of particles within a droplet is $\gg 1$. Outside such droplets the saturation pressure p_R is enhanced w.r.t. the value p_∞ for a planar surface, i.e., for $R \to \infty$.

For the *Helmholtz free energy* of a droplet we obtain

$$\mathrm{d}F = -p \cdot \mathrm{d}V + \sigma \cdot \mathrm{d}O - S \cdot \mathrm{d}T\,, \quad \text{with} \quad \mathrm{d}V = 4\pi R^2 \cdot \mathrm{d}R$$
$$\text{and} \quad \mathrm{d}O = 8\pi R \cdot \mathrm{d}R\,,$$

where $\mathrm{d}V$ is an infinitesimally small increment of volume and $\mathrm{d}O$ an increment of surface area of the droplet.

The above relation for dF defines the surface tension σ, an *energy per surface area*. This is best explained in the framework of the physics of soap bubbles.[7]

We thus obtain for the *Helmholtz free energy* of the droplet:

$$F_{\text{droplet}}(T, R) = N_{\text{droplet}} \cdot f(T, R) + 4\pi R^2 \sigma(T, R) \ .$$

Here $f(T, R)$ represents the volume part of the *Helmholtz free energy* per atom, and $\sigma(T, R)$ is the above-mentioned surface tension, which we now want to calculate. Thermal phase equilibrium gives

$$\mu^{\text{liquid droplet}} = \frac{\partial G}{\partial N} = f + p_R \cdot v_{\text{l}} + 4\pi\sigma \cdot \frac{\partial R^2}{\partial N} \overset{!}{=} \mu^{\text{vapor}} \ .$$

With

$$N \equiv \frac{4\pi R^3}{3 v_{\text{l}}} \ , \quad \text{hence} \quad dN = 4\pi R^2 \frac{dR}{v_{\text{l}}} \quad \text{and} \quad dR^2 = 2R \cdot dR \ ,$$

it follows that:

$$\mu^{\text{liquid}} = f + v_{\text{l}} \cdot \left(p_R + \frac{2\sigma}{R}\right) \overset{!}{=} \mu^{\text{vapor}} \equiv k_B T \cdot \ln \frac{p_R}{p_0} + B(T) \ , \quad [8]$$

where p_0 is an arbitrary unit pressure and $B(T)$ a temperature-dependent constant of physical dimension *energy per atom*.

For $R < \infty$ we thus have:

$$f + v_{\text{l}} \cdot \left(p_R + \frac{2\sigma}{R}\right) = k_B T \cdot \ln \frac{p_R}{p_0} + B(T) \ ,$$

and for $R = \infty$:

$$f + v_{\text{l}} \cdot p_\infty = k_B T \cdot \ln \frac{p_\infty}{p_0} + B(T) \ .$$

By subtracting the second equation from the first, we obtain:

$$k_B T \cdot \ln \frac{p_R}{p_\infty} = v_{\text{l}} \cdot \left(\frac{2\sigma}{R} + p_R - p_\infty\right) \ ,$$

and neglecting $(p_R - p_\infty)$[9]:

$$\frac{p_R}{p_\infty} \approx e^{\frac{v_{\text{l}}}{k_B T} \cdot \frac{2\sigma}{R}} \ . \tag{54.11}$$

---

[7] For the physics of soap bubbles see almost any textbook on basic experimental physics.

[8] The last term on the r.h.s. is reminiscent of the analogously defined *free enthalpy of mixing*: $\mu^{\text{vapor}} = k_B T \cdot \ln c_i + \ldots$, with the *partial pressure* $p_i = c_i \cdot p$.

[9] The result reminds us (not coincidentally) of the Clausius-Clapeyron equation.

The saturation vapor pressure outside a spherical droplet of radius R is thus greater than above a planar surface. With decreasing radius R the tendency of the particles to leave the liquid increases.

A more detailed calculation shows that a critical droplet radius R_c exists, such that smaller droplets, $R < R_c$, shrink, whereas larger droplets, $R > R_c$, increase in size. Only at R_c is there thermal equilibrium. For R_c the following rough estimate applies:

$$\sigma \cdot 4\pi R_c^2 \approx k_B T .$$

In this context the analogy with the Ising model (cf. (42.5)) is again helpful. For a cubic lattice with nearest-neighbor separation a and ferromagnetic nearest-neighbour interaction $J(> 0))$ in an external magnetic field $h_0(> 0)$ one has:

$$\text{Ising model:} \quad \mathcal{H} = -J \sum_{|r_l - r_m| = a} s_l s_m - h_0 \sum_l s_l , \qquad (54.12)$$

with s_l and $s_m = \pm 1$, corresponding to ↑ and ↓ spins.

We shall now consider the following situation: embedded in a ferromagnetic "lake" of ↓ spins is a single "island" (approximately spherical) of ↑ spins. The volume V of the 3d-"island" is assumed to be

$$V \approx \frac{4\pi R^3}{3} ,$$

while the surface area of the *island* is

$$O \approx 4\pi R^2 .$$

The energy of formation ΔE of the *island* in the *lake* is then given by

$$\Delta E \approx \frac{O}{a^2} \cdot 2J - \frac{V}{a^3} \cdot 2h_0 ,$$

as one can easily see from a sketch.

The first term on the r.h.s. of this equation ($\propto J$) corresponds to the surface tension, while the second term ($\propto h$) corresponds to the difference of the chemical potentials, $\mu^R - \mu^\infty$. From $\Delta E \overset{!}{=} 0$ it follows that

$$R_c = 3aJ/h_0 .$$

In this way one can simultaneously illustrate *nucleation processes* (e.g., nucleation of vapor bubbles and of condensation nuclei) in the context of overheating or supercooling, e.g., in the context of the *van der Waals equation*. The analogy is enhanced by the above lattice-gas interpretation. [10]

---

[10] A reminder: In the *lattice-gas interpretation* $s_l = \pm 1$ means that the site l is either *occupied*, (+), or *unoccupied*, (-).

55 Conclusion to Part IV

For all readers (not only those who understand German[1] or wish to practise their knowledge of the language) for all four parts of our compendium there are (translated) exercises to complement this book. These are separately documented on the internet, see [2], and purposely not integrated into this script.

In Thermodynamics and Statistical Physics both types of learning material, i.e., textbook and corresponding exercises, have been centered around

a) *Phenomenological Thermodynamics*, with the four quantities

$$F(T, V, N, \ldots) \quad (Helmholtz\ free\ energy)\,,$$
$$U(T, V, N, \ldots) \quad (internal\ energy)\,,$$
$$S(T, V, N, \ldots) \quad (entropy)$$

and the *absolute (or Kelvin) temperature T*, and
b) *Statistical Physics*, for historical reasons usually called *Statistical Mechanics*, although this name is unnecessarily restrictive.

a) and b) are closely interdependent, due essentially to the fundamental relation

$$F(T, V, N, \ldots) \equiv -k_B T \cdot \ln Z(T, V, N, \ldots)\,, \qquad (55.1)$$

where

$$Z(T, V, N, \ldots) = \sum_i e^{-\frac{E_i(V, N, \ldots)}{k_B T}} \qquad (55.2)$$

is the *partition function*.

This relation between a) and b), (55.1), applies for so-called *canonical ensembles*, i.e., when the particle number N is fixed and the heat bath only exchanges energy with the system considered – defining the reciprocal Kelvin temperature

$$\beta \left(= \frac{1}{k_B T} \right)$$

as conjugate parameter (conjugate to the energy).

---

[1] Sometimes even a partial understanding may indeed be helpful.

For a *grand canonical ensemble*, where the heat bath not only exchanges energy with the system, but also particles, such that the particle number in a volume element V *fluctuates* around the average $N(T, \mu, V)$, in addition to

$$\beta \left(= \frac{1}{k_B T} \right) ,$$

one obtains for the distribution a second parameter μ (the so-called *chemical potential*) for the analogous quantity to the *Helmholtz free energy*, i.e., for the *Gibbs grand canonical thermodynamic potential*:

$$\Phi(T, \mu, V, \ldots) = -k_B T \cdot \ln \mathcal{Z}(T, \mu, V, \ldots) ,$$

with the *grand canonical partition function*:

$$\mathcal{Z}(T, \mu, V, \ldots) = \sum_{i,j} e^{-\frac{E_i(V, N_j) - \mu N_j}{k_B T}} .$$

The mathematical relation between the *Helmholtz free energy* and the *Gibbs grand canonical potential* Φ is a Legendre transform, i.e.:

$$\Phi(T, \mu, V, \ldots) = F(T, V, N(T, \mu, V), \ldots) - \mu \cdot N(T, \mu, V) ,$$

similarly to the way the *internal energy* $U(T, V, N, \ldots)$ and the *enthalpy* I depend on each other:

$$I(T, p, N, \ldots) = U(T, V(T, p, N, \ldots), N, \ldots) + p \cdot V(T, p, N, \ldots) ,$$

with the *pressure* p as the conjugate Lagrange parameter regulating fluctuations in V.

These Legendre transformations are mathematically analogous to the transition from the *Lagrange function* $\mathcal{L}(\boldsymbol{v}, \ldots)$[2] to the *Hamilton function* $\mathcal{H}(\boldsymbol{p}, \ldots)$ in classical mechanics; incidentally (this may be used for mnemonic purposes!) the corresponding letters are similar, i.e., V (and $\boldsymbol{v}$) and p (and $\boldsymbol{p}$), although the meaning is completely different.

The relation between a) and b) can also be expressed as

$$U(T, V, N, \ldots) \equiv \langle \mathcal{H}(V, N, \ldots) \rangle_T .$$

Where $\left\langle \hat{A} \right\rangle_T$ is the thermodynamic expectation with the suitable *canonical* (and *microcanonical* and *grand canonical*) Boltzmann-Gibbs distribution, e.g.,

$$\left\langle \hat{A} \right\rangle_T = \sum_i p_i(T) \cdot \langle \psi_i | \hat{A} | \psi_i \rangle , \quad \text{with} \quad p_i(T) = \frac{e^{-\frac{E_i(V, N, \ldots)}{k_B T}}}{Z(T, V, N, \ldots)} ,$$

---

[2] Actually by the Legendre transformation of $-\mathcal{L}$.

and the Hermitian operator $\hat{A}$ represents an *observable* (i.e., a measurable quantity). The ψ_i represent the complete system of eigenfunctions of the Hamilton operator $\mathcal{H}$; the E_i are the corresponding eigenvalues.

Concerning the *entropy*: This is a particularly complex quantity, whose complexity should not simply be "glossed over" by simplifications. In fact, the entropy is a *quantitative* measure for complexity, as has been stated already. In this context one should keep in mind that there are at least three commonly used methods of calculating this quantity:

a) by differentiating the *Helmholtz free energy* with respect to T:

$$S(T, V, N, \ldots) = -\frac{\partial F(T, V, N, \ldots)}{\partial T} ;$$

b) from the *difference expression*

$$S = \frac{1}{T} \cdot (U - F)$$

following from the relation

$$F(T, V, N, \ldots) = U(T, V, N, \ldots) - T \cdot S(T, V, N, \ldots) ,$$

where the quantity $T \cdot S$ represents the *heat loss*. This formulation seems to be particularly useful educationally.

c) A third possibility of quantification follows directly from statistical physics

$$S = -k_B \cdot \sum_i p_i \ln p_i$$

(where this relation can even be simplified to $S = -k_B \cdot \langle \ln \hat{\varrho} \rangle_{\hat{\varrho}}$ in the above trace formalism.)

d) Shannon's informational entropy[3] is also helpful.

It should also have become clear that the Second Law (and even the Third Law) can be formulated without recourse to *entropy*. However, the notion of *absolute temperature* (Kelvin temperature) T is indispensable; it can be quantified via the efficiency of Carnot machines and constitutes a prerequisite for statistical physics.

Amongst other important issues, the *Maxwell relations* remain paramount. Here one should firstly keep in mind how these relations follow from a differential formulation of the First and Second Laws in terms of *entropy*; secondly one should keep in mind the special relation

$$\frac{\partial U}{\partial V} = T\frac{\partial p}{\partial T} - p$$

---

[3] This is essentially the same: k_B is replaced by 1, and the *natural logarithm* is replaced by the *binary logarithm*.

and the application to the *Gay-Lussac experiment*, and thirdly one should remember that one can always obtain important cross-relations by equating mixed second-order derivatives, e.g.,

$$\frac{\partial^2 F}{\partial x_i \partial x_j} \, ,$$

in the total differentials of the *Helmholtz free energy*,

$$\mathrm{d}F = -p\mathrm{d}V + \mu\mathrm{d}N + \ldots - S\mathrm{d}T \, ,$$

or analogous thermodynamic potentials.

In this book we have also stressed similarities between the four different parts. Therefore, looking back with a view on common trends, it seems that the systematic exploitation of "coherence properties" has a promising future, not only in optics (*holography*), but also in quantum mechanics (*quantum computing, etc.*). Unfortunately, as discussed above, *thermalization* also leads to *decoherence*. However, recent success in obtaining ultralow temperatures means that this barrier may become surmountable in the not-too-distant future.

References

1. W. Nolting: *Grundkurs theoretische Physik*, in German, 5th edn (Springer, Berlin Heidelberg New York 2002), 7 volumes.
   This is a good example of a series of textbooks which in seven volumes covers sometimes less but sometimes much more than the four parts of our compendium.
2. http://www.physik.uni-regensburg.de/forschung/krey

To Part I:

3. As a recommendable standard textbook on Classical Mechanics we recommend H. Goldstein, Ch.P. Poole, J. Safko: *Classical Mechanics*, 3rd edn (Addison-Wesley, San Francisco Munich 2002), pp 1–638
4. H. Hertz: *Die Constitution der Materie*, in: A. Fölsing (Ed.), Springer, Berlin Heidelberg New York, 1999, pp 1–171
5. A. Einstein: *Zur Elektrodynamik bewegter Körper*, Ann. der Physik **17**, 891 (1905)
6. R. von Eötvös, D. Pekár, E. Fekete: Ann. d. Physik **68**, 11 (1922)
7. S.M. Carroll: *Spacetime and Geometry. Introduction to General Relativity*, (Addison Wesley, San Francisco and elsewhere 2003), pp 1–513
8. L.D. Landau, E.M. Lifshitz: *Course of Theoretical Physics, volume 2, The Classical Theory of Fields*, 4th edn (Pergamon Press, Oxford New York and elsewhere, 1975), pp 1–402 (This book contains very readable chapters on relativity.)
9. H.G. Schuster, W. Just: *Deterministic Chaos*, 4th edn (Wiley-VCH, Weinheim 2005), pp 1–287

To Part II:

10. B.I. Bleaney, B. Bleaney: *Electricity and Magnetism*, 3rd edn (Oxford University Press 1976), pp 1–761
11. J.D. Jackson: *Classical Electrodynamics*, 3rd edn (Wiley, New York Weinheim Singapore 1999), pp 1–938
12. A. Heck: *Introduction to Maple*, 3rd edn (Springer, Berlin Heidelberg New York 2003), pp 1–828
13. L.D. Landau, E.M. Lifshitz: *Course of Theoretical Physics, volume 1, Mechanics*, 3rd edn (Pergamon Press, Oxford New York and elsewhere, 1976), pp 1–169, Chapt. 44 . (This chapter is useful for understanding the cross relations to the Fermat principle in geometrical optics.)
14. A. Sommerfeld: *Optics*, 4th edn (Academic Press, New York 1967), pp 1–383

15. F. Pedrotti, L. Pedrotti: *Introduction to optics*, 2nd edn (Prentice Hall, Upper Saddle River (NY, USA) 1993), pp 1–672

16. E. Hecht: *Optics*, 4th edn (Addison-Wesley, San Francisco New York 2002), pp 1–565

17. K. Bammel: Physik Journal, in German, (1) 42 (2005)

To Part III:

18. A. Einstein: *Über einen die Erzeugung und Verwandlung des Lichtes betreffenden heuristischen Gesichtspunkt*, Ann. der Physik **17**, 132 (1905)

19. W. Heisenberg: Z. Physik **33**, 879 (1925)

20. M. Born, W. Heisenberg, P. Jordan: Z. Pysik **35** 557 (1926)

21. C.I. Davisson, L.H. Germer: Nature **119**, 890 (1927); Phys. Rev. **30**, 705 (1927)

22. E. Schrödinger: Ann. Physik (4) **79**, 361; 489; 734 (1926); **80**, 109 (1926)

23. J. von Neumann: *Mathematische Grundlagen der Quantenmechanik*, in German, reprinted from the 2nd edn of 1932 (Springer, Berlin Heidelberg New York 1996), pp 1–262

24. W. Döring: *Quantenmechanik*, in German (Vandenhoek & Ruprecht, Göttingen 1962), pp 1–517

25. J. Bardeen, L.N. Cooper, J.R Schrieffer: Phys. Rev. **106**, 162 (1957); **108**, 1175 (1957)

26. D.D. Osheroff. R.C. Richardson, D.M. Lee, Phys. Rev. Lett. **28**, 885 (1972)

27. H. Börsch, H. Hamisch, K. Grohmann, D. Wohlleben: Z. Physik **165**, 79 (1961)

28. M. Berry: Phys. Today **43** 34 (12) (1990)

29. A. Einstein, B. Podolski, N. Rosen: Phys. Rev. **47**, 777 (1935)

30. J.S. Bell: Physics **1**, 195 (1964)

31. P. Kwiat, H. Weinfurter, A. Zeilinger: Spektrum der Wissenschaft **42**, (1) (1997)

32. A. Zeilinger: *Einsteins Schleier – die neue Welt der Quantenphysik*, in German (C.H. Beck, München 2003), pp 1–237

33. D. Loss, D.P. DiVincenzo: Phys. Rev. A **57**, 120 (1998)

34. F.H.L. Koppens, J.A. Folk, J.M. Elzerman, R. Hanson, L.H.W. van Beveren, I.T. Fink, H.P. Tranitz, W. Wegscheider, L.M.K. Vandersypen, L.P. Kouwenhoven: Science **309**, 1346 (2005)

To Part IV:

35. A. Einstein: *Über die von der molekularkinetischen Theorie der Wärme geforderte Bewegung von in ruhenden Flüssigkeiten suspendierten Teilchen*, Ann. der Physik **17**, 549 (1905)

36. P. Papon, J. Leblond, P.H.E. Meijer: *The Physics of Phase Transitions* (Springer, Berlin Heidelberg New York 2002), pp 1–397

37. C. Kittel: *Introduction to Solid State Physics*, 8th edn (Wiley, New York London Sidney Toronto 2005), pp 1–680

38. W. Gebhardt, U. Krey: *Phasenübergänge und kritische Phänomene*, in German (Vieweg, Braunschweig Wiesbaden 1980), pp 1–246

39. W. Döring: *Einführung in die theoretische Physik, Sammlung Göschen*, in German, 5 volumes, 3rd edn (de Gruyter, Berlin 1965), pp 1–125, 1–138, 1–117, 1–107, 1–114

40. A. Sommerfeld, H.A. Bethe: *Elektronentheorie der Metalle* (Springer, Berlin Heidelberg New York 1967)

41. H. Thomas: Phase transitions and critical phenomena. In: *Theory of condensed matter*, directors F. Bassani, G. Cagliotto, J. Ziman (International Atomic Energy Agency, Vienna 1968), pp 357–393. At some libraries this book, which has no editors, is found under the name E. Antoncik.

42. R. Sexl, H. Sexl: *White dwarfs – black holes*, 2nd edn (Springer, Berlin Heidelberg New York 1999), pp 1–540

43. R. Sexl, H.K. Urbantke: *Gravitation und Kosmologie*, in German, 3rd edn (Bibliograpisches Institut, Mannheim 1987), pp 1–399

44. C.W. Misner, K.S. Thorne, J.A. Wheeler: *Gravitation*, 25th edn (Freeman, New York 1003), pp 1–1279

45. D. Vollhardt, P. Wölfle, *The superfluid phases of helium 3*, (Taylor & Francis, London New York Philadephia 1990), pp 1–690

46. V.L. Ginzburg, L.D. Landau: J. Exp. Theor. Physics (U.S.S.R.) **20**, 1064 (1950)

47. A. Abrikosov: Sov. Phys. JETP **5**, 1174 (1957)

48. L.D. Landau, E.M. Lifshitz: *Course of Theoretical Physics, volumes 5 and 9* (= Statistical Physics, Part 1 and Part 2), 3rd edn, revised and enlarged by E.M. Lifshitz and L.P. Pitaevskii (Pergamon Press, Oxford New York and elsewhere, 1980), pp 1–544 and 1–387

49. J. de Cloizeaux: Linear response, generalized susceptibility and dispersion theory. In: *Theory of condensed matter*, directors F. Bassani, G. Cagliotto, J. Ziman (International Atomic Energy Agency, Vienna 1968), pp 325–354. At some libraries this book, which has no editors, is found under the name E. Antoncik.

50. N. Metropolis, A.W. Rosenbluth, M.N. Rosenbluth, A.H. Teller, E. Teller: J. Chem. Phys. **21**, 1087 (1953)

51. D.P. Landau, K. Binder: *A guide to Monte Carlo simulations in statistical physics*, 2nd edn (Cambridge University Press, Cambridge UK, 2000), pp 1–448

Index